Advances in

Laser Physics and Technology

Edited by

Man Mohan, Anil Kumar Maini,
Aranya B. Bhattacherjee and Anil K. Razdan

FOUNDATION
BOOKS

FOUNDATION
BOOKS

Foundation Books is an imprint of Cambridge University Press, India.

Cambridge House, 4381/4 Ansari Road, Daryaganj, New Delhi 110 002

www.cambridgeindia.org

First published 2015

Printed in India at AnVi Composers, New Delhi - 110063

ISBN 978-93-84463-41-0 Hardback

The book is dedicated to the memory of my parents, Late Shri Girja Dutt ji, Late Shrimati Dikeshwari Devi ji and my PhD supervisor Late Professor S. N. Biswas. Their loving memory will always remain in my heart for their constant encouragement and support.

Man Mohan

Contents

List of Figures

List of Tables

Introduction

There has been great advancement in the field of laser science and technology with wide applications which has led to several Noble Prize awards in recent years in Physics and Chemistry. Lasers are being developed from femto to atto-second range enabling scientists in the study of phenomena on the time scale of electron motion in atoms and molecules. Consequently, other interesting areas of scientific study include solid state, plasma physics, nano-science and defence technology. Lasers also permit us to reach exceedingly low temperatures often less than a billionth of a degree above absolute zero which includes the implications of Bose-Einstein condensation (BEC), fermionic superfluidity etc. Initially, the focus on BEC was on the study of basic properties such as superfluidity, structural properties, collective and elementary excitations, vortices, light matter interactions etc. But in recent years, the focus has shifted to some important applications such as detection of weak forces, cavity electrodynamics and opto-mechanics.

The book describes some of the latest advances in laser physics and technology and its applications. The book is divided into five major sections: Laser Processes Involving Short Pulses, Laser Spectroscopy, Laser Cooling and Trapping, Quantum Optics and Laser Applications.

The first section deals with the laser processes in atoms and molecules involving short pulse lasers. It contains topics from high-harmonic generation (HHG) and strong-field non-sequential double ionization (NSDI), with emphasis on atto-second orbital imaging to the field of molecular quantum computing including a description of all current theoretical and experimental results. Chapter 1, Augstein et al., addresses high harmonic generation (HHG) and strong-field non-sequential double ionization (NSDI), with emphasis on atto-second orbital imaging. In Chapter 2, Alex Brown et al., a general introduction is given to the field of molecular quantum computing including a description of all current theoretical and experimental results. These chapters enumerate the laser requirements and molecular properties that govern the implementation of high-fidelity quantum gates.

The effect of a low energy atto-second pulse sustaining only a few-cycles is studied in Chapter 3, on a simple two-level system for different pulse envelope profiles. An atto-second pulse, due to its very short duration, can be sustained only by wavelengths of hundreds of angstroms. Such pulses are highly energetic and can cause molecular dissociation. Thus, atto-second pulses cannot be used to study molecular processes. The ability to sustain only a few cycles in the pulse is a characteristic unique to the atto-second domain. To capture this quintessential feature in molecular processes, the authors use the idea of few-cycle pulses to know what happens when a few-cycle pulse interacts with matter. In their study of the interaction of few-cycle pulses with ideal two-level systems, the approximation of smoothly varying envelope function breaks down. They observed that despite the overwhelming importance of the oscillatory nature of a few-cycle pulse, the signatures of the envelope profile still persist in the interactions.

The next six chapters belong to sub-topic Laser Spectroscopy. In Chapter 4, the group led by Manoj Mishra deals with the time dependent investigation for photodynamic control of selective dissociation of ^{18}O–^{16}O and ^{16}O–^{16}O bonds in isotopic ozone ($^{18}O^{16}O^{16}O$) molecule using simple Gaussian UV pulses applied to the ground vibrational state ($|0,0>$) of the ground electronic state are presented and discussed. Results obtained from their investigation are analyzed using time evolution plots and flux profiles on both ground and excited states and shows a ^{18}O–^{16}O + $^{16}O/^{18}O$ + ^{16}O–^{16}O flux branching ratio $\sim$ 1.8 and thereby demonstrates a possible route to selective photo dissociation in $^{18}O^{16}O^{16}O$ molecule. In Chapter 5, the possibility of enhancement and suppression of chemical reaction by CW laser is discussed and numerically demonstrated by Yoshiaki Teranishi, Hiroki Nakamura and Sheng H. Lin. The study demonstrates that non-adiabatic transitions induced by laser play a crucial roles and the Zhu-Nakamura theory can be usefully utilized. The study of the creation of doubly charged ions upon the multiphoton ionization of alkaline-earth metals by the visible and infrared laser radiation is reviewed by V. V. Suran and I. I. Bondar in Chapter 6. Their results show that the features of the creation of doubly charged ions depend on the spectral range of the ionizing laser radiation. In the ionization of the alkaline-earth atoms by the visible laser radiation, the cascade creation of the doubly charged ions is predominantly implemented when the ions result from the multiphoton ionization of singly charged ions created in the presence of the same laser pulse. In the multiphoton ionization of the alkaline-earth atoms by the infrared (IR) laser radiation, the two-electron process occurs when the doubly charged ions are directly created from neutral atoms owing to the ionization via induced resonances with strongly perturbed states of these atoms. In Chapter 7, Man Mohan's group discusses the application of the non-perturbative Floquet theory for studying non-linear multiphoton processes in nanostructures. Chapter 8 by Mainuddin et al., describes optical based singlet oxygen measurement technique which is an essential diagnostics for state of the art chemical oxygen iodine laser (COIL). Optical emission spectroscopy at 1.27 μm has been used for on line estimation of concentration of singlet oxygen which is the pumping medium of COIL. The details of calibration of the diagnostic set up, its characterization and the obtained experimental results along with analysis have been discussed in the article. Fundamentals of Raman theory for silicon is discussed in Chapter 9 by A. K. Shukla et al. First-order and second-order Raman scatterings are discussed at length which will help the researchers to learn Raman characterization of their samples.

Chapters 10 and 11 belong to sub-topic Laser Cooling and Trapping. Ultra-stable coherent matter-wave gyroscope exploiting quantized vortices in Bose-Einstein condensates generated from orbital angular momentum beams of light is discussed by Kishore T. Kapale, Sulakshana Thanvanthri, and Jonathan P. Dowling in Chapter 11. They review the methods for coupling of orbital angular momentum of light with trapped Bose-Einstein Condensates (BEC). Also, there's a discussion on the application of these methods to coherent generation of superpositions of quantized vortex states in BEC, and applications of these superpositions for the development of a sensitive gyroscope that has no moving parts. In Chapter 11, the authors have developed a new tool for controlling the superfluid properties of a condensate loaded in an optical lattice inside a cavity. They have studied the opto-mechanical effects (effects of moving one of the cavity mirror) on the Bloch energy, the effective mass, the Bogoliubov excitation spectrum, superfluid fraction and the mott-superfluid phase diagram of a Bose-Einstein condensate.

Chapters 12 to 14 belong to Quantum Optics section. In Chapter 12, S. Aithal et al., describe their observation about low-intensity optical phase-conjugation in DASPB dye in PMMA–PA polymer matrix using a degenerate four-wave mixing setup, employing 633 nm light radiation from a He-Ne laser. Optical filters find its applications in transmission links, optical instruments, spectroscopy which is described in Chapter 13. Here, the authors have proposed a tunable optical filter based on one–dimensional photonic crystal structure. In Chapter 15, Tarun Kumar et al., discuss the recent work on quantum mechanical system in quantum regime in the field of cavity quantum electrodynamics. Various properties of a Fabry-perot cavity are discussed by showing that a non-linear kerr medium present inside a cavity can be used as a tuneable quantum mechanical device by photon blockade mechanism. They also discuss various aspects of cooling of a micromirror in two optically coupled micromirrors and the process of normal mode splitting.

The following chapters belong to Laser Applications. In Chapter 15, A. K. Gupta et al., discuss the generation of mid-infrared (IR) and THz radiation using nonlinear optical techniques. They describe different methods and materials available for mid-IR and THz generation such as quasi-phase matched (QPM) materials. In this chapter they describe the fabrication of periodically poled lithium niobate (PPLN) using E-field poling technique followed by design and development of PPLN based mid-IR tunable source and DAST for generating THz radiation through the process of difference frequency generation (DFG)/ terahertz parametric generation (TPG).

Chapter 16 by Anil Kumar Maini et al., elaborates on a new test system design for comprehensive characterization of Non-imaging laser IR guided missiles. Electro optically guided munitions, which include IR guided and laser guided munitions, are the most widely used precision strike weapons on a variety of military platforms. To ensure 100 percent hit accuracy for these weapons, it is imperative that these are tested comprehensively at regular intervals. Comprehensive testing of these weapons is in addition to the pre-flight functionality check, also known as serviceability check, usually performed just before a mission without dismantling the weapon from the launch platform. This chapter presents a new design concept that can be used to build a test system capable of carrying out comprehensive characterization of IR guided missiles employing non-imaging type single colour or dual colour infrared seekers. The chapter begins with a brief outline of the requirements of comprehensive characterization and then discusses the proposed design concept along with its implementation in experimental hardware. Test results obtained with prototype hardware are also presented in this chapter.

Chapter 17 discusses the highly advanced laser and optoelectronics technologies for low intensity conflict (LIC) applications. This chapter briefly describes both established and emerging applications of laser and optoelectronics technologies in homeland security in terms of salient features, potential usage and international developments. Chapter 18 by A. K. Varshney et al., describes the development of highly advanced laser cavity and resonator for high power chemical oxygen iodine laser developed at Laser Science and Technology Centre (LASTEC). In Chapter 19, Mukesh Kumar and K. B. Singh have reviewed the theory of laser induced plasma with the micro-Raman scattering spectroscopy under the condition of the atmospheric domain of ionosphere. The aim of their work is to study the effect of micro-Raman scattering spectroscopy to generate the transparent dusty plasma. In Chapter 20, laser-cluster interaction is studied in the sub-cycle pulse regime, for the first time, with three dimensional particle-in-cell simulation and with a model, by P. K. Kaw's group. Significant amount of laser energy coupling with the cluster nanoplasma is found due to the linear and nonlinear resonance absorption process. Most recent work on Light Detection and Ranging (LIDAR) system is described by A. K. Razdan et al., in the last chapter. They use highly sensitive *intense pulsed lasers* which can detect extremely low concentrations of different pollutants, chemical or biological species present in the atmosphere at distances of several kilometers with very high degree of discrimination among molecules of different species. Their work will be useful against the threat of use of chemical, biological and explosive (CBE) materials by rogue countries against their adversaries.

This book will be useful for scientists, teachers, graduate and postgraduate students, especially dealing in laser physics, atomic and molecular physics, chemical physics and nano-technology, defence technology and condensed matter. In view of the very wide range of laser applications and people working on multi-disciplinary subjects, there is need for those other than physicists to have a working knowledge of laser phenomenon in different systems. The knowledge in this book will hopefully enable them to make informed judgments on the purchase of appropriate lasers and associated components. Such people might include mechanical, electronic and telecommunication engineers, chemists, life scientists, surgeons and military personnel.

Man Mohan

Acknowledgments

We are thankful to Professor Dinesh Singh, the Vice-Chancellor of University of Delhi, Shri Avinash Chander, Scientific Advisor to Defence Minister, Secretary Department of Defence R&D and DG DRDO, New Delhi, Shri H. B. Srivastava, Director, LASTEC, DRDO, New Delhi, Professor R. P. Tandon and Professor Amitabha Mukherjee (H.O.D.) Department of Physics and Astrophysics, University of Delhi for their encouragement.

Man Mohan is thankful to Professor T. W. Hansch (Nobel Laureate) and Late Professor H. Walther for giving him the opportunity to work in Max Planck Institute, Garching. He is also thankful to Professor N. P. Bigelow, the Head of Quantum Optics and BEC Divison, University of Rochester, USA for the constant encouragement. Many thanks are due to Dr P. K. Jha, Dr A. K. Singh, Dr Sanjay Tyagi, Dr Vinay Singh, Dr Nupur Verma, Professor R. Ramnathan, Professor R. Seth, Professor R. K. Sharma,

Dr Rinku Sharma, Dr N. Singh, Dr Rachna Joshi, Dr Kriti Batra, Dr Alok Jha, Dr Monica, Dr Tarun Kumar, Jagjit Singh, Siddhartha Lahon, Priyanka Verma, Sunny Aggarwal, Sonam Mahajan and Mr. Manoj Malik for helping in compiling this book. Man Mohan would like to express his gratitude towards wife Sneh Mohan, daughter Anu, son Anurag and daughter-in-law Skita for their constant patience and encouragement in completing this project.

Man Mohan is also thankful to DST and UGC for UGC-BSIR National Fellowship award.

I

Laser Processes Involving Short Pulses

1 High-order Harmonic Generation and Non-sequential Double Ionization in Atoms and Molecules: Quantum Interference, Causality and the Role of Multiple Orbitals

B. B. Augstein, J. Wu, T. Shaaran and C. Figueira de Morisson Faria

1.1 Introduction

In the past decade, the phenomena occurring when matter interacts with short laser fields, whose intensity is of the order of 10^{13} W/cm^2 or higher, have been looked at as powerful attosecond (10^{-18} s) imaging tools[1, 2, 3, 4]. Recent examples include the reconstruction of the highest occupied molecular orbital (HOMO) of N_2[5] and the imaging of attosecond electron–hole dynamics[6] employing high-order harmonic generation, or the velocity-map imaging of atomic orbitals employing above-threshold ionization[7, 8]. This is a direct consequence of the fact that these phenomena result from the laser-induced rescattering or recombination of an electron with its parent ion, or molecule[9]. Recombination of the active electron with a bound-state leads to high-order harmonic generation (HHG), while rescattering may cause above-threshold ionization (ATI) or laser-induced non-sequential double ionization (NSDI). These processes occur within a fraction of a driving field cycle, which for typical parameters used in experiments is roughly $\tau =$ 2.7 fs. Hence, their duration comprises hundreds of attoseconds. Furthermore, upon recollision, the active electron will probe the structure of the target, and this will leave imprints in the high-harmonic or photoelectron spectra.

The above-mentioned imprints may be static or dynamic. Widely investigated structural effects include interference fringes due to recollision at spatially separated centers, and the influence of the geometry of the molecular orbitals involved, such as nodal planes or orbital orientation, on the spectra. These effects have been mostly studied by considering the core to be static, and assuming the HOMO to be the sole active orbital. Recent studies, however, have shown that the dynamics of the core, in particular electron–electron correlation, core excitation and relaxation are expected to play an important role[6]. Indeed, when the active electron is removed, a hole will be created in the core, which, potentially, will migrate. Apart from that, the electric field may polarize the core, or induce excitation and relaxation processes, especially if one electron has been removed[10, 11].

This chapter is a brief review of our recent study on HHG and NSDI, both in atoms and molecules, together with novel results or aspects which have not been brought up before. Throughout, particular emphasis is placed on attosecond imaging. In this context, it is worth mentioning that the vast majority of the studies performed in literature have focused on HHG and ATI, while NDSI is, comparatively, much less explored as an attosecond imaging tool. We have devoted a great deal of attention to this phenomenon, however, for two main reasons. First, in recent years, there has been experimental evidence that NSDI electron momentum distributions from molecules depend strongly on the target geometry[12, 13]. This is not surprising, as NSDI also results from the recollision of an electron with its parent ion. The main difference is that, in ATI, elastic recollision with the core occurs, while in NSDI, the returning electron gives part of its kinetic energy to the core so that a second electron is freed. Second, in principle, NSDI electron momentum distributions contain more detailed information about dynamic processes than ATI or HHG spectra. In fact, the shape of such distributions is extremely sensitive to the type of interaction by which the second electron is dislodged, the electron bound states involved and the type of collision through which NSDI occurs (for a recent example of correlated attosecond imaging using NSDI see[14]).

Two main themes permeate this study: quantum interference and electron orbits. Quantum interference may occur spatially, or temporally. Spatial interference occurs when the returning electron, for instance, recollides with spatially separated centers in a molecule. This leads to well-defined maxima and minima, whose position will depend both on the internuclear separation and on the symmetry of the orbitals involved. For homonuclear molecules, this interference has been looked at as a microscopic "double-slit experiment". Another example of spatial quantum interference is when the electrons involved in the rescattering or recombination interact with more than one molecular orbital. This may happen if the system is prepared in a coherent superposition of states, if electrons occupying different shells tunnel with comparable probability or if excitation occurs. Temporal interference occurs because, for a given return energy, an electron may return to the core along several trajectories. Quantum mechanically, the corresponding transition amplitudes may be added coherently. For a brief historical backround on quantum interference effects in HHG, see our review[15].

Electron orbits are presented from a more methodological viewpoint, as we employ approaches which allow a direct connection to the physical picture of a returning electron which rescatters or recombines with its parent ion. Here, we employ mainly the strong field approximation (SFA), which is one of the most widespread approaches in strong field physics. In the SFA, the influence of the laser field is neglected when the active electron is bound, and the binding potential is neglected when it is in the continuum. This means that, in the SFA, the continuum is approximated by field-dressed plane waves, i.e., by Volkov states. It simplifies the problem considerably, and allows it to be treated analytically to a very great extent. This is in particular true if the steepest descent method is used[16]. Apart from the SFA, we also discuss preliminary results obtained using the coupled-coherent state (CCS) approach. In this approach, the time-dependent Schrödinger equation is solved using a coherent state basis which is guided by classical trajectories[17]. In contrast to the SFA, this method fully incorporates the residual binding potential when the electron is in the continuum.

The present chapter is organized in two main parts. Section 1.2 and Section 1.3 presents selected examples of our work on HHG and NSDI, respectively. The HHG studies have been mostly based on the SFA, which has been modified in order to account for more than one active electron or orbital. These modifications are reviewed in Section 1.2.1, together with their role in the spectra. In Section 1.2.2, preliminary spectra obtained using the coupled coherent state approach are introduced, with emphasis on the temporal quantum interference that leads to a plateau. The NSDI studies focus on the recollision-excitation with subsequent ionization (RESI) rescattering mechanism, in which the first electron, upon return, promotes the second electron to an excited state. This mechanism is expected to be important for molecular targets. Section 1.3.1 presents the specific SFA transition amplitude associated with RESI. Subsequently, we analyze the momentum constraints delimited by these transition amplitudes, and provide selected examples of RESI distributions, for atoms and diatomic molecules (Section 1.3.2 and 1.3.3, respectively). Finally, in Section 1.4, an overall summary of our results is provided and the results been placed in a broader context.

1.2 High-order Harmonic Generation

High-order harmonic generation is a coherent process which occurs when atoms or molecules are exposed to a strong laser field, causing the target system to emit multiples of the driving laser field frequency. Much in contrast to the monotonic decrease in the intensity of harmonics which is expected from time-dependent perturbation theory with regard to the field, a typical harmonic spectrum exhibits a sharp decline, followed by a plateau of roughly constant intensity, and finally an abrupt cutoff at higher harmonic order, found universally at,

$$n = \frac{(E_0 + 3.17U_p)}{\omega} \tag{1.1}$$

where n is the harmonic order, ω is the frequeny of the driving laser field, E_0 is the ionization potential of the system, and U_p is the ponderomotive energy, which is the cycle averaged quiver energy for an electric field of strength ε_0. For a monochromatic field, $U_p = (\varepsilon_0^2/4\omega^2)$.

In order to calculate the high-order harmonic spectrum, we must find the Fourier transform of the expectation value of the dipole operator. This requires calculating the wavefunction of the system, $|\Psi_{(t)}\rangle$. Solving the time-dependent Schrödinger equation (TDSE), which, apart from not having an exact analytical solution, is a highly demanding task for systems with more than one active electron.[1]

Employing the strong field approximation (SFA), neglecting depletion of the ground state, and continuum–continuum transitions leads to the transition amplitude[19, 20],

$$M(\Omega) = -i\int_{-\infty}^{\infty} dt \int_{-\infty}^{t} dt' \int d^3\mathbf{k}a^*_{\text{rec}}(\mathbf{k}+\mathbf{A}(t))a_{\text{ion}}(\mathbf{k}+\mathbf{A}(t'))\times\exp[iS(t,t',\Omega,\mathbf{k})]+c.c., \tag{1.2}$$

[1] In general, the fully numerical solution of the TSDE suffers from the so-called "exponential wall", i.e., the numerical effort increases exponentially with the number of degrees of freedom involved. Only recently has this equation been solved for the simplest existent two-electron system, namely helium, in a realistic three-dimensional model[18].

where

$$a_{\mathrm{rec}}(\mathbf{k}) = \langle \mathbf{k} \mid \hat{\mathbf{r}} \mid \Psi_0 \rangle, \tag{1.3}$$

and

$$a_{\mathrm{ion}}(\mathbf{k}) = \langle \mathbf{k} \mid H_I(t') \mid \Psi_0 \rangle, \tag{1.4}$$

are referred to as the recombination and ionization dipole matrix elements, respectively, $\mid \Psi_0 \rangle$ is the ground state wave function, $\Omega = n\omega$ is the harmonic order, $H_I(t')$ is the interaction Hamiltonian, which will be chosen to be in the length gauge ($H_I(t') = \hat{\mathbf{r}}\cdot\mathbf{E}(t')$), where $\mathbf{E}(t')$ is the instantaneous laser field at the instant of ionization and

$$S(t,t',\Omega,\mathbf{k}) = -\frac{1}{2}\int_{t'}^{t} [\mathbf{k}+\mathbf{A}(\tau)]^2 d\tau - E_0(t-t') + \Omega t, \tag{1.5}$$

is the semi-classical action.

The multi-dimensional integral in Eq. (1.2) can be solved by using the saddle-point approximation (SPA), which looks for solutions while the action in Eq. (1.5) is stationary. This implies that $\partial S(t,t',\mathbf{k})/\partial t' = 0$, $\partial S(t,t',\mathbf{k})/\partial t = 0$ and $\partial S(t,t',\mathbf{k})/\partial \mathbf{k} = \mathbf{0}$. Explicitly,

$$\frac{[\mathbf{k}+\mathbf{A}(t')]^2}{2} + E_0 = 0, \tag{1.6}$$

$$\int_{t}^{t'} d\tau [\mathbf{k}+\mathbf{A}(\tau)] = 0, \tag{1.7}$$

and

$$\frac{[\mathbf{k}+\mathbf{A}(t)]^2}{2} + E_0 = \Omega. \tag{1.8}$$

These equations may be interpreted as follows. Equation (1.6) represents the conservation of energy while the electron is tunneling. This equation only has complex solutions, which corresponds to the quantum mechanical nature of tunneling ionization at time t'. Equation (1.7) constrains the electron to return to the site of its ionization. Finally, Eq. (1.8), represents the conservation of energy when the returning electron recombines with the orbital of energy E_0, giving rise to a harmonic of frequency Ω[21, 22].

The SFA outlined for HHG so far can be applied to any one-electron ground state wavefunction $\mid \Psi_0 \rangle$. For the specific case of homonuclear diatomic molecules, the ground state can be represented by using a linear combination of atomic orbitals (LCAO),

$$\Psi(\mathbf{r}) = \sum_{\alpha} c_\alpha \psi_\alpha \left(\mathbf{r} + \frac{\mathbf{R}}{2}\right) + (-1)^{l_\alpha - m_\alpha + \lambda_\alpha} c_\alpha \psi_\alpha \left(\mathbf{r} - \frac{\mathbf{R}}{2}\right), \tag{1.9}$$

where $\mathbf{R}$ is the internuclear distance, l_α is the orbital quantum number, m_α is the magnetic quantum number and λ_α determines the orbital symmetry, with $\lambda = m$ corresponding to gerade (even) and $\lambda_\alpha = m_\alpha + 1$ corresponding to ungerade (odd). In literature, the wavefunctions ψ_α at each ion have been expanded in both exponential[23] and Gaussian-type orbitals. The latter type of orbitals considerably simplifies the computations, and hence have been used by us. The LCAO coefficients c_α, and the expansion coefficients and exponents may be computed using a quantum-chemistry code, such as GAMESS-UK[24].

The first theoretical calculations for harmonic generation from H_2^+ at arbitrary alignment angles[25, 26, 27] predicted an angle-dependent minimum in the harmonic spectrum. This effect has been investigated for more complicated molecular orbitals, such as the highest occupied molecular orbital (HOMO) in N_2, where s-p mixing must be considered. An interference condition was derived within the framework of the SFA[23, 21] and also for heteronuclear diatomic systems[28].

1.2.1 SFA models beyond the single-active orbital approximation

We now provide two examples where we investigate the role of multiple molecular orbitals in HHG. Firstly, we employ a coherent superposition of molecular orbitals within the single-active electron approximation (SAE), and, subsequently, we take a multielectron state.

The explicit expressions for $|\Psi_0\rangle$ in both cases read

$$|\Psi_0\rangle = \sum_j C_j |\Psi_0^{(j)}\rangle, \tag{1.10}$$

where C_j is the relative weight of each contributing orbital, and

$$|\Psi_0\rangle = \frac{1}{\sqrt{N!}}\hat{A}\,|\Psi_0^{(1)}, \Psi_0^{(2)}, \ldots, \Psi_0^{(N)}\rangle, \tag{1.11}$$

respectively, where $\hat{A}$ is the anti-symmetrization operator and $|\Psi_0^{(j)}\rangle$, $j = 1, \ldots, N$ are the wavefunctions related to each of the active electrons, i.e., the active orbitals. Throughout, spin degrees of freedom are neglected. In the latter model, we also consider the N-particle dipole operator $\hat{\mathbf{r}} = \sum_{j=1}^{N} \mathbf{r}^{(j)}$, and construct a multi-electron time evolution operator such that, if an electron is ionized from an orbital, its time evolution is described by the SFA, while the bound electrons are propagated using the field-free time evolution operator. In other words, it is assumed that the evolution of the electron that is freed by tunnel ionization and that of the bound electrons is decoupled, i.e., we do not allow for excitation or core relaxation to occur.

Taking the coherent superposition for the ground state wavefunction in Eq. (1.10), the SFA transition amplitude reads

$$M = \sum_{j,\nu} M_{j,\nu}^{(1)} + c.c, \tag{1.12}$$

where $c.c.$ is the complex conjugate,

$$M_{j,\nu}^{(n)} = -iC_j^* C_\nu \int_{-\infty}^{\infty} dt \int_{-\infty}^{t} dt' \int d^3\mathbf{k}_n a_{\text{rec}}^{*j}[\mathbf{k}_n + \mathbf{A}(t)]$$
$$\times\, a_{\text{ion}}^{j}[\mathbf{k}_n + \mathbf{A}(t')] \exp[iS_{j\nu}(t, t', \mathbf{k}_n)], \tag{1.13}$$

and

$$S_{j\nu}(t,t',\mathbf{k}_n) = S(t,t',\mathbf{k}_n) - (E_j t - E_\nu t'). \tag{1.14}$$

Here, E_j and E_ν are the ionization energies of orbitals j and ν. The number of electrons is denoted by n, and for coherent superposition, $n = 1$. We note that, crucially, Eq. (1.13) allows the possibility of ionization and recombination to different orbitals.

If the multi-electron ground state wavefunction in Eq. (1.11) is taken, the SFA transition amplitude becomes,

$$M = \sum_{j,n} M^{(n)}_{j,j} + c.c.. \tag{1.15}$$

We see from Eq. (1.15) that the electron must ionize and recombine to the same orbital[21]. This is due to the restrictions placed upon the time evolution operator, which means that the core remains static and that, due to the Pauli exclusion principle, the electron may only return to the orbital from which it came.

For a coherent superposition, we consider a target system of N_2^+, and a coherent superposition between the HOMO, which is a $3\sigma_g$ orbital, and the LUMO, which is a $1\pi_g$ orbital. Both orbitals are negatively bound, which implies that ionization proceeds by tunneling. Vibrational degrees of motion can be neglected[29]. We restrict the calculation to two dimensions, although it is easily extended to three[21], and the results are very similar. Taking the weights C_j to be equal the wavefunction for N_2^+ reads,

$$|\Psi_0\rangle = \frac{1}{\sqrt{2}}(|3\sigma_g\rangle + |1\pi_g\rangle). \tag{1.16}$$

The diatomic molecule is aligned at an angle, θ_L, relative to the laser field polarization and the harmonic spectrum is calculated. These results are displayed in Fig. 1.1, where we focus on the contributions of particular ionization and recombination channels to the overall harmonic spectrum. Immediately, one observes dramatic differences in the overall harmonic spectrum Fig. 1.1(a) which considers all contributing channels, compared to the contributions from the $3\sigma_g$ molecular orbital only Fig. 1.1(f). The two-center interference observed in the latter spectrum is completely washed out in Fig. 1.1(a), up to around harmonic $\Omega = 60\omega$, after which it is visible again.

Figure 1.1(b) shows the spectrum arising from processes in which the electron returns to the $3\sigma_g$ orbital, but may ionize from either the $3\sigma_g$ orbital or the $1\pi_g$ orbital. A clear two-center interference minimum is seen over all angles and harmonics up to the cutoff, and as the recombination step determines the interference, one can conclude that the blurring seen in Fig. 1.1(a) is due to the contribution of channels which return to the $1\pi_g$ orbital. At angles $\theta_L = 0°$ and $\theta_L = 90°$, there is a strong suppression in the HHG yield. This is due to ionization from the $1\pi_g$ orbital, which has nodal planes aligned along the laser field polarization at these angles. Thus, as a consequence of the geometry of the orbital, tunneling ionization is greatly reduced. However, because the potential barrier of the $1\pi_g$ orbital is smaller than that of the $3\sigma_g$ orbital, at other angles, the harmonic yield is higher. In comparison to Fig. 1.1(c), which considers processes which return to the $1\pi_g$ orbital, the cutoff is higher in Fig. 1.1(a), at around $\Omega = 80\omega$. This is due to recombination to the $3\sigma_g$ orbital, which has a higher ionization potential. As the $1\pi_g$ orbital has a lower ionization potential, we see a lower cutoff in Fig. 1.1(c). There is also no interference minimum in the spectrum from channels which return to the $1\pi_g$ orbital. This is because this minimum occurs at a harmonic frequency beyond the cutoff.

When observing Fig. 1.1(d), which displays the contributions from processes starting at the $3\sigma_g$ molecular orbital, one immediately sees a drop in yield. This is due to the higher ionization potential in comparison to the $1\pi_g$ orbital. There is also a clear interference minimum occurring after $\Omega = 60\omega$, which has been washed out at lower harmonics due to recombination to the $1\pi_g$ orbital. There are no cuts at $\theta_L = 0°$ and $\theta_L = 180°$ due to the process of ionization and recombination to the $3\sigma_g$ orbital. This washes out the cuts in the spectrum due to ionization from the $3\sigma_g$ orbital and recombination to the $1\pi_g$, which one would expect to see due to the nodal planes in the $1\pi_g$ orbital.

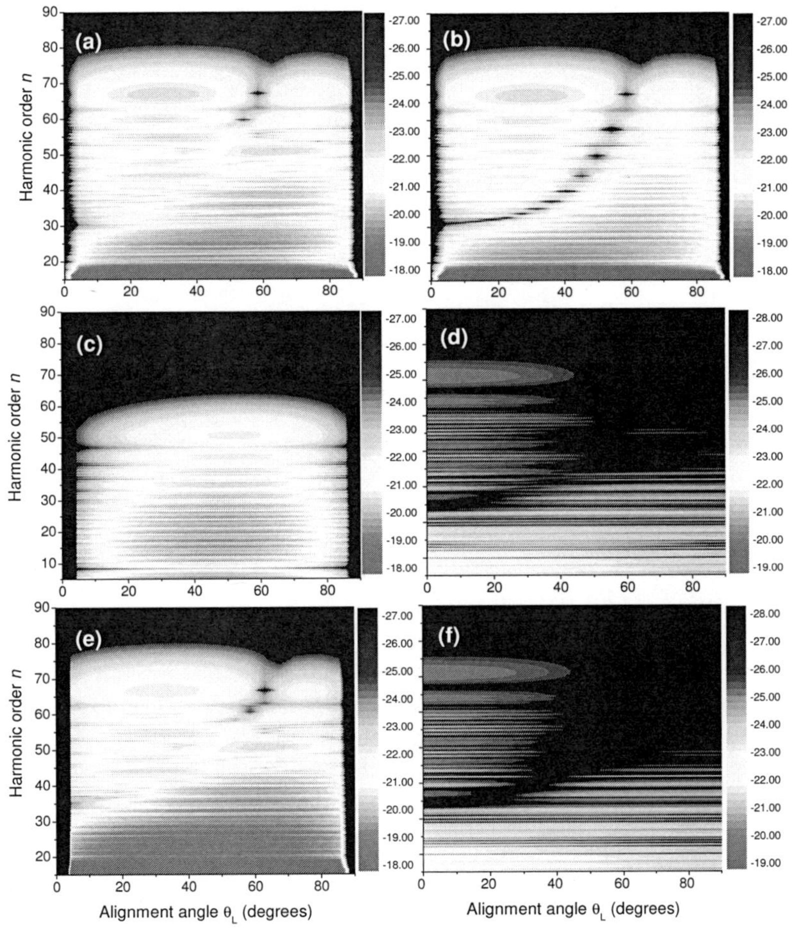

Figure 1.1 Harmonic spectrum as a function of alignment angle, for an N_2^+ molecule prepared in a coherent superposition of the $3\sigma_g$ HOMO and $1\pi_g$ LUMO, subject to a linearly polarized laser field of frequency $\omega = 0.057$ a.u. and intensity $I = 4\times10^{14}$ Wcm^{-2}. The energies of the orbitals are $E_{3\sigma_g} = 1.12657012$ a.u. and $E_{1\pi_g} = 0.26871290$ a.u., respectively, with the internuclear distance taken as $\mathbf{R} = 2.113$ a.u. Panel (a): the coherent sum of all contributions; panel (b) the contributions finishing at the $3\sigma_g$ orbital; panel (c) those finishing at the $1\pi_g$ orbital; panel (d) those starting at the $3\sigma_g$ orbital; panel (e) those starting from the $1\pi_{gx}$ orbital; panel (f): those considering the HOMO only. Source: Figure 1.5 in[21]

In Fig. 1.1(e), where only processes starting at the $1\pi_g$ orbital are considered, we see a spectrum which is very similar to Fig. 1.1(a). Thus, we conclude that in the overall spectrum, ionization from the $1\pi_g$ orbital is dominant, as expected due to the lower ionization potential. The main difference is that at angles $\theta_L = 0°$ and $\theta_L = 90°$, there is a slight distortion when all channels are considered, which is due to the lack of nodal structure in the $3\sigma_g$ orbital. More information on the importance of nodal structures and their effect on HHG spectra can be found in[30, 6, 23, 28].

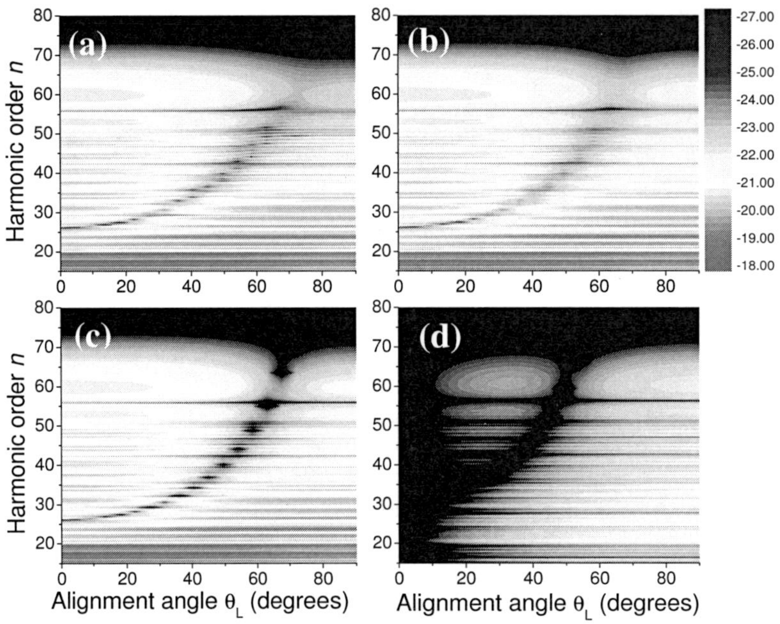

Figure 1.2 Contribution of different processes to the high-harmonic spectra, as functions of the alignment angle θ_L, for an N_2 molecule subject to a linearly polarized laser field of frequency $\omega = 0.057$ a.u. and intensity $I = 4 \times 10^{14}$ Wcm^{-2}. The orbital energies are, $E_{3\sigma_g} = 0.63485797$ a.u and $E_{1\pi_u} = 0.65087981$ a.u., and the internuclear distance is $\mathbf{R} = 2.068$ a.u. Panel (a): coherent sum $|M_{\sigma\sigma} + M_{\pi_x\pi_x}|^2$; panel (b): incoherent sum $|M_{\sigma\sigma}|^2 + |M_{\pi_x\pi_x}|^2$; panel (c): processes involving the $3\sigma_g$ orbital, i.e., $|M_{\sigma\sigma}|^2$; panel (d): processes involving the $1\pi_{ux}$ orbital, i.e., $|M_{\pi_x\pi_x}|^2$.
Source: Figure 1.3 in[21]

The spectrum obtained using the multi-electron wavefunction including the $3\sigma_g$ HOMO and $1\pi_u$ HOMO-1 of N_2 is presented in Fig. 1.2. We compare Fig. 1.2(a) and 1.2(c), which correspond to the two-electron wavefunction, where the HOMO and HOMO-1 are considered, and the one electron case, which considers only the HOMO. We see that the main effect of the HOMO-1 orbital is to blur the interference minimum in the spectrum, with the blurring becoming more pronounced at higher harmonic orders. The lack of influence from the HOMO-1 at lower harmonic orders can be understood by observing Fig. 1.2(d), which shows the harmonic spectrum when considering the HOMO-1 only. We see that, in this frequency region, the intensity of the spectrum is around two orders of magnitude lower than that of the spectrum from the HOMO only and that there is vanishing harmonic signal at $\theta_L = 0°$. The slightly higher ionization potential of the $1\pi_u$ orbital has the effect of reducing the overall ionization probability. However, it is the orbital geometry of the π_u orbital which has the main effect on the harmonic spectrum, as seen for the coherent-superposition example.

The nodal plane along the internuclear axis means that when the molecule is aligned parallel to the laser field, ionization is greatly reduced. In contrast, the $3\sigma_g$ orbital is entirely localized along the internuclear axis. This implies that the overall harmonic yield from the HOMO will decrease with increasing alignment angle.

The influence of the nodal plane in the $1\pi_u$ orbital however will give rise to the opposite behavior. Therefore, in the overall spectrum, displayed in Fig. 1.2(a), the influence of the HOMO-1 would be expected to be more pronounced as the alignment angle increases. Finally, we note the increased substructure in the spectrum of the coherent sum in Fig. 1.2(a) compared to Fig. 1.2(b). Both coherent and incoherent sums, however, are quite similar.

1.2.2 The coupled coherent state approach

The work presented up to this stage has been based upon the strong field approximation (SFA), which is a very powerful approach, but, as we have seen, makes a series of physical approximations that are not always well justified. This leads to some important electron trajectories not being incorporated into SFA calculations[31], causing the SFA to misinterpret certain experimental features where such trajectories are important, such as the so-called "ionization surprise"[32, 33]. Moreover, dynamic multi-electron effects are extremely important in strong-field phenomena[6, 11], whereas the SFA assumes a frozen core.

The coupled coherent state (CCS) approach[34, 35, 17] is a fully quantum, trajectory based method, which fully includes the Coulomb potential, but is numerically not too expensive. Semi-classical approaches, such as the Herman–Kluk propagator[36], have been shown to accurately reproduce the HHG cutoff and plateau, and fully include the Coulomb potential. However, so far, they do not simulate the recollision picture, because ionization is neglected[37, 38].

The CCS method exploits the properties of Gaussian wavepackets, which are guided by classical trajectories governed by Hamilton's equations, as a basis for solving quantum mechanical equations. The initial state may be chosen by randomly selecting trajectories in phase space using Monte–Carlo sampling. This facilitates very favorable scaling with dimensionality compared to solving the TDSE numerically without approximation. Therefore, this scheme is suitable for modeling large systems.

Under these approximations, the wavefunction in the coherent state, $|\,\mathbf{z}\rangle$, basis is written as,

$$\langle \mathbf{z} \,|\, \Psi \rangle = C(\mathbf{z},t)\exp[iS_z], \tag{1.17}$$

where,

$$S_z = \int \left(\frac{i}{2}(\mathbf{z}^* \frac{d\mathbf{z}}{dt} - \frac{d(\mathbf{z})^*}{dt}\mathbf{z}) - H_{\text{ord}}(\mathbf{z},(\mathbf{z})^*) \right) dt, \tag{1.18}$$

and the coefficients $C(\mathbf{z},t)$ can be calculated by solving,

$$\frac{dC(\mathbf{z},t)}{dt} = -i \int \langle \mathbf{z} \,|\, \mathbf{z}' \rangle \delta^2 H_{\text{ord}}(\mathbf{z}^*,\mathbf{z}') C(\mathbf{z}'(t)) \exp[i(S_{z'} - S_z)] \frac{d^2\mathbf{z}'}{\pi} \tag{1.19}$$

where $H_{\text{ord}}(\mathbf{z}^*,\mathbf{z}')$ is the ordered Hamiltonian[17].

Equation (1.19) is the working equation of the CCS method and includes coupling between different coherent states. For coherent states that are close, $\mathbf{z} = \mathbf{z}'$, $\delta^2 H_{\text{ord}}(\mathbf{z}^*,\mathbf{z})$ will vanish[17] whereas for coherent states which are far apart, the overlap $\langle z \mid z' \rangle$ will also cause Eq. (1.19) to vanish which means that the coupling between coherent states is always sparse, making the method numerically less expensive.

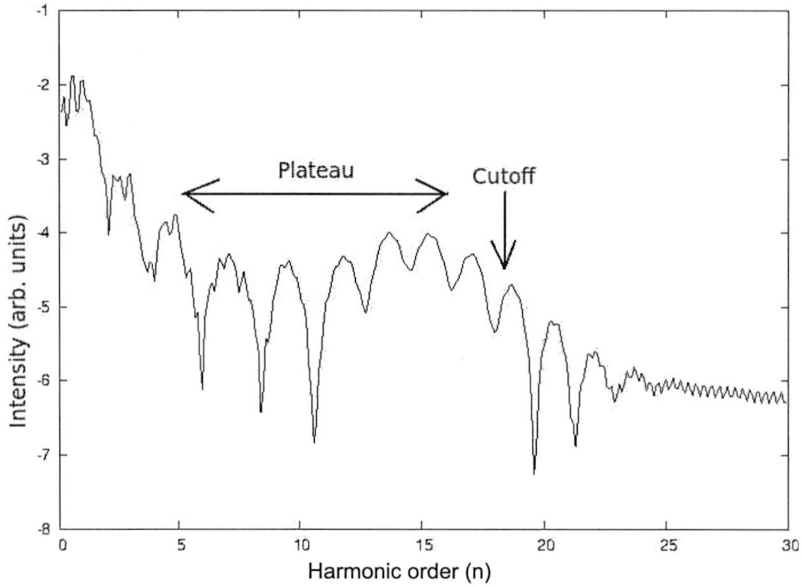

Figure 1.3 Harmonic spectrum from a sample of CCS trajectories distributed at the quiver distance using the acceleration form of the dipole operator with an electric field with a maximum strength of $\varepsilon_0 = 0.5$ a.u. and frequency $\omega = 0.05$ a.u. We consider a trapezium shaped pulse with one cycle ramp up, one cycle ramp down and three cycles of constant intensity

Equation (1.19) can be solved using Monte–Carlo sampling. An accurate choice of initial conditions is extremely important in obtaining reliable converged results. Much semi-classical work has placed the initial wavepacket at the quiver distance, thus neglecting the ionization step of the harmonic generation process[38, 37, 39]. Although in principle tunneling can be modeled by the CCS approach, starting trajectories around the quiver distance, which is the distance an electron travels due to the influence of the laser field in half a field cycle, allows us to study the recombination step in detail. In this instance, we expect the cutoff harmonic to be located at harmonic order $n_{\max} = (E_0 + 2U_p)/\omega$.

We present results for atomic hydrogen, of ionization potential $E_0 = 0.5$ a.u., exposed to a trapezoidal laser field. We use a sample of 3000 trajectories, of which five trajectories are selected due to their interaction with the core. All the other trajectories show little or no Coulombic interaction and will therefore oscillate with the field frequency and not give rise to higher harmonics. The electron is started from an unbound location, situated approximately at the quiver distance in the z-direction, which is the direction of polarization of the laser field. Performing a one-dimensional calculation, we obtain the spectrum in Fig. 1.3.

We see a clear cutoff and plateau, with odd harmonic peaks. There is splitting in the lower order harmonic peaks, which occurs due to interference between many contributing trajectories. Although these results are very promising, numerical instabilities in the method lead to problems as the number of trajectories is increased.

1.3 Excitation in Non-sequential Double Ionization

Non-sequential double ionization is a strong-field phenomenon for which electron-electron correlation is huge. In fact, while the rough features encountered in HHG and ATI spectra, such as the plateau and the cutoff, are well reproduced employing the single-active electron approximation, for double ionization, the predictions of sequential models, in which each electron is removed independently of the other, fail completely in a wide intensity range. Striking examples are (i) a deviation of at least two orders of magnitude from the predictions of sequential models, observed in the total double ionization yield as a function of the driving-field intensity[40, 41], and (ii) electron-momentum distributions peaked at non-vanishing momentum components $p_{n\|}$, $(n = 1, 2)$ parallel to the laser-field polarization, extensively observed for rare gases such as Ar and Ne in the early 2000s[42, 43][2]. Both effects are very well reproduced by the rescattering model, in which NSDI is related to the laser-induced inelastic recollision of an electron with its parent ion.

Several features observed in the NSDI distributions suggest its great potential as an attosecond-imaging tool. For instance, the shapes of the electron momentum distributions are extremely sensitive to the type of interaction by which the electron is dislodged or excited. This has been shown by several theoretical studies, using a wide range of methods such as the strong-field approximation[44, 45, 46, 47, 48, 49], classical-trajectory methods[50] and the fully numerical solution of the time-dependent Schrödinger equation[18, 51]. Such studies have shown that a contact-type interaction favors equal parallel momenta while a Coulomb-type interaction leads to a V-shaped structure. This structure has been identified in both theoretical[44, 46, 47, 50, 52, 48, 53] and experimental[54, 55] investigations. Apart from that, the initial bound states in which the electrons are located are also expected to influence such distributions considerably[12, 13, 47, 49, 56].

Most studies, however, focus on the direct double ionization pathway, in which the first electron, upon rescattering, provides the second electron with enough energy so that it can overcome the second ionization potential and reach the continuum by electron-impact ionization. Both electrons leave the target simultaneously. However, in recent years, time-delayed pathways have become the object of intensive scrutiny, due to several experiments in the so-called "below-threshold regime", in which the driving-field intensity is too low for electron-impact ionization to occur[57, 58]. Furthermore, there is increasing experimental evidence that RESI (recollision-excitation with subsequent ionization) is prominent for molecular targets[13]. For a detailed discussion on this issue, we refer to the recent review[59]. In this context, a very important rescattering mechanism is recollision-excitation with subsequent ionization (RESI). In this mechanism, the first electron is freed by tunneling, propagates in the continuum and is driven by the external field towards its parent ion, with which it recollides. Upon recollision, it gives enough energy to the core so that a second electron is excited. This electron then reaches the continuum at a later time, by tunneling or above-the-barrier ionization. Most theoretical investigations of RESI so far use purely numerical methods (mostly classical-trajectory computations). If, on the one hand, these models have the advantage of making very few physical approximations, on the other hand, they resemble numerical experiments, in the sense that they contain several rescattering mechanisms whose effect may be difficult to disentangle.

---

[2]The experimental electron-momentum distributions exhibited peaks at parallel momentum components $p_{1\|} = p_{2\|} = \pm 2\sqrt{U_p}$, while sequential models predicted distributions centered at $p_{1\|} = p_{2\|} = 0$.

In our studies, we have concentrated on the RESI mechanism from the outset. We have constructed the transition amplitude corresponding to this specific process, within the SFA, which is then solved employing the steepest-descent method.

1.3.1 Transition amplitudes and saddle-point equations

Explicitly, the RESI transition amplitude reads[49]

$$M(\mathbf{p}_1,\mathbf{p}_2) = \int_{-\infty}^{\infty} dt \int_{-\infty}^{t} dt' \int_{-\infty}^{t'} dt'' \int d^3k \times V_{\mathbf{p}_2 e} V_{\mathbf{p}_1 e,\mathbf{k}g} V_{\mathbf{k}g} \exp[iS(\mathbf{p}_1,\mathbf{p}_2,\mathbf{k},t,t',t'')], \tag{1.20}$$

where

$$S(\mathbf{p}_1,\mathbf{p}_2,\mathbf{k},t,t',t'') = E_1^{(g)}t'' + E_2^{(g)}t' + E_2^{(e)}(t-t') - \int_{t''}^{t'} \frac{[\mathbf{k}+\mathbf{A}(\tau)]^2}{2} d\tau - \int_{t'}^{\infty} \frac{[\mathbf{p}_1+\mathbf{A}(\tau)]^2}{2} d\tau - \int_{t}^{\infty} \frac{[\mathbf{p}_2+\mathbf{A}(\tau)]^2}{2} d\tau \tag{1.21}$$

denotes the semiclassical action and

$$V_{\mathbf{kg}} = \langle \tilde{\mathbf{k}}(t'') | V | \psi_1^{(g)} \rangle = \frac{1}{(2\pi)^{3/2}} \int d^3r_1 V(\mathbf{r}_1) e^{-i\tilde{\mathbf{k}}(t'')\cdot \mathbf{r}_1} \psi_1^{(g)}(\mathbf{r}_1), \tag{1.22}$$

$$V_{\mathbf{p}_1 e,\mathbf{kg}} = \langle \tilde{\mathbf{p}}_1(t'), \psi_2^{(e)} | V_{12} | \tilde{\mathbf{k}}(t'), \psi_2^{(g)} \rangle = \frac{1}{(2\pi)^3} \int\int d^3r_2 d^3r_1 \exp[-i(\mathbf{p}_1-\mathbf{k})\cdot \mathbf{r}_1] \times V_{12}(\mathbf{r}_1,\mathbf{r}_2)[\psi_2^{(e)}(\mathbf{r}_2)]^* \psi_2^{(g)}(\mathbf{r}_2), \tag{1.23}$$

and

$$V_{\mathbf{p}_2 e} = \langle \tilde{\mathbf{p}}_2(t) | V_{\text{ion}} | \psi_2^{(e)} \rangle = \frac{1}{(2\pi)^{3/2}} \int d^3r_2 V_{\text{ion}}(\mathbf{r}_2) e^{-i\,\tilde{\mathbf{p}}_2(t)\cdot \mathbf{r}_2} \psi_2^{(e)}(\mathbf{r}_2) \tag{1.24}$$

are prefactors related to the ionization of the first electron, the recollision of the first electron with excitation of the second electron, and tunnel ionization of the second electron [Eqs (1.22), (1.23) and (1.24), respectively]. Within the SFA, they contain all information on the initial bound states of the first electron, and of the second electron, prior and subsequent to excitation, and on the type of interaction by which the second electron was excited.

Physically, Eq. (1.20) describes a process in which the first electron tunnel ionizes at time t'' from its initial bound-state $|\psi_1^{(g)}\rangle$ with energy $E_1^{(g)}$ into a Volkov state $|\tilde{\mathbf{k}}(t'')\rangle$. Subsequently, from the time t'' to the time t', it propagates in the continuum with intermediate momentum $\mathbf{k}$ under the sole influence of the laser field. At time t', it rescatters inelastically with the core and, through the interaction V_{12}, excites the second electron from the ground state $|\psi_2^{(g)}\rangle$ of the singly ionized species, whose energy is $E_2^{(g)}$, to the state $|\psi_2^{(e)}\rangle$, whose energy is $E_2^{(e)}$. Immediately after rescattering, the first electron reaches the detector with final momentum $\mathbf{p}_1$, while the second electron remains bound until a subsequent time t. At this time, the second electron tunnels from $|\psi_2^{(e)}\rangle$ into a Volkov state $|\tilde{\mathbf{p}}_2(t)\rangle$, and reaches the detector with final momentum $\mathbf{p}_2$. Here, $\tilde{\mathbf{k}}(\tau) = \mathbf{k} + \mathbf{A}(\tau)$ and $\tilde{\mathbf{p}}_n(\tau) = \mathbf{p}_n + \mathbf{A}(\tau)$ $(\tau = t, t', t'')$ in the length gauge, and $\tilde{\mathbf{k}}(\tau) = \mathbf{k}$ and $\tilde{\mathbf{p}}_n(\tau) = \mathbf{p}_n$ in the velocity gauge.

The above-stated transition amplitude is solved by means of steepest descent methods. This implies that we seek values of t'', $\mathbf{k}$, t' and t so that the action shown in Eq. (1.21) is stationary, i.e., so that its partial derivatives with respect to these variables vanish. This gives the saddle-point equations

$$[\mathbf{k} + \mathbf{A}(t'')]^2 = -2E_1^{(g)}, \tag{1.25}$$

$$\mathbf{k} = -\frac{1}{t' - t''} \int_{t''}^{t'} d\tau \mathbf{A}(\tau), \tag{1.26}$$

$$\left[\mathbf{p}_1 + \mathbf{A}(t')^2\right] = [\mathbf{k} + \mathbf{A}(t')]^2 - 2(E_2^{(g)} - E_2^{(e)}), \tag{1.27}$$

and

$$[\mathbf{p}_2 + \mathbf{A}(t)]^2 = -2E_2^{(e)}, \tag{1.28}$$

respectively.

Similar to the saddle-point equations obtained for HHG, Eqs (1.25) and (1.28) give the conservation of energy for the first and second electron, respectively, upon tunneling ionization, and Eq. (1.26) gives the return condition for the first electron. Equation (1.27) gives the energy conservation upon excitation of the first electron. Instead of being fully converted into high-frequency light, as in HHG, now the kinetic energy that the first electron acquired from the field is partly given to the core, in an inelastic collision. Since these two equations are coupled with Eq. (1.25), which has no real solution, the rescattering time t' and the intermediate momentum $\mathbf{k}$ will also be complex.

One should note that the saddle-point equations related to the first and the second electron are decoupled. This is a consequence of the fact that Eq. (1.21) is factorizable into $S(\mathbf{p}_1, \mathbf{p}_2, \mathbf{k}, t, t', t'') = S_1(\mathbf{p}_1, \mathbf{k}, t', t'') + S_2(\mathbf{p}_2, t)$. This does not mean, however, that the transition amplitude (Eq. (1.20)) may be written as the product of one-electron transition amplitudes. In fact, correlation comes via time ordering, i.e., the second electron cannot tunnel ionize from an excited state before the first electron recollides with the core. Hence, causality must be considered when computing RESI electron momentum distributions[60].

1.3.2 Momentum–space constraints

An important issue is what regions in the $p_{1\|}p_{2\|}$ plane will be populated by the RESI mechanism. Traditionally, RESI has been associated with the probability density observed in the second and fourth quadrant of the parallel-momentum plane. This interpretation has been supported by classical arguments, which state that, due to the time delay between rescattering of the first electron and ionization of the second electron, the momenta of the first and the second electron should have opposite directions[61, 62]. These constraints, however, are mostly related to another mechanism, namely electron-impact ionization with a time delay[59]. The saddle-point equations (1.25)–(1.27), together with arguments based on the above-threshold ionization, lead to a markedly different set of constraints for RESI. These constraints have been explained in detail in[59, 63], and will be briefly recalled here.

If rewritten in terms of the electron-momentum components—parallel and perpendicular to the laser-field polarization—and after a few manipulations, Eq. (1.27) states that the final momentum component $p_{1\|}$ must satisfy the condition

$$-A(t') - \sqrt{2\triangle E} \leq p_{1\|} \leq -A(t') + \sqrt{2\triangle E}, \tag{1.29}$$

where $\triangle E = E_{\text{kin}}(t',t'') - (E_2^{(g)} - E_2^{(e)} + p_{1\perp}^2/2)$ is the energy difference between the kinetic energy $E_{\text{kin}}(t',t'')$ of the first electron upon return and an effective excitation energy, which increases with perpendicular momentum $p_{1\perp}$. The above-stated equation delimits a region in the $p_{1\|}p_{2\|}$ plane centered at $p_{1\|} = -A(t')$, which is the most probable momentum the first electron may have upon return. Its extension depends on the driving-field intensity, the energy difference between the ground and the excited bound states for the second electron, and the perpendicular momentum $p_{1\perp}$. If the maximal kinetic energy of the electron upon return is larger than the effective excitation energy, $\triangle E > 0$ and Eq. (1.29) delimits a region in the parallel momentum plane within which rescattering has a classical counterpart. For monochromatic fields, $A(t') = \pm 2\sqrt{U_p}$ and the maximal kinetic energy of the first electron upon return is $E_{\text{kin}}^{(\text{max})}(t',t'') = 3.17U_p$, while for a few-cycle pulse, this value may vary.

Similarly, the momentum constraints upon the second electron can be determined from Eq. (1.28). This gives

$$[p_{2\|} + A(t)]^2 = -2E_2^{(e)} - p_{2\perp}^2. \tag{1.30}$$

The above-stated equation delimits a region centered at $p_{2\|} = -A(t)$ whose width, in terms of U_p, will not vary with the driving-field intensity. In fact, because the saddle-point of Eq. (1.30) is formally equivalent to that describing direct ATI, the maximal kinetic energy of the second electron will be dictated by the direct ATI cutoff. For a monochromatic field, this cutoff is at $2U_p$ so that $|p_{2\|} + A(t)| \leq 2\sqrt{U_p}$. Note that according to Eq. (1.30), a non-vanishing transverse momentum $p_{2\perp}$ effectively widens the potential barrier through which the second electron must tunnel. This will lead to a suppression in the yield. Furthermore, since the barrier is the narrowest when the field is at its maximum (i.e., when $A(t) \simeq 0$), the distributions, as functions of $p_{2\|}$, will be peaked at $p_{2\|} = 0$.

Figure 1.4 shows a schematic representation of the above-mentioned constraints for three intensities of a monochromatic driving field employing the information obtained above.

As we are interested in an upper bound for the momentum regions to be populated, we considered $p_{n\perp} = 0$. We have also considered symmetrization with regard to electron exchange, as both the first and the second electron are indistinguishable. As an overall feature, the distributions exhibit maxima at $(p_{j||}, p_{\nu||}) = (\pm 2\sqrt{U_p}, 0)$, with $j \neq \nu$, $j = 1,2$ and $\nu = 1,2$. As the driving-field intensity increases, the region delimited by the constraints (1.29) becomes more extensive, while the region delimited by Eq. (1.30) will always be determined by the direct ATI cutoff. Consequently, the electron-momentum distributions will become increasingly more elongated, shifting from ring- to cross-shaped, as shown in Fig. 1.4 panels (a) to (c).

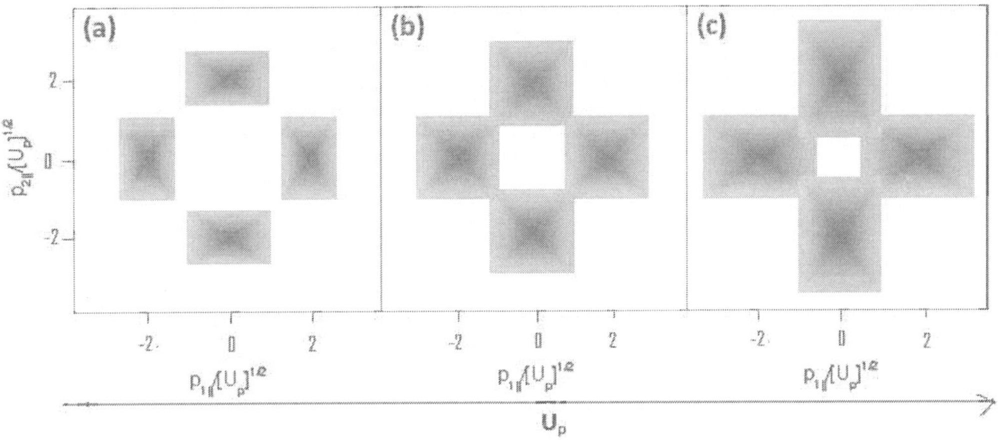

Figure 1.4 Schematic representation of the regions of the parallel momentum plane populated by the recollision–excitation–tunneling –ionization mechanism, highlighted as the rectangles in the figure. As the driving-field intensity increases, the momentum distributions change from ring-shaped to cross-shaped [panels (a) to (c), respectively]. Different sets of trajectories, whose start and recollision times are separated by half a cycle of the field, and symmetrization $\mathbf{p}_1 \leftrightarrow \mathbf{p}_2$ with respect to the indistinguishability of the two electrons were considered. Vanishing transverse momenta were considered., so that the constraints provided constitute an upper bound for this region

This estimate is in excellent agreement with the electron-momentum distributions obtained for a monochromatic field if the prefactors $V_{\mathbf{p}_1 e,\mathbf{k}g}$ and $V_{\mathbf{p}_2 e}$ are kept constant and if the transverse momentum components are integrated over. An example is shown in Fig. 1.5, for helium. The three driving-field intensities in the figure have been chosen such that (i) $E_2^{(g)} - E_2^{(e)} \simeq 3.17U_p$, i.e., the maximum kinetic energy of the returning electron is just enough to promote the second electron to an excited state (panel (a)); (ii) $E_2^{(g)} - E_2^{(e)} < 3.17U_p$, but $E_2^{(g)} > 3.17U_p$, i.e., even if RESI is allowed to occur, the driving-field intensity is not enough to release the second electron by electron-impact ionization (panel (b)); (iii) $E_2^{(g)} - E_2^{(e)} \ll 3.17U_p$ and $E_2^{(g)} \lesssim 3.17U_p$, i.e., the intensity chosen is far above the threshold for RESI and slightly above the threshold for electron-impact ionization (panel (c)). The shapes of the distributions mirror the above-mentioned constraints, and move from four well-localized maxima at $(p_{1||}, p_{2||}) = (\pm 2\sqrt{U_p}, 0)$ and $(p_{1||}, p_{2||}) = (0, \pm 2\sqrt{U_p})$ to crosses whose axis lie at $p_{n||} = 0$, $(n = 1,2)$.

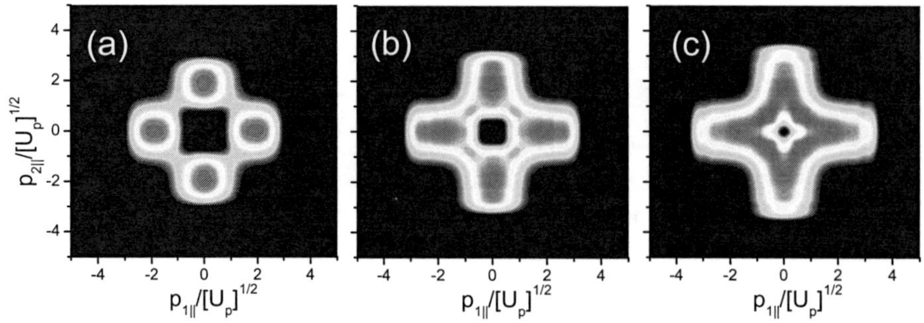

Figure 1.5 Electron-momentum distributions as functions of the electron momentum components parallel to the laser field polarization, for helium in a linearly polarized, monochromatic field of frequency $\omega = 0.057$ a.u. and intensities $I = 2.16 \times 10^{14}$ W/cm^2, $I = 2.5 \times 10^{14}$ W/cm^2 and $I = 3 \times 10^{14}$ W/cm^2 (panels (a), (b) and (c), respectively). The transverse momentum components have been integrated over and the contour plots have been normalized to the maximum probability in each panel.
Source: Figure 1.2 in[49]

1.3.3 Bound-state influence in atoms and molecules

Apart from the momentum–space constraints, the shapes of the RESI electron momentum distributions are strongly influenced by the prefactors (1.23) and (1.24). They will introduce an additional momentum bias in the electron momentum distributions, which will be target-specific. In order to facilitate the comparison with existing experiments, the transverse momentum components have been integrated over in all distributions that follow.

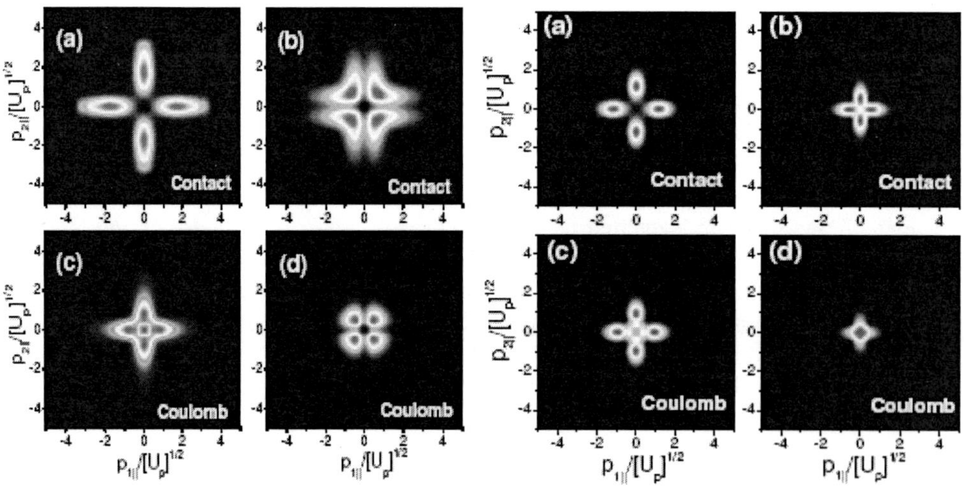

Figure 1.6 The left and the right panels in the figure exhibit electron-momentum distributions as functions of the parallel momentum components $p_{1\|}$, $p_{2\|}$ for helium ($E_1^{(g)} = 0.97$ a.u., $E_2^{(2g)} = 2$ a.u. and $E_e^{(2)} = 0.5$ a.u.) and argon ($E_1^{(g)} = 0.58$ a.u., $E_2^{(g)} = 1.02$ a.u. and $E_2^{(e)} = 0.40$ a.u.) in a linearly polarized monochromatic field of frequency $\omega = 0.057$ a.u. For helium, we considered the laser field intensity to be $I = 3 \times 10^{14}$ W/cm^2, while, for Argon, we chose $I = 9 \times 10^{13}$ W/cm^2 and $I = 1.5 \times 10^{14}$ W/cm^2 in the upper (panels (a) and (c)) and lower (panels (b) and (d)) panels, respectively. The type of interaction V_{12} taken is indicated in the figure. The contour plots have been normalized to the maximum probability in each panel.
Source: The left and right panels are Figs 4 and 7 in[49], respectively

Figure 1.6 illustrates this issue, for helium and argon (left and right panels, respectively). The explicit expressions for these prefactors are provided in[49]. For helium, if we assume that the electron is excited from the $1s$ to the $2s$ state (panels (a) and (c) on the left-hand side, respectively), the distributions will be localized along the axes $p_{n\|} = 0$. This holds regardless of whether excitation occurs via a Coulomb or contact interaction ($V_{12}(\mathbf{r}_2, \mathbf{r}_1) =$ or $V_{12}(\mathbf{r}_2, \mathbf{r}_1) = \delta(\mathbf{r}_2 - \mathbf{r}_1)$, respectively). In fact, the main effect of the Coulomb-type interaction is to enhance the yield in the low-momentum regions. However, it does not alter the shape of the distributions considerably. Similar cross-shaped distributions are observed for argon, for which we consider a $3p$ to $4s$ excitation, regardless of the driving-field intensity or of the electron–electron interaction, as shown in the panels (a) to (d) on the right-hand side. These examples are evidence that tunnel ionization from an s state favors events close to the axes $p_{n\|} = 0$. Physically, this can be understood in two ways. First, an s state is spatially isotropic. This implies that the distributions obtained if the second electron is excited to an s state are not expected to be significantly different from those determined by the momentum constraints alone, i.e., when the prefactors are constant. Any differences will come from the spatial range of such prefactors, or from the presence of radial nodes. Second, the spatial probability density related to an s wavefunction is non-vanishing at the region occupied by the core, i.e., near $\mathbf{r} = 0$. This means that the corresponding prefactors, which according to Eqs (1.23) and (1.24) depend on this wavefunction, will be non-vanishing in the region around $\mathbf{p}_n = 0$ as well.

In contrast, the shapes of the distributions are markedly different if one assumes that the second electron is excited to a p state, such as for the helium electron-momentum distributions in which excitation from $1s$ to $2p$ has been considered (panels (b) and (d) on the left-hand side). In this case, there is a strong suppression along the axes, and the distributions are peaked at $p_{1\|} = \pm p_{2\|}$. This suppression is caused by the fact that, spatially, the probability density associated with a p wavefunctions disappears at $\mathbf{r} = 0$. It will lead to vanishing prefactors around $\mathbf{p}_n = 0$. This suppression at the axes is also present for diatomic molecules, if the state to which the second electron is excited shares some of the p characteristics, such as vanishing probability density at $\mathbf{r} = 0$. In Fig. 1.7, we display the outcome of a RESI computation for N_2, performed in our recent publication[56], for three alignment angles between the molecular axis and the laser-field polarization. In this computation, we assume that the first electron tunnels out from the HOMO, which is a $3\sigma_g$ orbital, and, upon return, recollides with the singly ionized molecule N_2^+. Upon this recollision, a second electron is excited from the HOMO to the LUMO, which is a $1\pi_g$ orbital, as seen in Section 1.2.1. In position space, a $1\pi_g$ orbital has two nodal planes, along and perpendicular to the internuclear axis, and vanishing probability at the geometric center of the molecule. For all alignment angles, the shapes of the electron-momentum distributions are very similar to those obtained for helium if the $1s \to 2p$ excitation is considered. Finally, apart from the features discussed here, there is also two-center interference. The maxima and minima caused by this interference, however, will only be important for small alignment angles, at least if the transverse momentum components are integrated over. A comprehensive discussion of this interference, the explicit expressions for the molecular prefactors and specific details on our molecular RESI model are provided in[56].

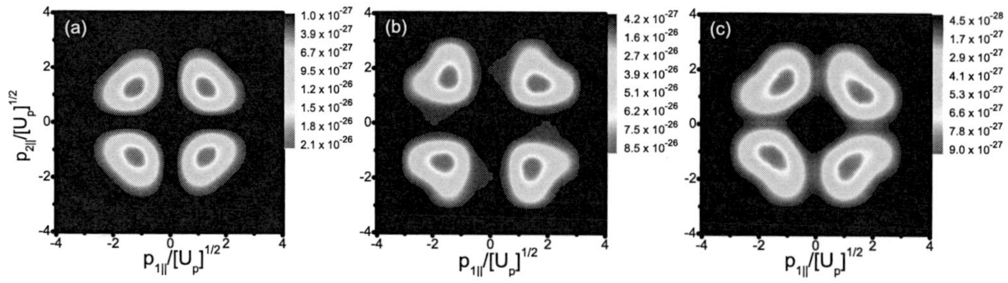

Figure 1.7 (Color online) RESI electron-momentum distributions as functions of the parallel momentum components $p_{1\|}, p_{2\|}$ for N_2 (bound-state energies $E_1^{(g)} = 0.63486$ a.u., $E_2^{(g)} = 1.12657$ a.u., and $E_2^{(e)} = 0.26871290$ a.u. and equilibrium internuclear distance $R = 2.11$ a.u.) in a linearly polarized field monochromatic field of intensity $I = 1.25 \times 10^{14}$ W/cm^2 and frequency $\omega = 0.057$ a.u. A contact-type interaction has been assumed throughout. Panel (a), (b) and (c) correspond to the alignment angles $\theta_L = 0, 45$ and 90 degrees, respectively. Note that the false color scales are different for each panel.
Source: Figure 1.6 in[56]

1.4 Summary and Conclusions

The studies reported here lead us to conclude that the dynamics of the core are very important for the attosecond imaging of matter. This has been evident both in our HHG and NSDI investigations. The outcome of our HHG studies, in which multiple molecular orbitals and electrons have been incorporated, but in which the dynamics of the continuum and bound electrons have been decoupled, is very similar to that obtained if a single-active electron, single-active orbital approximation is used. In the specific example of N_2, the presence of the HOMO-1 in such a model causes at most a blurring in the two-center minimum, which is determined by the HOMO. This is in agreement with further investigation for which multi-electron effects have been incorporated using many-body perturbation theory around the HOMO[22]. These multi-electron corrections were very small, except near the two-center minima. On the other hand, our one-electron model in which a coherent superposition of orbitals has been included shows that the processes in which the active electron tunnels from one orbital and recombines with the other leave a significant imprint in the spectra. In the model considered, this could be taken into consideration as the HOMO and LUMO were not completely filled. In a multi-electron model incorporating the HOMO and lower-lying orbitals, the energy shells are filled and vacancies are created when an electron ionizes. Hence, if the electron leaves from a specific orbital, unless relaxation, excitation or hole migration occurs, Pauli exclusion principle dictates that it can only return to the orbital from where it was freed. In conclusion, a huge effect in the spectra can only be expected if one allows for the core dynamics, i.e., if multiple channels are allowed to occur. These conclusions are strenghtened by recent results in HHG of molecules[6].

It is, however, not trivial to incorporate such dynamics. The model used in[6] is based on the key assumptions that there are spatial regions in which either the core or the laser field are dominant. These assumptions allow one to consider the field perturbatively for the multi-electron core, and the binding potential as a perturbation around the SFA when the laser field is dominant. Such a procedure is very effective in the description of small molecules, for which such a distinction can be made. For large molecules, however, they are expected to break down, or at least to be of limited applicability. For that reason, it is desirable to employ an orbit based method such as the CCS method, in which a clear distinction between "bound" and "continuum" is not necessary. The CCS method, however, poses several challenges. First, since the CCS method employs a coherent-state basis, which can be associated to Gaussians in phase space, it is expected to work well if the binding potential of the system in question resembles that of a harmonic oscillator. A Coulomb potential, however, is highly anharmonic. In the presence of the external laser field, this leads to large number of irregular trajectories, which, depending on the situation, may cause numerical instabilities or noise in the spectra. This is particularly critical near the core, and makes the treatment of tunneling ionization rather difficult. Furthermore, HHG is a highly coherent process, so that the phase differences between the trajectories included in the computation play a major role. Hence, the effect of numerical noise tends to be amplified in the spectra. It may require a large degree of ensemble averaging in order to make the spectra robust with regard to the initial sampling. This will be discussed in detail in a future publication[64]. Nevertheless, the present results show that high-harmonic spectra can be obtained from a CCS computation.

The importance of the core dynamics, in particular, excitation and electron-electron correlation, can also be seen in NSDI, if the recollision-excitation with subsequent ionization (RESI) mechanism is considered. We have provided a few selected examples that showed how the RESI NSDI electron-momentum distributions depend on the bound-state to which the second electron is excited. This illustrates the potential of RESI for attosecond imaging, especially in view of several experimental findings. For instance, in the experimental results reported in[57, 58] for argon below the threshold, the shapes of the electron-momentum distributions indicate that excitation occurred from an s to a p state. This can be inferred from the suppression of the probability densities at the axes $p_{n\|} = 0$ and the peaks along $p_{1\|} = \pm p_{2\|}$. If, on the other hand, much shorter pulses are taken, such as in the recent experiments in[65] for RESI with few-cycle pulses, the cross-shaped distributions obtained are a strong evidence that excitation to an s state has occurred. Cross-shaped distributions have also been reported in[53], in comparison to existing experiments[54][3]. This is evidence that the driving pulse shape plays an important role in determining to which bound-state the electron is most probably excited, and opens a wide range of possibilities for attosecond imaging of correlated electron emission.

Bibliography

1. M. Lein. 2007. 'Molecular imaging using recolliding electrons.' *J. Phys. B.* 40: R135.
2. F. Krausz and M. Yu. Ivanov. 2009. 'Attosecond physics.' *Rev. Mod. Phys.* 81: 163.
3. M. F. Kling and M. J. J. Vrakking. 2008. 'Attosecond electron dynamics.' *Annu. Rev. Phys. Chem.* 59: 463.
4. S. Haessler, J. Caillat and P. Salières. 2011. 'Self-probing of molecules with high harmonic generation.' *J. Phys. B.* 44: 203001.
5. J. Itatani, J. Levesque, D. Zeidler, H. Niikura, H. Pépin, J. C. Keiffer, P. B. Corkum and D. M. Villeneuve. 2004. 'Tomographic imaging of molecular orbitals.' *Nature* 432: 867.
6. O. Smirnova, Y. Mairesse, S. Patchkovskii, N. Dudovich, D. M. Villeneuve, P. B. Corkum and M. Yu. Ivanov. 2009. 'High harmonic interferometry of multi-electron dynamics in molecules.' *Nature* 460: 972.

[3]The authors of[53] attribute this shape, however, to a physical mechansim distinct from RESI, in disagreement with the constraints in Section 1.3.2.

7. Y. Huismans, A. Rouzée, A. Gijsbertsen, J. H. Jungmann, A. S. Smolkowska, P. S. W. M. Logman, F. Lépine, C. Cauchy, S. Zamith, T. Marchenko, J. M. Bakker, G. Berden, B. Redlich, A. F. G. van der Meer, H. G. Muller, W. Vermin, K. J. Schafer, M. Spanner, M. Yu. Ivanov, O. Smirnova, D. Bauer, S. V. Popruzhenko and M. J. J. Vrakking. 2011. 'Time-resolved holography with photoelectrons.' *Science* 331: 61.

8. M. F. Kling, J. Rauschenberger, A. J. Verhoef, E. Hasović, T. Uphues, D. B. Milošević, H. G. Muller and M. J. J. Vrakking. 2008. 'Imaging of carrier-envelope phase effects in above-threshold ionization with intense few-cycle laser fields.' *New J. Phys.* 10: 025024.

9. P. B. Corkum. 1993. 'Plasma perspective on strong field multiphoton ionization.' *Phys. Rev. Lett.* 71: 1994.

10. S. Sukiasyan, C. McDonald, C. Destefani, M. Yu. Ivanov and T. Brabec. 2009. 'Multi-electron correlation in high-harmonic generation: A 2D model analysis.' *Phys. Rev. Lett.* 102: 223002.

11. S. Sukiasyan, S. Patchkovskii, O. Smirnova, T. Brabec and M. Yu. Ivanov. 2010. 'Exchange and polarization effect in high-order harmonic imaging of molecular structures.' *Phys. Rev. A.* 82: 043414.

12. E. Eremina, X. Liu, H. Rottke, M. G. Schätzel, A. Dreischuch, G. G. Paulus, H. Walther, R. Moshammer and J. Ullrich. 2004. 'Influence of molecular structure on double ionization of N_2 and O_2 by high intensity ultrashort laser pulses.' *Phys. Rev. Lett.* 92: 173001.

13. D. Zeidler, A. Staudte, A. B. Bardon, D. M. Villeneuve, R. Dörner and P. B. Corkum. 2005. 'Controlling attosecond double ionization dynamics via molecular alignment.' *Phys. Rev. Lett.* 95: 203003.

14. N. Camus, B. Fischer, M. Kremer, V. Sharma, A. Rudenko, B. Bergues, M. Kübel, N. G. Johnson, M. F. Kling, T. Pfeifer, J. Ullrich and R. Moshammer. 2012. 'Attosecond correlated dynamics of two electrons passing through a transition state.' *Phys. Rev. Lett.* 108: 073003.

15. B. B. Augstein and C. Figueira de Morisson Faria. 2012. 'High-order harmonic generation in diatomic molecules: Quantum interference, nodal structures and multiple orbitals.' *Mod. Phys. Lett. B.* 26: 1130002.

16. P. Salières, B. Carré, L. Le Déroff, F. Grasbon, G. G. Paulus, H. Walther, R. Kopold, W. Becker, D. B. Milošević, A. Sanpera and M. Lewenstein. 2001. 'Feynman's path-integral approach for intense-laser-atom interactions.' *Science* 292: 902.

17. D. V. Shalashilin and M. S. Child. 2004. 'The phase space CCS approach to quantum and semi-classical molecular dynamics for high-dimensional problems.' *J. Chem. Phys.* 304: 103.

18. J. S. Parker, B. J. S. Doherty, K. T. Taylor, K. D. Schultz, C. I. Blaga and L. F. DiMauro. 2006. 'High-energy cutoff in the spectrum of strong-field non-sequential double ionization.' *Phys. Rev. Lett.* 96: 133001.

19. M. Lewenstein, Ph. Balcou, M. Yu. Ivanov, A. L. Huillier and P. B. Corkum. 1994. 'Theory of high-harmonic generation by low-frequency laser fields.' *Phys. Rev. A.* 49: 2117.

20. W. Becker, A. Lohr, M. Kleber and M. Lewenstein. 1997. 'A unified theory of high-harmonic generation: Application to polarization properties of the harmonics.' *Phys. Rev. A.* 56: 645.

21. C. Figueira de Morisson Faria and B. B. Augstein. 2010. 'Molecular high-order harmonic generation with more than one active electron: Quantum-interference effects.' *Phys. Rev. A*. 81: 043409.

22. B. B. Augstein and C. Figueira de Morisson Faria. 2011. 'Multi-electron corrections in molecular high harmonic generation for different formulations of the strong-field approximation.' *J. Mod. Opt*. 58: 1173.

23. S. Odžak and D. B. Milošević. 2009. 'Interference effects in high-order harmonic generation by homonuclear diatomic molecules.' *Phys. Rev. A*. 79: 023414.

24. M. F. Guest, I. J. Bush, H. J. J. van Dam, P. Sherwood, J. M. H. Thomas, J. H. van Lenthe, R. W. A. Havenith and J. Kendrick. 2005. 'The GAMESS-UK electronic structure package: Algorithms, developments and applications.' *Mol. Phys*. 103: 719.

25. R. Kopold, W. Becker and M. Kleber. 1998. 'Model calculations of high-harmonic generation in molecular ions.' *Phys. Rev. A*. 58: 4022.

26. M. Lein, N. Hay, R. Velotta, J. P. Marangos and P. L. Knight. 2002. 'Role of intramolecular phase in high-harmonic generation.' *Phys. Rev. Lett*. 88: 183903.

27. N. Hay, R. Velotta, M. Lein, R. de Nalda, E. Heesel, M. Castillejo and J. P. Marangos. 2002. 'High-order harmonic generation in laser-aligned molecules.' *Phys. Rev. A*. 65: 053805.

28. B. B. Augstein and C. Figueira de Morisson Faria. 2011. 'Influence of asymmetry and nodal structures on high-order harmonic generation in heteronuclear molecules.' *J. Phys. B*. 44: 055601.

29. C. B. Madsen and L. B. Madsen. 2007. 'Theoretical studies of high-order harmonic generation: Effects of symmetry, degeneracy and orientation.' *Phys. Rev. A*. 76: 043419.

30. B. K. McFarland, J. P. Farrell, P. H. Bucksbaum and M. Gühr. 'High harmonic generation from multiple orbitals in N2.' *Science* 322: 1232.

31. Tian-Min Yan, S. V. Popruzhenko, M. J. J. Vrakking and D. Bauer. 2010. 'Low-energy structures in strong field ionization revealed by quantum orbits.' *Phys. Rev. Lett*. 105: 253002.

32. C. I. Blaga, F. Catoire, P. Colosimo, G. G. Paulus, H. G. Muller, P. Agostini and L. F. DiMauro. 2009. 'Strong field photoionization revisited.' *Nature Physics* 5: 335.

33. W. Quan, Z. Lin, M. Wu, H. Kang, H. Liu, J. Chen, J. Liu, X. T. He, S. G. Chen, H. Xiong, L. Guo, H. Xu, Y. Fu, Y. Cheng and Z. Z. Xu. 2009. 'Classical aspects in above-threshold ionization with a mid-infrared strong laser field.' *Phys. Rev. Lett*. 103: 093001.

34. D. V. Shalashilin and M. S. Child. 2000. 'Time dependent quantum propagation in phase space.' *J. Chem. Phys*. 113: 10028.

35. D. V. Shalashilin and M. S. Child. 2001. 'Description of tunneling with the help of coupled frozen Gaussians.' *J. Chem. Phys*. 114: 9296.

36. M. F. Herman and E. Kluk. 1984. 'A semiclassical justification for the use of non-spreading wavepackets in dynamics calculations.' *Chem. Phys*. 91: 27.

37. G. van de Sand and J. M. Rost. 1999. 'Irregular orbits generate higher harmonics.' *Phys. Rev. Lett*. 83: 524.

38. M. Protopapas, D. G. Lappas, C. H. Keitel and P. L. Knight. 1996. 'Recollisions, bremsstrahlung, and attosecond pulses from intense laser fields.' *Phys. Rev. A*. 53: R2933.

39. G. van de Sand and J. M. Rost. 2000. 'Semi-classical description of multiphoton processes.' *Phys. Rev. A*. 62: 053403.
40. G. Petite, P. Agostini and H. G. Muller. 1988. 'Intensity dependence of non-perturbative above-threshold ionization spectra: Experimental study.' *J. Phys. B*. 21: 4097.
41. B. Walker, B. Sheehy, L. F. DiMauro, P. Agostini, K. J. Schafer and K. C. Kulander. 1994. 'Precision measurement of strong field double ionization of helium.' *Phys. Rev. Lett*. 73: 1227.
42. Th. Weber, M. Weckenbrock, A. Staudte, L. Spielberger, O. Jagutzki, V. Mergel, F. Afaneh, G. Urbasch, M. Vollmer, H. Giessen and R. Dörner. 2000. 'Momentum distributions of Nen+ ions created by an intense ultrashort laser pulse.' *Phys. Rev. Lett*. 84: 443.
43. R. Moshammer, B. Feuerstein, W. Schmitt, A. Dorn, C. D. Schröter, J. Ullrich, H. Rottke, C. Trump, M. Wittman, G. Korn, K. Hoffmann and W. Sandner. 2000. *Phys. Rev. Lett*. 84: 447.
44. S. P. Goreslavskii, S. V. Propuzhenko, R. Kopold and W. Becker. 2001. 'Electron-electron correlation in laser-induced non-sequential double ionization.' *Phys. Rev. A*. 64: 053402.
45. A. Becker and F. H. M. Faisal. 2002. 'S-matrix analysis of coincident measurement of two-electron energy distribution for double ionization of He in an intense laser field.' *Phys. Rev. Lett*. 89: 193003.
46. C. Figueira de Morisson Faria, H. Schomerus and X. Liu and W. Becker. 2004. 'Electron-electron dynamics in laser-induced non-sequential double ionization.' *Phys. Rev. A*. 69: 043405.
47. C. Figueira de Morisson Faria and M. Lewenstein. 2005. 'Bound-state corrections in laser-induced non-sequential double ionization.' *J. Phys. B*. 38: 3251.
48. D. I. Bondar, W. K. Liu and M. Yu. Ivanov. 2009. 'Two-electron ionization in strong laser fields below intensity threshold: Signatures of attosecond timing in correlated spectra.' *Phys. Rev. A*. 79: 023417.
49. T. Shaaran, M. T. Nygren and C. Figueira de Morisson Faria. 2010. 'Laser-induced non-sequential double ionization at and above recollision-excitation-tunneling threshold.' *Phys. Rev. A*. 81: 063413.
50. D. F. Ye, X. Liu and J. Liu. 2008. 'Classical trajectory diagnosis of a fingerlike pattern in the correlated electron momentum distribution in strong field double ionization of helium.' *Phys. Rev. Lett*. 101: 233003.
51. S. Baier, C. Ruiz, L. Plaja and A. Becker. 2006. 'Nonsequential double ionization of the hydrogen molecule in a few-cycle laser pulse.' *Phys. Rev. A*. 74: 033405.
52. A. Emmanouilidou. 2008. 'Recoil collisions as a portal to field-assisted ionization at near-uv frequencies in the strong-field double ionization of helium.' *Phys. Rev. A*. 78: 023411.
53. Zhangjin Chen, Yaqiu Liang and C. D. Lin. 2010. 'Quantitative rescattering theory of correlated two-electron momentum spectra for strong-field non-sequential double ionization of helium.' *Phys. Rev. A*. 82: 063417.
54. A. Staudte, C. Ruiz, M. Schöffler, S. Schllösser, D. Zeidler, Th. Weber, M. Meckel, D. M. Villeneuve, P. B. Corkum, A. Becker and R. Dörner. 2007. 'Binary and recoil collisions in strong field double ionization of helium.' *Phys. Rev. Lett*. 99: 263002.

55. A. Rudenko, V. L. B. de Jesus, Th. Ergler, K. Zrost, B. Feuerstein, C. D. Schröter, R. Moshammer and J. Ullrich. 2007. 'Correlated two-electron momentum spectra for strong-field non-sequential double ionization of He at 800 nm.' *Phys. Rev. Lett.* 99: 263003.

56. T. Shaaran, B. B. Augstein and C. Figueira de Morisson Faria. 2011. 'Excitation, two-center interference and the orbital geometry in laser-induced non-sequential double ionization of diatomic molecules.' *Phys. Rev. A.* 84: 013429.

57. E. Eremina, X. Liu, H. Rottke, W. Sandner, A. Dreischuh, F. Lindner, F. Grasbon, G. G. Paulus, H. Walther, R. Moshammer, B. Feuerstein and J. Ullrich. 2003. 'Laser-induced non-sequential double ionization investigated at and below the threshold for electron impact ionization.' *J. Phys. B.* 36: 3269.

58. Y. Liu, D. Ye, J. Liu, A. Rudenko, S. Tschuch, M. Dürr, M. Siegel, U. Morgner, Q. Gong, R. Moshammer and J. Ullrich. 2010. 'Multiphoton double ionization of Ar and Ne close to threshold.' *Phys. Rev. Lett.* 104: 173002.

59. C. Figueira de Morisson Faria and X. Liu. 2011. 'Electron-electron correlation in strong laser fields.' *J. Mod. Opt.* 58: 1076.

60. T. Shaaran, C. Figueira de Morisson Faria and H. Schomerus. 2012. 'Causality and quantum interference in time-delayed laser-induced non-sequential double ionization.' *Phys. Rev. A.* 85: 023423.

61. Y. Liu, S. Tschuch, A. Rudenko, M. Dürr, M. Siegel, U. Morgner, R. Moshammer and J. Ullrich. 2008. 'Strong-field double ionization of Ar below the recollision threshold.' *Phys. Rev. Lett.* 101: 053001.

62. S. L. Haan, L. Breen, A. Karim and J. H. Eberly. 2006. 'Variable time lag and backward ejection in full-dimensional analysis of strong-field double ionization.' *Phys. Rev. Lett.* 97: 103008.

63. T. Shaaran and C. Figueira de Morisson Faria. 2010. 'Laser-induced non-sequential double ionization: Kinematic constraints for the recollision-excitation-tunneling mechanism.' *J. Mod. Opt.* 57: 11.

64. Jie Wu. 'Novel Approaches for Matter in Strong Laser Fields.' PhD Thesis, University College London, 2014.

65. Boris Bergues, Matthias Kübel, Nora G. Johnson, Bettina Fischer, Nicolas Camus, Kelsie J. Betsch, Oliver Herrwerth, Arne Senftleben, A. Max Sayler, Tim Rathje, Thomas Pfeifer, Itzik Ben-Itzhak, Robert R. Jones, Gerhard G. Paulus, Ferenc Krausz, Robert Moshammer, Joachim Ullrich and Matthias F. Kling. 2012. 'Attosecond tracing of correlated electron-emission in non-sequential double ionization'. *Nature Comm.* 3: 813.

2 Current and Future Progress Towards Quantum Computation using Diatomic Molecules and Shaped Laser Pulses

Alex Brown and Ryan R. Zaari

2.1 Introduction

The use of tailored laser fields to control quantum dynamics has been of much theoretical and experimental interest over the last 25 years, (see the recent review[1] and references therein). In general, many of the theoretical problems are posed in the form of optimal control theory (OCT)[2, 3] where monotonically convergent iterative techniques are used to find the laser field. Alternatively, and as often done in experimental implementation, closed loop feedback control is utilized, where genetic algorithms (GA) are used to determine the appropriate laser field[4]. In 2002, Tesch and de Vivie-Riedle proposed the use of a molecule's vibrational modes for quantum computation[5] and demonstrated the utility on a model of acetylene using OCT to find the laser field required to implement quantum gates. Concurrently, the work of Apkarian and co-workers appeared[6, 7], suggesting the manipulation of ro-vibronic states for quantum computing applications using time–frequency resolved coherent anti-Stokes Raman scattering or four-wave mixing. Within molecular quantum computing using OCT, various studies on polyatomics emerged incorporating: (i) quantum gates[5, 8–13], (ii) quantum algorithms[14, 15] and experimental limitations of OCT concerned with (iii) frequency constraints[16, 17, 18] and (iv) mask functions[13].

In the present chapter, the focus is on quantum computing using the internal state structure of diatomic molecules. The research involving polar diatomic molecules on a 1D array coupled by dipole–dipole interactions, as suggested by DeMille,[19] is not included in the present discussion. Table 2.1 details the theoretical and experimental studies specifically concerning *diatomic quantum computing* using OCT or GA shaped laser pulses. The list only includes two experimental implementations and in both cases there was no feedback loop setup. The vast majority of calculations done exploring molecular and diatomic quantum computing implement the OCT algorithm to shape laser pulses, which in general do not reflect experimental pulse shaping conditions. One of the papers by Hosaka[20] proposes an experimental setup based on available experimental techniques and explores its feasibility via numerical simulations. However, in general, there is a disconnect between computational simulations and experimental realization that needs to be bridged if further progress in molecular quantum computation is to be achieved. In this paper, we suggest areas lacking in theoretical progress that we have or are currently attempting to provide useful insights into with respect to diatomic quantum computing.

Table 2.1 *Current list and key point summary of theoretical and experimental studies in diatomic quantum computing. For details regarding the difference between population ($\bar{P}$) and fidelity (F) optimization see Section 2.3.4*

Molecular model	Optimization method	Qubit states	Logic gates/ algorithm	$\bar{P}$ or F optimization	Ref
$^{14}N^{16}O$	OCT	rovibrational	Deutsch-Jozsa	fidelity	[21]
$^{12}C^{16}O$	OCT	rovibrational	NOT,Had,$CNOT_1$	fidelity	[22]
$^{12}C^{16}O$	OCT	rovibrational	Deutsch-Jozsa	fidelity	[23]
$^{12}C^{16}O$	GA	rovibrational	NOT,Had,$CNOT_1$	fidelity	[24]
I_2	none	vibronic	Deutsch-Jozsa	fidelity	[7]
Li_2	OCT	vibronic	Deutsch-Jozsa	fidelity	[25]
LiH	none	rovibrational	NOT,Had	population	[26]
Na_2	none	rovibronic	NOT,$\frac{\pi}{8}$,Had	population	[27]
Na_2 / Li_2	OCT	vibronic	Deutsch-Jozsa	fidelity	[28]
NaX–NaY$^{(a)}$	OCT	rotational	Deutsch-Jozsa	fidelity	[29]
NaCs	GA/OCT	rovibronic	$CNOT_1$	fidelity	[30]
OH	OCT	vibrational	NOT,$\frac{\pi}{8}$,Had	fidelity	[31]
OH/D($\omega_e\chi_e$)$^{(b)}$	OCT	vibrational	NOT,Had	population	[32]
ortho-N_2 -D_2	none	rovibrational	NOT,CNOT	population	[33]
I_2	none	vibrational	$CNOT_1$,QFT	population	[34]
I_2	OCT	vibronic	Grovers	fidelity	[35]
I_2	experiment	vibrational	QFT	–	[20]
Li_2	experiment	rovibrational	Deutsch-Jozsa	–	[36]
$^{12}C^{16}O$	GA	rovibrational	NOT,Had CNOT,ACNOT	fidelity	[37]
D($\omega_e\chi_e$,B_e)$^{(c)}$	OCT/GA	rovibrational	$ACNOT_1$,NOT_2	fidelity	[38]

$^{(a)}$ X,Y = Cl or Br
$^{(b)}$ D($\omega_e\chi_e$): model diatomic using vibrational anharmonicity
$^{(c)}$ D($\omega_e\chi_e$,B_e): model diatomic using vibrational anharmonicity and rotational constant

2.2 Diatomic Quantum Computing with Shaped Laser Pulses

In order for any physical system to be a viable candidate for a quantum computer, it must meet five criteria as proposed by DiVincenzo[39]. Diatomics can in general satisfy all the five conditions given below.

1. Scalable system with well characterized qubits: Rovibrational states as the qubits with scalability being satisfied through the inclusion of more rovibrational qubit states
2. Initialization: Rovibrational state preparation through current experimental techniques[36, 20], e.g., jet-cooling
3. Long decoherence times: Rovibrational states in diatomic molecules exhibit long decoherence times on the order of a thousand laser pulse (quantum gate) applications before significant information loss
4. Universal set of quantum gates: Laser pulses are shaped to represent universal gates (e.g, Hadamard, NOT, phase, CNOT, ACNOT)
5. Measurement: Rovibrational states can be projected onto other molecular states for computational read-out[36, 20]

In experimental methods for control, a genetic algorithm (GA) is used in a closed loop feedback setup to perform the laser pulse shaping. A metric that describes the ability of the current laser pulse to carry out the required operation (e.g., dissociation, rovibrational state control, fluorescence) is fed back to the GA. Adjustments are then made to the applied laser pulse to increase the desired operation, hence, the term *laser pulse shaping*. This type of experimental setup was first suggested by Judson and Rabitz[4]. A theoretically analogous procedure can be implemented for laser pulse shaping towards quantum computation on diatomic molecules using the GA and was initially investigated for carbon monoxide (CO) by Tsubouchi and Momose[24].

2.3 Theory

2.3.1 Qubits: Rovibrational states

In order for quantum calculations to be done, a set of quantum bits (qubits) analogous to the classical bit, are chosen. In our scheme for 2 qubit operations, 4 rovibrational states (v, J) of a diatomic are chosen to act as the qubits namely: $|00\rangle \equiv (1,2), |01\rangle \equiv (0,1), |10\rangle \equiv (2,1), |11\rangle \equiv (1,0)$. The choice of qubit representation is arbitrary but it may have some unapparent benefits or drawbacks. For the case considered here, the choice of qubits allows for single photon excitation only between rovibrational states constituting a quantum gate[37, 38]. Typical rovibrational qubits chosen for OCT calculations are those analogous to the qubit representation, namely $|q_1q_2\rangle \equiv (v, \mathrm{J})$=(0,0), (0,1), (1,0) and (1,1). While it is advantageous to have control of single qubits (q_1 or q_2), in these cases, using such a qubit basis results in fields that contain both microwave *and* infrared frequencies[21, 22, 23]—a result that represents a significant, if not insurmountable, challenge to experimental realization. Other studies[31, 32, 34] have used only vibrational states for qubits, ignoring any rotational effects.

2.3.2 Quantum gates

The qubits undergo specific transformations as determined by the quantum gate operation. Any quantum algorithm can be constructed from a set of universal quantum gates. In the present case, the CNOT, ACNOT, NOT and Hadamard gates of which a possible universal set of quantum gates includes the Hadamard, CNOT and phase gates (not shown here) were studied[40]. In order to implement a specific quantum gate, a laser pulse is shaped to represent the action of the gate itself through the required qubit (rovibrational state) excitations. Table 2.2 details the operation of each of the above mentioned quantum gate.

Table 2.2 *Quantum gate operations studied*[37]. *A set of quantum gates includes the Hadamard (Had), CNOT and phase gate (not described here). The subscript number following the NOT and Hadamard gates refers to the qubit that the gate operates on. The subscript number following the CNOT and ACNOT gates refers to the qubit that is the control qubit*

NOT_1 :	$\lvert 00\rangle \leftrightarrow \lvert 10\rangle$	NOT_2 :	$\lvert 00\rangle \leftrightarrow \lvert 01\rangle$
	$\lvert 01\rangle \leftrightarrow \lvert 11\rangle$		$\lvert 10\rangle \leftrightarrow \lvert 11\rangle$
Had_1 :	$\lvert 00\rangle \leftrightarrow \frac{1}{\sqrt{2}}(\lvert 00\rangle + \lvert 10\rangle)$	Had_2 :	$\lvert 00\rangle \leftrightarrow \frac{1}{\sqrt{2}}(\lvert 00\rangle + \lvert 01\rangle)$
	$\lvert 01\rangle \leftrightarrow \frac{1}{\sqrt{2}}(\lvert 01\rangle + \lvert 11\rangle)$		$\lvert 01\rangle \leftrightarrow \frac{1}{\sqrt{2}}(\lvert 00\rangle - \lvert 01\rangle)$
	$\lvert 10\rangle \leftrightarrow \frac{1}{\sqrt{2}}(\lvert 00\rangle - \lvert 10\rangle)$		$\lvert 10\rangle \leftrightarrow \frac{1}{\sqrt{2}}(\lvert 10\rangle + \lvert 11\rangle)$
	$\lvert 11\rangle \leftrightarrow \frac{1}{\sqrt{2}}(\lvert 01\rangle - \lvert 11\rangle)$		$\lvert 11\rangle \leftrightarrow \frac{1}{\sqrt{2}}(\lvert 10\rangle - \lvert 11\rangle)$
CNOT_1 :	$\lvert 00\rangle \rightarrow \lvert 00\rangle$	CNOT_2 :	$\lvert 00\rangle \rightarrow \lvert 00\rangle$
	$\lvert 01\rangle \rightarrow \lvert 01\rangle$		$\lvert 10\rangle \rightarrow \lvert 10\rangle$
	$\lvert 10\rangle \leftrightarrow \lvert 11\rangle$		$\lvert 01\rangle \leftrightarrow \lvert 11\rangle$
ACNOT_1 :	$\lvert 10\rangle \rightarrow \lvert 10\rangle$	ACNOT_2 :	$\lvert 01\rangle \rightarrow \lvert 01\rangle$
	$\lvert 11\rangle \rightarrow \lvert 11\rangle$		$\lvert 11\rangle \rightarrow \lvert 11\rangle$
	$\lvert 00\rangle \leftrightarrow \lvert 01\rangle$		$\lvert 00\rangle \leftrightarrow \lvert 10\rangle$

2.3.3 Laser pulse field

The laser pulse is represented in the frequency-domain on a discretized grid according to:

$$\epsilon(\nu_j) = \epsilon_0 \sqrt{A(\nu_j)} \exp\left[-2\ln 2\left(\frac{\nu_j - \nu_0}{\Delta\nu}\right)^2\right] \exp\left[i\phi(\nu_j)\right], \tag{2.1}$$

where ϵ_0 is the peak field strength, ν_0 is the central frequency, and ν_j represents the discrete frequencies at which the field is shaped. The amplitude and phase at each discretized frequency are chosen from the range $0 \leq A(\nu_j) \leq 1$ and $0 \leq \phi(\nu_j) \leq 2\pi$, respectively. The transform limited (TL) pulse, which is the input pulse for experimental pulse shaping, defines the pulse structure which is to be shaped. The TL-pulse is defined by $A(\nu_j) = 1$ and $\phi(\nu_j) = 0$ in Eq. (2.1). In the shaping procedure, the resulting frequency domain laser pulse is Fourier transformed into the time domain in order to solve the time-dependent Schrödinger equation to determine the laser pulse effectiveness at carrying out the chosen quantum gate operation. An example of the connection between the frequency- and time-domain representations of a laser pulse is shown in Fig. 2.1 for $^{12}C^{16}O$ using binary pulse shaping—when only two choices of amplitude and phase are considered for any frequency component ν_j, i.e, $A(\nu_j) = 0$ or 1 and $\phi(\nu_j) = 0$ or π.

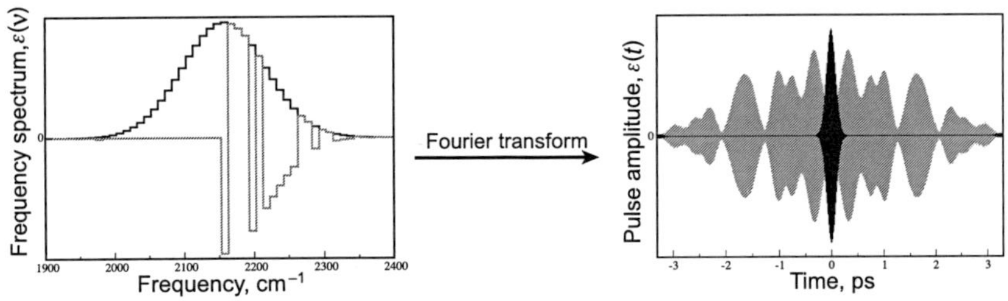

Figure 2.1 A sample shaped laser pulse in the frequency-domain (left) and the corresponding Fourier transformed time-domain laser pulse (right). Black lines denote the TL-pulse shape and the grey lines denote the shaped laser pulse. Negative values of $\epsilon(\nu)$ denote, in the case of binary pulse shaping, values of phase $\phi(\nu_j) = \pi$. The time-domain TL-pulse (right;black) has been scaled by a factor of 0.3

2.3.4 Optimization with genetic algorithms

A genetic algorithm (GA) is a heuristic search space optimization routine that uses the biological aspects of natural selection and survival of the fittest. In the context of laser pulse shaping, the GA determines the optimal combinations of amplitudes ($A(\nu_j)$) and phases ($\phi(\nu_j)$) at each frequency component (ν_j) that gives rise to the optimal pulse shape. In all of our GA optimizations, we used the driver developed by Carroll[41].

Early GA optimization laser pulse shaping computations used the average population, $\bar{P}$, as the target for appropriate quantum gate operation construction:

$$\bar{P} = \frac{1}{N}\sum_{k=1}^{N} |\langle\Psi_k(T)|\Phi_k\rangle|^2. \tag{2.2}$$

The function is a summation of the overlap of the resulting state after laser pulse interaction, $\Psi_k(T)$, to the target final state, Φ_k, over N qubit operations. Later, it was realized that the global phase acquired among the qubit transformations must be aligned[15, 31], and this is carried out by targeting the fidelity, F,

$$F = \frac{1}{N^2}\left|\sum_{k=1}^{N} \langle\Psi_k(T)|\Phi_k\rangle\right|^2 \tag{2.3}$$

rather than the population. The task of the GA is to determine the combinations of amplitude and phase at each frequency component (ν_j), see Eq. 2.1, that maximizes the fidelity. The fidelity function is a maximum value ($F = 1$) when each of the required excitations in the quantum gate operation is satisfied and all the resultant qubit states are globally phase aligned. Global phase alignment is a requirement for appropriate quantum gate application. It implies that upon interaction with the optimized quantum gate laser pulses, each qubit should undergo the same phase rotation, thereby making all the qubit states phase aligned. Qubits which are not aligned appropriately will undergo a different phase change when subsequent laser pulses are applied. This will lead to an ineffective quantum gate operation, and, hence, the rapid failure of any quantum algorithm requiring the application of multiple quantum gates.

When using the GA, the fidelity can be optimized directly. However, when using the standard iterative OCT approach[3], one only optimizes the population. None the less, within this OCT approach, there is a simple method for incorporating global phase alignment by optimizing an auxiliary transition[15]. Along with the 4 qubit transformations, there is an extra transition to optimize and enforce global phase alignment, Ψ_5:

$$[|\Psi_{00}\rangle + |\Psi_{01}\rangle + |\Psi_{10}\rangle + |\Psi_{11}\rangle]_{t=0} \longrightarrow \left[(|\Psi_{00}\rangle + |\Psi_{01}\rangle + |\Psi_{10}\rangle + |\Psi_{11}\rangle)e^{i\theta_5}\right]_{t=T} \tag{2.4}$$

2.4 Current Insights

Many of the theoretical investigations of molecular quantum computing provide a plausible system without identifying useful insights that one would need in order to apply it to experimental realization. The fields generated may be too complex in structure and in frequency to produce experimentally; the candidate molecules may be only theoretically ideal; or the data reported may not be specific to experimental practice. Table 2.1 provides a summary of all computational and experimental studies utilizing diatomic molecules for quantum computing that have been carried out. By analysis of Table 2.1 and other related studies on polyatomic molecules, the following important ideas can be garnered:

1. Binary pulse shaping: What is the effect of restricting the number of amplitude and phase parameters [Eq. (2.1); A and ϕ] on the resulting quantum gate fidelity? Is a large parameter space of amplitude and phase variation essential for high fidelity control of diatomics?
2. Choice of diatomic: Are there inherent properties (anharmonicity, rotational constant, selection rules) that make some diatomics more promising candidates than others for quantum computing?
3. Further pulse shape variations: What is the effect of reducing the number of frequency components and/or increasing the frequency resolution on the resulting quantum gate fidelity? Is it necessary to use the entire range of discretized frequencies? Is the discretized frequency width, and, therefore, the resulting pulse duration, of importance?

2.4.1 Binary pulse shaping

One method for shaping laser pulses experimentally, which is analogous to our current theoretical investigations, is to use a liquid crystal spacial light modulator (LC-SLM). The device is a pixelated liquid crystal thin layer in which each pixel can be modulated by varying the individual voltages. In turn, the laser pulse frequency components striking a particular pixel can be varied by amplitude and phase. For example, a typical LC-SLM device can be divided into 128 pixels for frequency with 256 amplitude/phase voltage variations, thus producing a parameter space of $(256\times256)^{128}$ laser pulse combinations. Such a parameter space size can be difficult for the GA to search within and will increase the time in finding the optimal pulse shape. In a first attempt to minimize the parameter space, using 51 frequency components each with a width of 10 cm^{-1}, as chosen by Tsubouchi and Momose[24], we utilized only 2 amplitude and 2 phase variations, namely *binary pulse shaping*. A binary shaped laser pulse also represents the minimum fidelity laser pulse since it is a subset belonging to non-binary pulse shapes. For example, a choice of 2 amplitudes and 2 phases ($A = 0,1$ and $\phi = 0,\pi$) is a subset of 4 amplitudes and 4 phases ($A = 0,\frac{1}{3},\frac{2}{3},1$ and $\phi = 0,\frac{\pi}{4},\pi,\frac{3\pi}{4}$). Our reduced parameter space is only $(2\times2)^{51}$ laser pulse combinations in size. The results for 8 gates tested on the diatomic $^{12}C^{16}O$ and previously reported[37] are shown in Table 2.3. The ACNOT and CNOT gates have fidelities at and above 80%, while the Hadamard and NOT gates have significantly lower fidelities (60% and lower). The decline in fidelities for the Hadamard and NOT gates is attributed to non-optimal choice of the initial pulse energy which significantly hinders the excitation. Also shown in Table 2.3 are results by Tsubouchi and Momose[24] using 51 frequency components, 64 amplitude and 64 phase components for CO. Although we implemented a comparatively different qubit basis, we were still able to achieve a greater fidelity while using only binary pulse shaping. Having a smaller search space also has the added benefit of increased certainty of finding the greatest fidelity since the GA is a heuristic search optimization routine.

2.4.2 Choice of diatomic

From an experimental and theoretical perspective, it is important to choose a diatomic that will exhibit good control for the universal set of quantum gates being optimized. Thus far, only 12 species have been tested theoretically, see Table 2.1. Babikov[32] provided a study using a general model diatomic but it was implemented using OCT on the vibrational states (i.e., no rotational structure) without global phase alignment. Though the choice of diatomics used may be limited, there is significant system variety, namely: (i) closed shell (e.g., CO) and open shell (e.g., NO), (ii) homonuclear (e.g., I_2) and heteronuclear (e.g., CO), and (iii) isolated diatomics (e.g., CO) and dipole coupled diatomics (e.g., NaCl–NaBr). Unfortunately, the majority of systems were studied using OCT to determine the laser pulses required for quantum gate operations. Even though fidelities nearing 100% were obtained, the resulting OCT pulses contain both microwave and infrared frequencies.

Table 2.3 *Central frequency ν_0, the resulting laser pulse energy and peak intensity, and fidelities for the quantum gate operations studied in Table 2.2. Also shown are results for a similar study by Tsubouchi and Momose[24]. The central frequencies of Ref.[24] differ from the present study due to a different choice of qubit states*

Quantum gate	Central frequency, ν_0 cm^{-1}	Pulse energy μJ	Peak intensity TW/cm^2	Fidelity, F
CNOT$_1$	2121	2.20	0.0466	0.8113
CNOT$_2$	2139	4.55	0.0896	0.7962
ACNOT$_1$	2151	4.07	0.0901	0.9729
ACNOT$_2$	2109	0.77	0.033	0.8416
Had$_1$	2124	13.21	0.229	0.6172
Had$_2$	2136	11.71	0.381	0.5075
NOT$_1$	2124	10.72	0.264	0.4921
NOT$_2$	2136	12.20	0.293	0.5118
CNOT$_1$[24]	2115	NR	0.592	0.6877
ACNOT$_2$[24]	2147	NR	0.232	0.6374

NR - Not reported

In our investigation[38], we used a GA to optimize the ACNOT$_1$ and NOT$_2$ quantum gates using the rovibrational states of a general diatomic (oscillating rotor) with globally phase aligned qubits, restricting ourselves to binary pulse shaping (see Section (2.4.1)). The parameters used to construct the model diatomic were from a study by Irikura[42] and taken from a plot in our previous study[38]. Figure 2.2 shows the resulting fidelities for the ACNOT$_1$ and NOT$_2$ gates as a function of the anharmonicity and rotational constant (i.e., each point represents a particular diatomic). A population only optimization (no global phase alignment) is also plotted. The difference between the fidelity and population only curves of ACNOT$_1$ in Fig. 2.2 suggests that the appropriate choice of diatomic is important when performing ACNOT$_1$ quantum gate operations. Moreover, the required global phase alignment dictates whether some diatomics are better than others for shaping laser pulses to represent the ACNOT$_1$ gate. If this were not the case, then the optimization with global phase alignment would be expected to produce a curve similar to that obtained through population only optimization (grey line; Fig. 2.2). Figure 2.2 also shows similar calculations for the NOT$_2$ gate. As stated previously in Section 2.4.1, there is a non-optimal choice for the NOT$_2$ initial pulse energy. This leads to both ineffective control of both the population and the resulting fidelities.

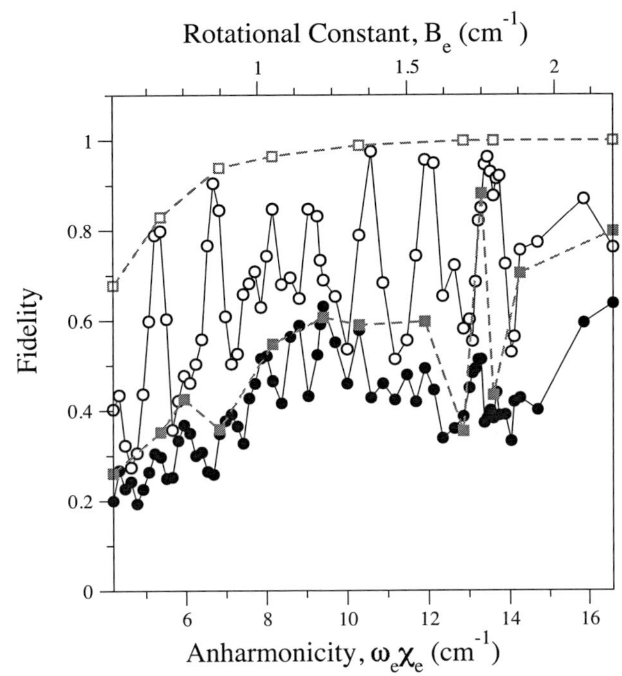

Figure 2.2 Resulting fidelities of optimized laser pulses for the $ACNOT_1$ (hollow black dots/lines) and NOT_2 (solid black dots/lines) for values of anharmonicity ($\omega_e\chi_e$) and rotational constants (B_e). Also plotted are respective optimizations without global phase alignment (average population) for the $ACNOT_1$ (hollow grey squares/dashed line) and NOT_2 (solid grey squares/dashed line) gates. (Reprinted with permission from Ryan R. Zaari and Alex Brown, Journal of Chemical Physics, Vol. 135, Page 044317, (2011). Copyright 2011, American Institute of Physics)

2.4.3 Further pulse shape variations

As has been alluded to in Sections 2.4.1 and 2.4.2, simple binary pulse shaping can produce high fidelities for the ACNOT and CNOT gates, and the choice of an appropriate diatomic for quantum computation is important. Future investigations will consider the effect of further variations in the laser pulse parameters on the resulting fidelity. As illustrated in Fig. 2.3, the ideas are as follows:

i. Increasing the frequency resolution, $d\nu$: In our previous calculations, a frequency width of 10 cm^{-1} was chosen because within CO, it is smaller than the overlap between excitation frequencies. What effect does increasing the resolution (decreasing the frequency width, and, hence, increasing the total pulse length in time) have on the resulting quantum gate fidelity?

ii. Reducing the number of frequency components, ν_j: In our previous calculations, 51 frequency components were used but which components are most influential in producing high fidelity quantum gates?

iii. Increasing the number of amplitude and phase components, A and ϕ: In our previous calculations, our optimization was restricted to binary pulse shaping ($A = 0, 1$ and $\phi = 0, \pi$). What effect does the choice of amplitude and phase variation have on the resulting quantum gate fidelity?

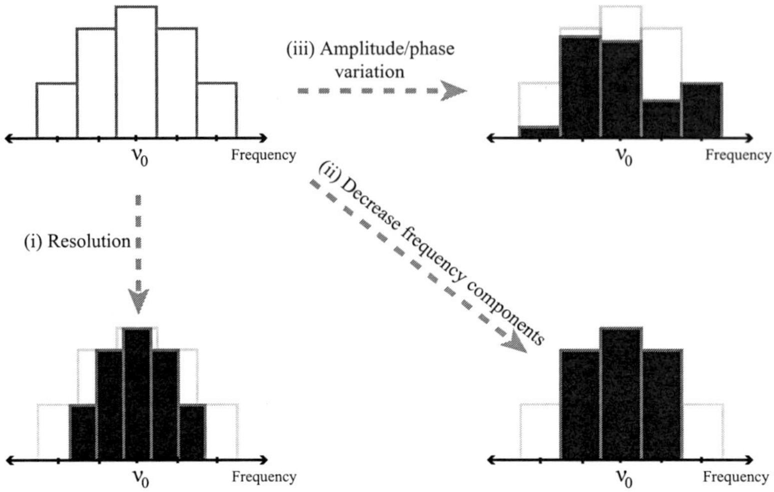

Figure 2.3 An illustration describing key ideas currently being investigated as described in Section 2.4.3

2.5 Concluding Remarks

The detailed investigation of diatomic molecular quantum computing involves several important elements: the choice of rovibrational (or rovibronic) states for the qubits, the quantum gates to be implemented, and the algorithm used to perform the laser field optimization. According to DiVincenzo[39], there are five essential requirements of a quantum computer and we discuss how a diatomic molecule fulfills them. In the present work, the current theoretical and experimental investigations that have been undertaken focusing specifically on diatomic quantum computing have been summarized, (see Table 2.1). Current theoretical laser pulse optimizations for determining the fields required for quantum gate or quantum algorithm implementation focus on either the use of iterative optimal control theory or a genetic algorithm. Each has their respective strengths and weaknesses. However, the use of a genetic algorithm produces laser fields that more closely resemble those that can be determined experimentally. While there are currently only two experimental implementations of a quantum algorithm, both studies do not use a closed loop feedback setup to determine optimal laser pulses. The simulations discussed demonstrate the critical requirement of global phase alignment needed for quantum gate operations. It is clear that this requirement can lead to a strong variation in gate fidelity as a function of molecular parameters (vibrational anharmonicity and rotational constant). Hence, the choice of molecule can be crucial. Moreover, expansion to beyond 2 qubits within the rovibrational manifold of a single diatomic molecule may prove extremely difficult. Therefore, the possibility of coupled polar diatomic molecules on a 1D array[19, 43] needs to be explored further. Whatever approach is taken, insight can be provided to experimentalists through simulations optimizing both laser field and molecular parameters.

Bibliography

1. C. Brif, R. Chakrabarti and H. Rabitz. 2010. *New J. Phys.* 12: 075008.
2. A. P. Peirce, M. A. Dahleh and H. Rabitz. 1988. *Phys. Rev. A*. 37: 4950.
3. W. Zhu, J. Botina and H. Rabitz. 1998. *J. Chem. Phys.* 108: 1953.
4. R. Judson and H. Rabitz. 1992. *Phys. Rev. Lett.* 68: 1500.
5. C. Tesch, L. Kurtz and R. de Vivie-Riedle. 2001. *Chem. Phys. Lett.* 343: 633.
6. R. Zadoyan, D. Kohen, D. A. Lidar and V. A. Apkarian. 2001. *Chem. Phys.* 266: 323.
7. Z. Bihary, D. R. Glenn, D. A. Lidar and V. A. Apkarian. 2002. *Chem. Phys. Lett.* 360: 459.
8. C. Tesch and R. de Vivie-Riedle. 2002. *Phys. Rev. Lett.* 89: 157901.
9. U. Troppmann, C. Tesch and R. de Vivie-Riedle. 2003. *Chem. Phys. Lett.* 378: 273.
10. S. Suzuki, K. Mishima and K. Yamashita. 2005. *Chem. Phys. Lett.* 410: 358.
11. M. Schroeder and A. Brown. 2009. *J. Chem. Phys.* 131: 034101.
12. B. M. R. Schneider, C. Gollub, K. L. Kompa and R. de Vivie-Riedle. 2007. *Chem. Phys.* 338: 291.
13. B. M. R. Korff, U. Troppman and R. de Vivie-Riedle. 2005. *J. Chem. Phys.* 123: 244509.
14. L. Bomble, D. Lauvergnat, F. Remacle and M. Desouter-Lecomte. 2010. *Phys. Chem. Chem. Phys.* 12: 15628.
15. C. M. Tesch and R. de Vivie-Riedle. *J. Chem. Phys.* 121: 12158.
16. C. Gollub, M. Kowalewski and R. de Vivie-Riedle. 2008. *Phys. Rev. Lett.* 101: 073002.
17. M. Schroeder and A. Brown. 2009. *New J. Phys.* 11: 105031.
18. M. Lapert, R. Tehini, G. Turinici and D. Sugny. 2009. *Phys. Rev. A*. 79: 063411.
19. D. DeMille. 2002. *Phys. Rev. Lett.* 88: 067901.
20. K. Hosaka, H. Shimada, H. Chiba, H. Katsuki, Y. Teranishi, Y. Ohtsuki and K. Ohmori. 2010. *Phys. Rev. Lett.* 104: 180501.
21. K. Mishima and K. Yamashita. 2010. *Chem. Phys.* 367: 63.
22. J. L. Chen, C. M. Ling, C. C. Hwang and Y. H. Ho. *J. Chem. Phys.* 134: 134103.
23. K. Shioya, K. Mishima and K. Yamashita. 2007. *Mol. Phys.* 105: 1283.
24. M. Tsubouchi and T. Momose. 2008. *Phys. Rev. A*. 77: 052326.
25. Y. Ohtsuki. 2005. *Chem. Phys. Lett.* 404: 126.
26. W. Z. Cao, L. J. Tian, H. J. Jiang and C. Li. 2008. *Int. J. Quant. Info.* 6: 1223.
27. C. Menzel-Jones and M. Shapiro. 2007. *Phys. Rev. A*. 75: 052308.
28. K. Mishima, K. Tokumo and K. Yamashita. 2008. *Chem. Phys.* 343: 61.
29. K. Mishima and K. Yamashita. 2009. *Chem. Phys.* 361: 106.
30. D. Sugny, L. Bomble, T. Ribeyre, O. Dulieu and M. Desouter-Lecomte. 2009. *Phys. Rev. A*. 80: 042325.
31. M. Zhao and D. Babikov. *J. Chem. Phys.* 125: 024105.
32. D. Babikov. 2004. *J. Chem. Phys.* 121: 7577.

33. E. A. Shapiro, I. Khavkine, M. Spanner and M. Y. Ivanov. 2003. *Phys. Rev. A*. 67: 013406.
34. Y. Teranishi, Y. Ohtsuki, K. Hosaka, H. Chiba, H. Katsuki and K. Ohmori. 2006. *J. Chem. Phys*. 124: 114110.
35. Y. Ohtsuki. 2010. *New J. Phys*. 12: 045002.
36. J. Vala, Z. Amitay, B. Zhang, S. R. Leone and R. Kosloff. 2002. *Phys. Rev. A*. 66: 062316.
37. R. R. Zaari and A. Brown. 2010. *J. Chem. Phys*. 132: 014307.
38. R.R. Zaari and A. Brown. *J. Chem. Phys*. 135: 044317.
39. D. P. DiVincenzo and D. Loss. 1998. *Super lattices and Microstructures* 23: 419.
40. M. A. Nielsen and I. L. Chuang. 2000. *Quantum Computation and Quantum Information*. Cambridge University Press, New York.
41. D. L. Carroll. 2004. *Genetic Algorithm Driver* 1.7a. http://www.cuaerospace.com/carroll/ga.html.
42. K. K. Irikura. 2007. *J. Phys. Chem*. Ref. Data 36: 389.
43. L. D. Carr, D. DeMille, R.V. Krems and J. Ye. 2010. *New J. Phys*. 11: 055049.

3 Investigating the Science of Few-cycle Pulses on Simple Model Systems

Amartya Bose and Debabrata Goswami

3.1 Introduction

As of today, attosecond pulses are in the forefront of state-of-the-art technology[1, 2]. An attosecond pulse, due to its very short duration, can be sustained only by wavelengths of hundreds of angstroms. Such pulses are highly energetic and can cause molecular dissociation. Thus, attosecond pulses cannot be used to study molecular processes.

The ability to sustain only a few cycles in the pulse is a characteristic unique to the attosecond domain. To capture this quintessential feature on molecular processes, the idea of few-cycle pulses at more viable energy ranges, which can represent femtosecond or sub-femtosecond duration was used. In this chapter, we will observe what happens when a few-cycle pulse interacts with matter.

A few-cycle pulse, even on a near-IR wavelength, would be of a temporal duration of the order of femtoseconds. Hence, any general two-level system can, under the interaction of such pulses, be treated as ideal and non-relaxing, since normal molecular processes relax at much longer time scales. In this chapter, we, therefore, study the interaction of few-cycle pulses with ideal two-level systems, where the approximation of a smoothly varying envelope function breaks down. We also see that despite the overwhelming importance of the oscillatory nature of a few-cycle pulse, the signatures of the envelope profile still persists in the interactions.

3.2 Breakdown of Rotating Wave Approximation

For many-cycle pulses, the smoothly varying envelope function approximation is valid. The approximation states that the internal oscillations, because of their comparatively high number, can be ignored and only the envelope function plays a role. This is also called the rotating wave approximation (RWA)[3]. When we move away from the many-cycle pulses domain to the case of few-cycles, the smoothly varying function approximation is invalidated, and hence RWA breaks down[4, 5, 6, 7].

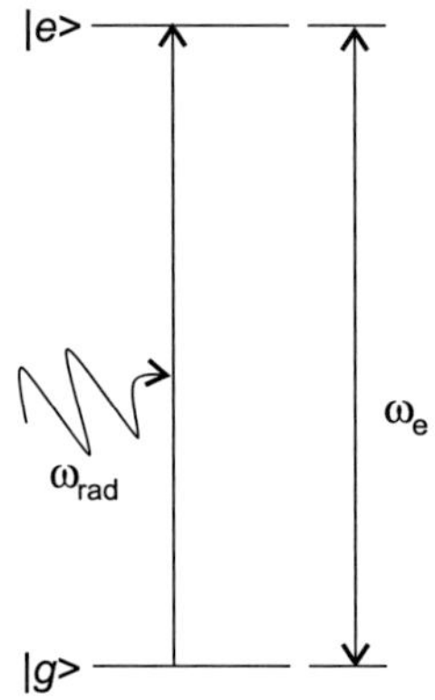

Figure 3.1 Two-level system

Let us consider an ideal 2-level system, with energy difference ω_e Fig. 3.1. The central frequency of the exciting laser pulse is $\omega_{\rm rad}$. The rotating wave approximation holds under the following condition [8]:

$$|\omega_e - \omega_{\rm rad}| \ll \omega_e + \omega_{\rm rad}. \tag{3.1}$$

Assume that the laser pulse is given by:

$$E(x,t) = e^{-\frac{(x-ct)^2}{2\sigma^2}} e^{i(kx-\omega t)},$$

where σ^2 is very small—the pulse is very short.

Consider the Fourier transform of $E(x,t)$ in the time frequency domain,

$$E(x,t) \leftrightarrow E(x,\omega)$$

$$\Rightarrow E(x,\omega) = \frac{\sigma}{c} e^{ikx} + \frac{(\sigma-\omega)[-2icx+\sigma^2(\omega-\sigma)]}{2c^2}.$$

Therefore, keeping the x coordinate fixed, we see that the standard deviation (width) of the Fourier domain pulse is inversely proportional to σ. Thus, the shorter the temporal duration of the pulse, the more spreadout its spectrum would be.

Now, if we have a very spreadout spectrum, as would be for a few-cycle pulse, though Eq. (3.1) might hold for the central part of the spectrum, it would fail for the extremities of the spectral range of the pulse.

$$|\omega_e - \omega_{\text{rad}}| \ll \omega_e + \omega_{\text{rad}}$$

$$\Rightarrow \sqrt{(\omega_e - \omega_{\text{rad}})^2} \ll \omega_e + \omega_{\text{rad}}$$

$$\Rightarrow \sqrt{\left(\frac{\omega_e}{\omega_{\text{rad}}} - 1\right)^2} \ll \frac{\omega_e}{\omega_{\text{rad}}} + 1 \tag{3.2}$$

or

$$\sqrt{\left(1 - \frac{\omega_{\text{rad}}}{\omega_e}\right)^2} \ll \frac{\omega_{\text{rad}}}{\omega_e} + 1. \tag{3.3}$$

Let us look at the last two equations at the spectral extremities. For a two-level system, the energy gap is well-defined. Therefore, the two extremities can be modeled as follows.

1. If $\frac{\omega_e}{\omega_{\text{rad}}} \to 0$, the condition in Eq. (3.2) fails.
2. If $\frac{\omega_{\text{rad}}}{\omega_e} \to 0$, Eq. (3.3) fails.

This is the reason for the breakdown of RWA, which is most prominent in the interaction of few-cycle pulses with matter. Therefore, the full Schrödinger equation has to be solved. A certain envelope-dependent pattern in the breakdown of the RWA was observed.

3.3 Formulating the Measure

3.3.1 Simulation

The interaction between a few-cycle laser pulse and an ideal, non-relaxing two-level system was simulated using the Euler forward time integration method. The width of the pulse is directly proportional to the number of cycles sustained in the same. The width of a pulse was defined as $2 \times l$, where if t_0 is the time coordinate of the peak of the pulse, then the area under the pulse in the interval $t_0 \pm l > 99\%$ of the total area. Gaussian, cosine squared, hyperbolic secant and von Mises ($k = 2$ and $k = 10$) profiles with varying widths, and consequently varying number of cycles per pulse was used.

3.3.2 Analysis of inversion contours

The constant area theorem at zero detuning fail on reaching the higher areas. Therefore, we define a threshold function for each type of profile as:

$$\text{area}(j) = \text{area of the } j\text{th peak at zero detuning}$$
$$\chi(n, \text{profile}) = \min\{\text{area}(j) : \text{area}(j+1) - \text{area}(j) > 2\pi\},$$

where n is the number of cycles, and the minimum is taken over the inversion contours of the corresponding profile. Figures 3.2–3.10 show how $\chi(n)$ varies for the different profiles considered.

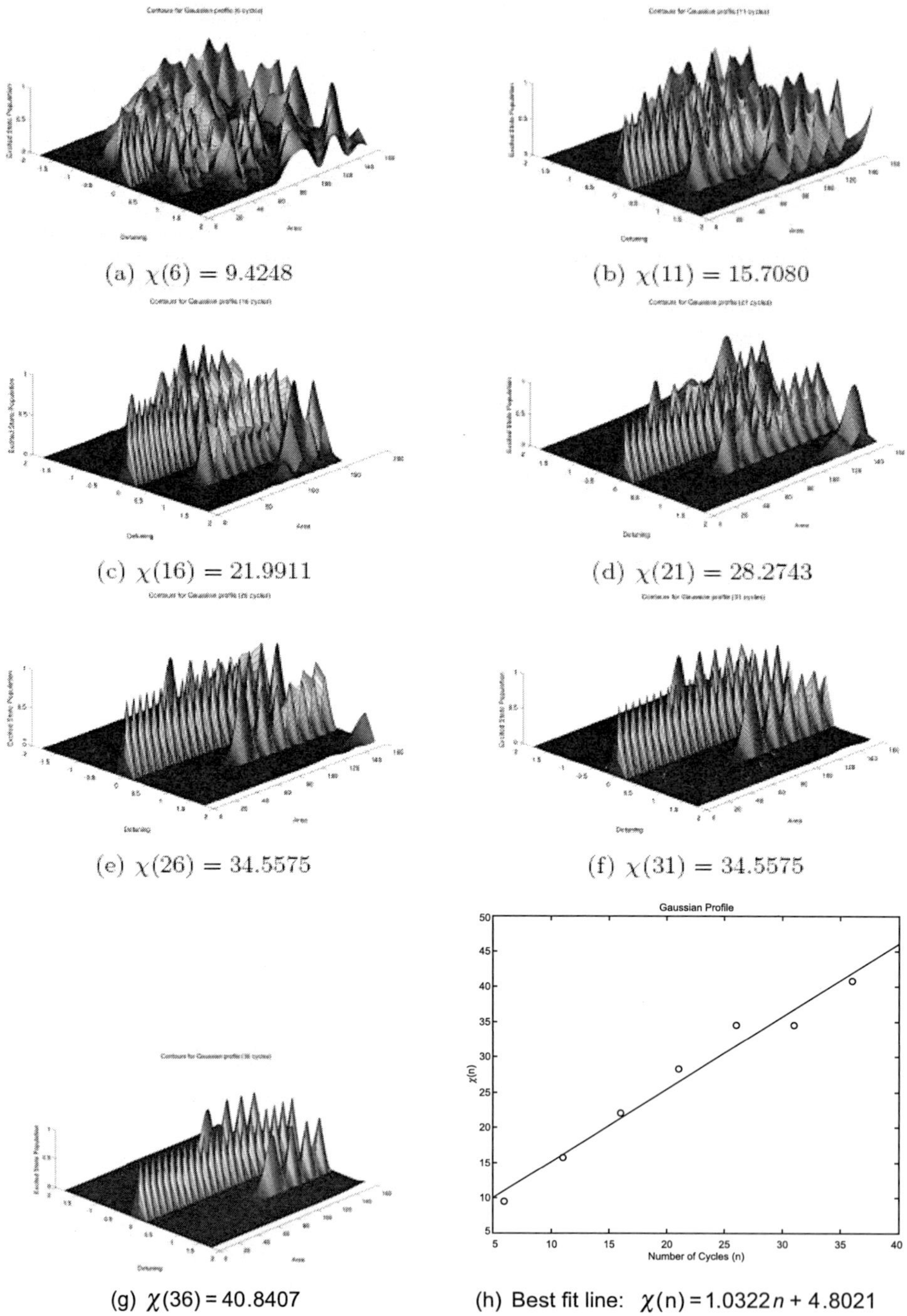

(a) $\chi(6) = 9.4248$

(b) $\chi(11) = 15.7080$

(c) $\chi(16) = 21.9911$

(d) $\chi(21) = 28.2743$

(e) $\chi(26) = 34.5575$

(f) $\chi(31) = 34.5575$

(g) $\chi(36) = 40.8407$

(h) Best fit line: $\chi(n) = 1.0322n + 4.8021$

Figure 3.2 Actual inversion profile for a Gaussian profile

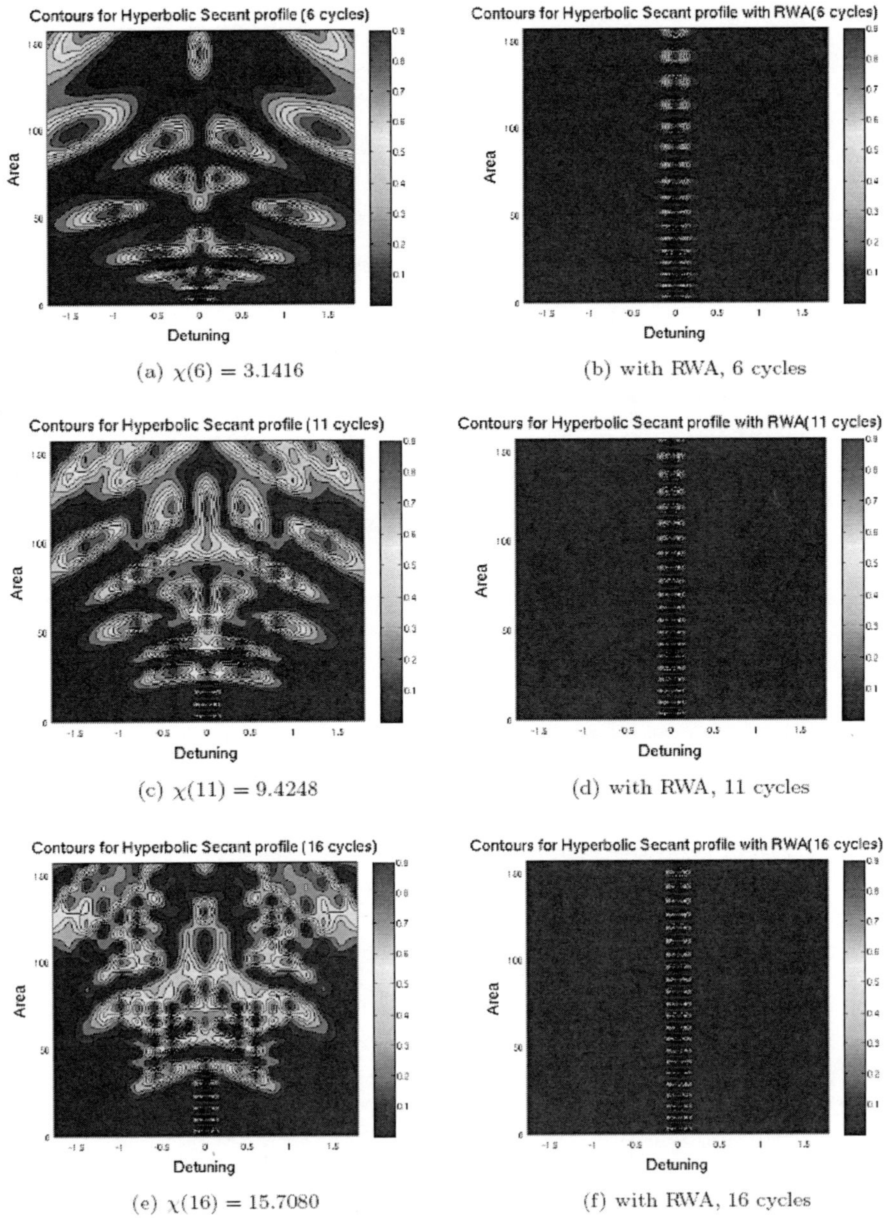

(a) $\chi(6) = 3.1416$

(b) with RWA, 6 cycles

(c) $\chi(11) = 9.4248$

(d) with RWA, 11 cycles

(e) $\chi(16) = 15.7080$

(f) with RWA, 16 cycles

Figure 3.3 Actual inversion profile for a hyperbolic secant profile vis-à-vis RWA case

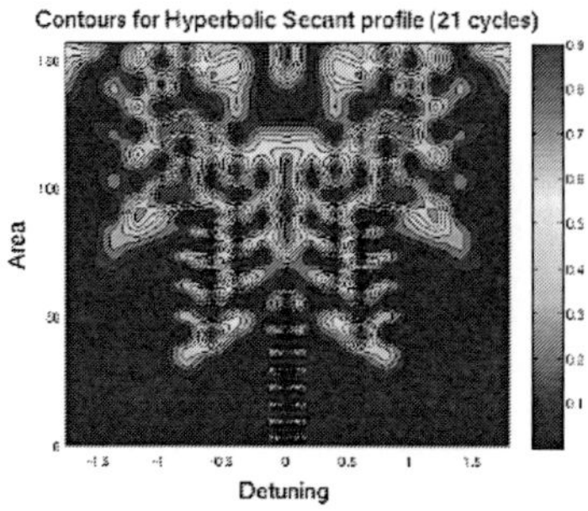

(g) $\chi(21) = 15.7080$

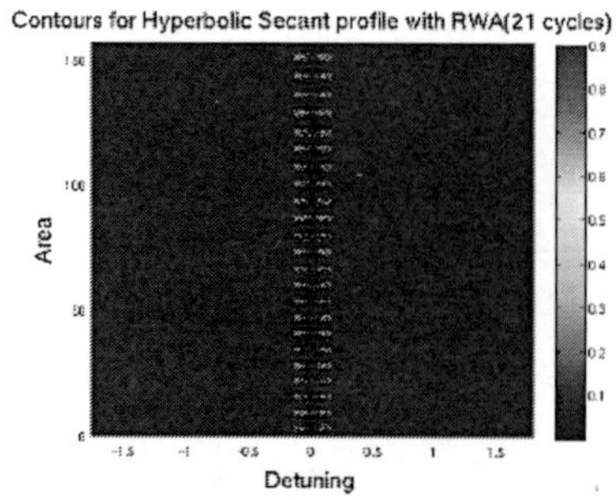

(h) with RWA, 21 cycles

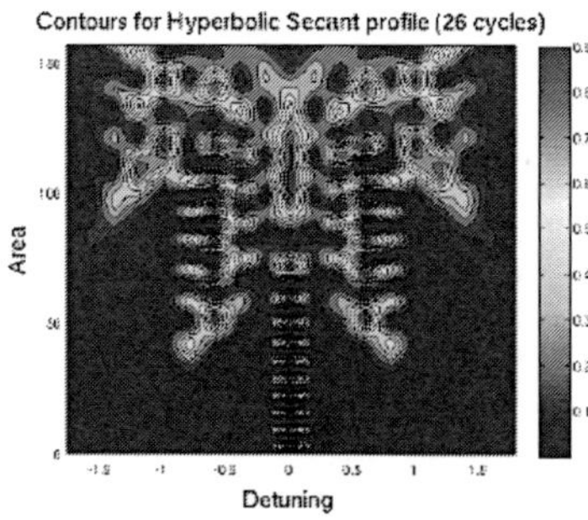

(i) $\chi(26) = 21.9911$

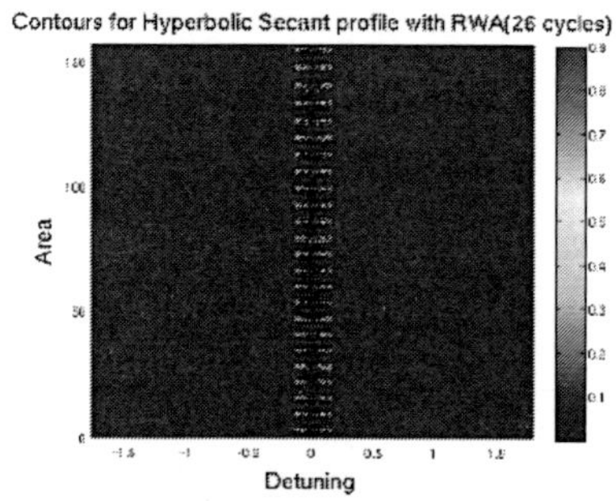

(j) with RWA, 26 cycles

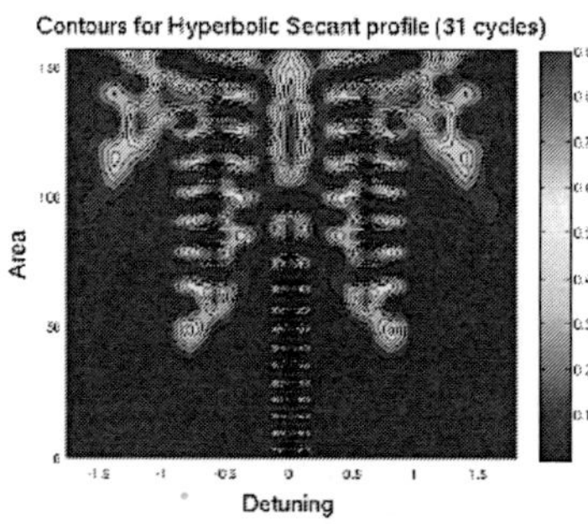

(k) $\chi(31) = 21.9911$

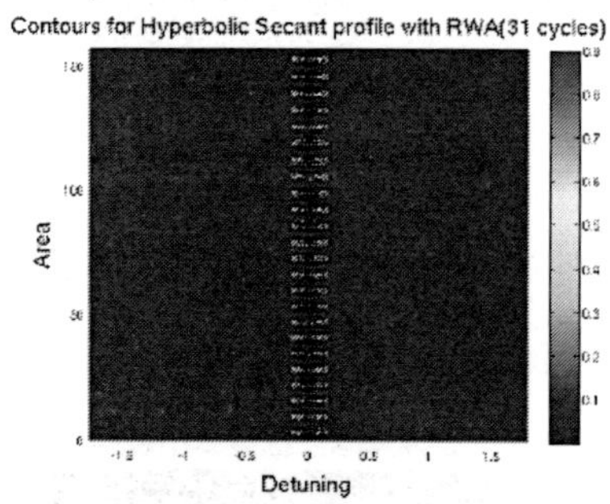

(l) with RWA, 31 cycles

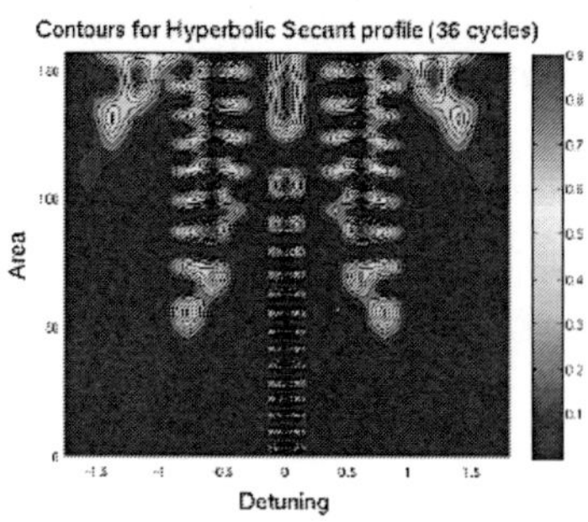

(m) $\chi(36) = 28.2743$

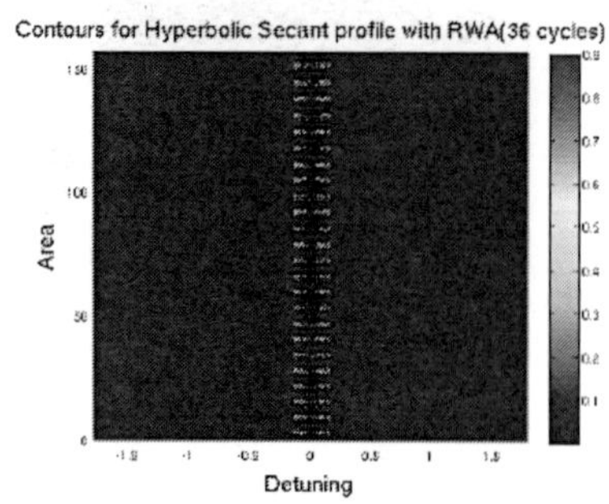

(n) with RWA, 36 cycles

Figure 3.3 contd...

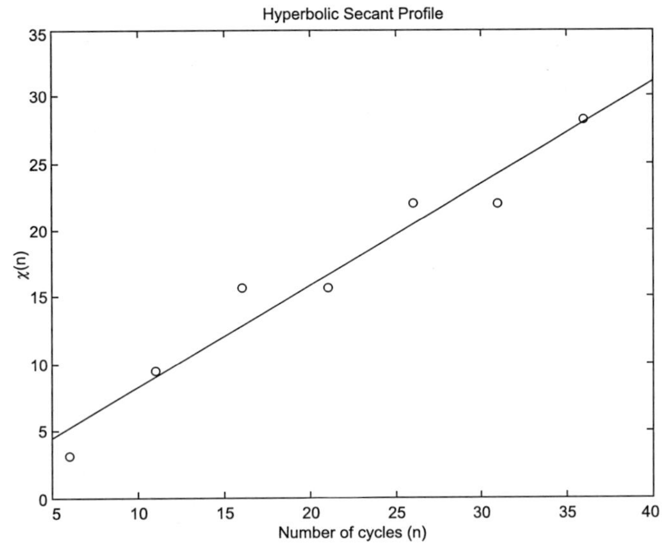

Figure 3.4 Best fit line: $\chi(n) = 0.7630\,n + 0.5834$

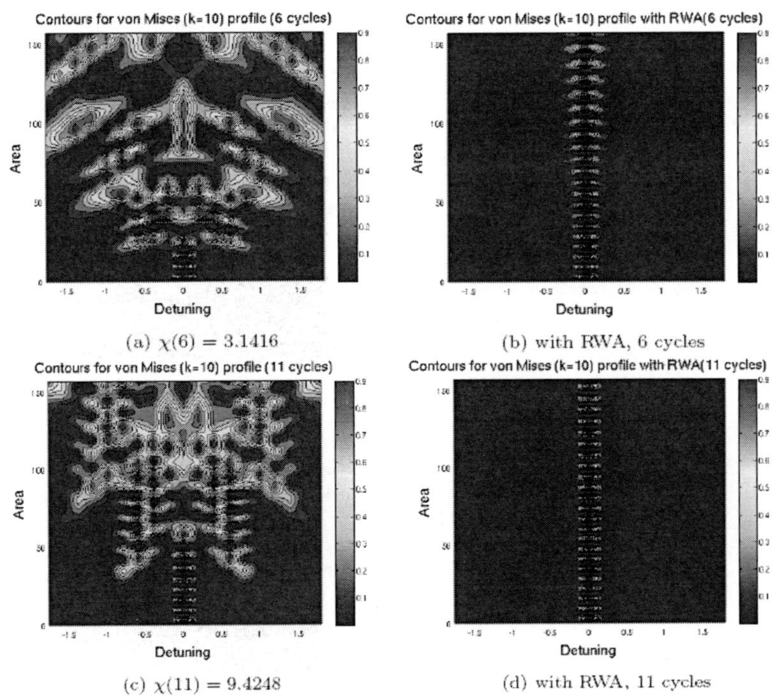

Figure 3.5 Actual inversion profile for a von Mises ($k = 10$) profile vis-à-vis RWA case

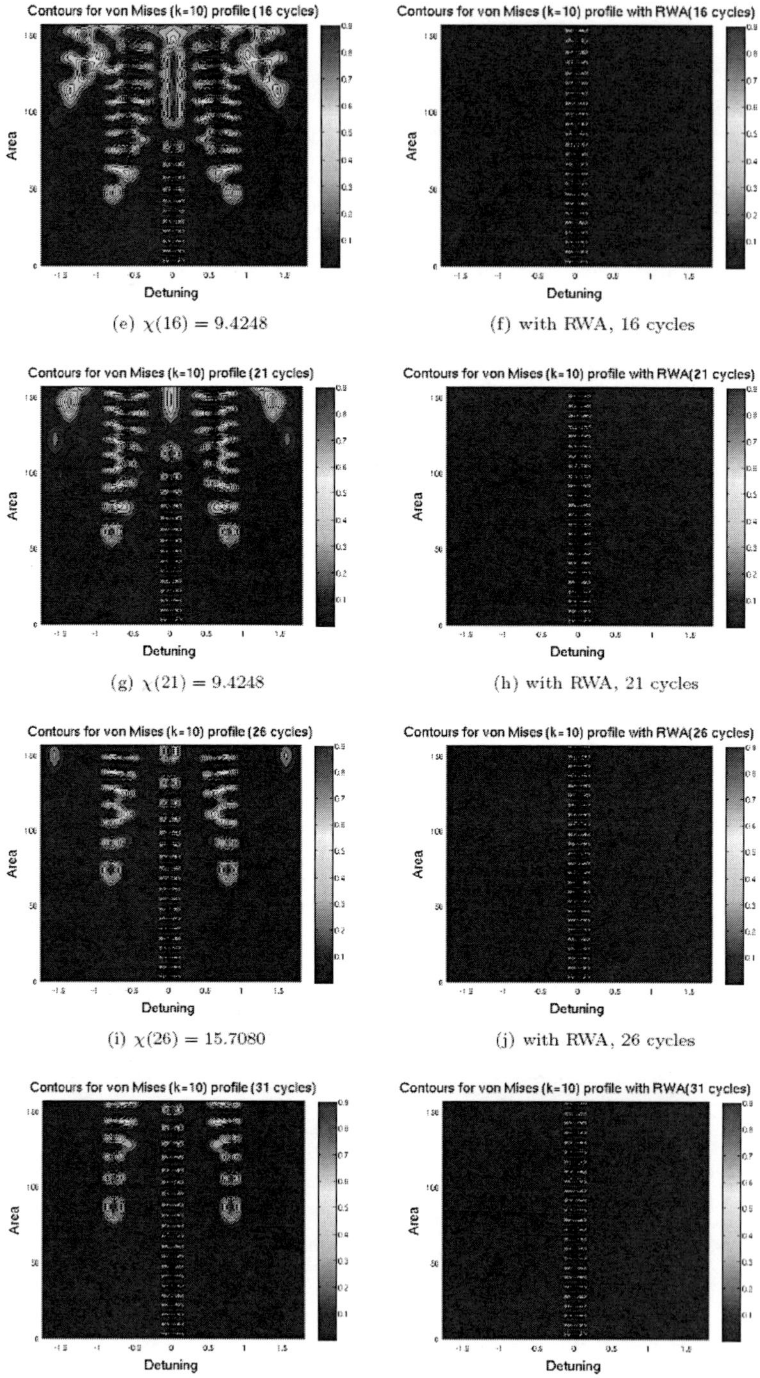

(e) $\chi(16) = 9.4248$

(f) with RWA, 16 cycles

(g) $\chi(21) = 9.4248$

(h) with RWA, 21 cycles

(i) $\chi(26) = 15.7080$

(j) with RWA, 26 cycles

(k) $\chi(31) = 15.7080$

(l) with RWA, 31 cycles

Figure 3.5 contd...

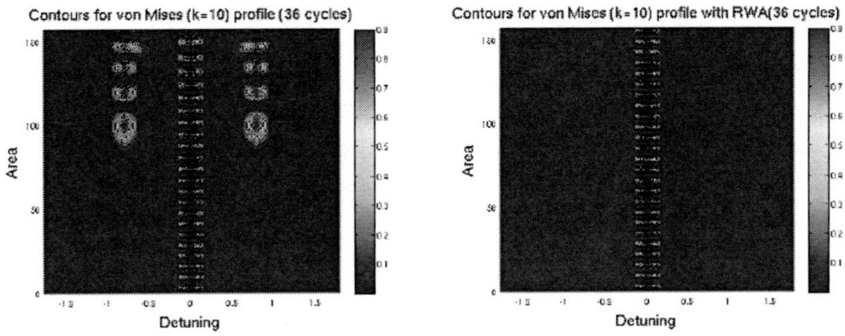

(m) $\chi(36) = 21.9911$

(n) with RWA, 36 cycles

Figure 3.5 contd...

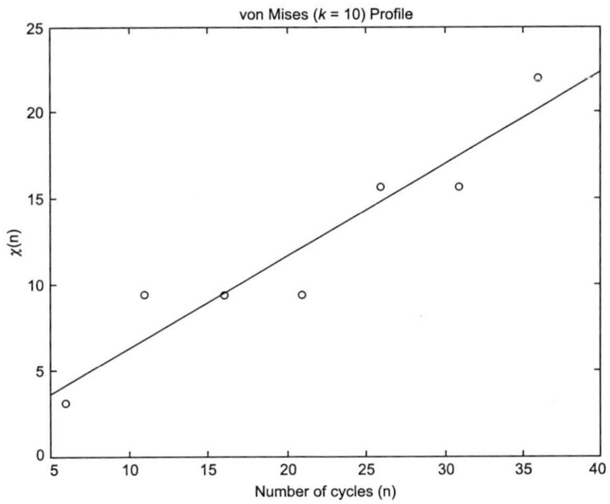

Figure 3.6 Best fit line: $\chi(n) = 0.5386n + 0.8078$

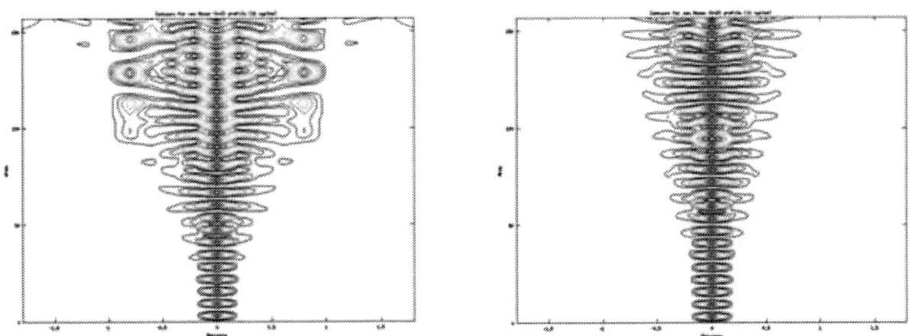

(a) $\chi(11) = 40.8407$

(b) with RWA, 11 cycles

Figure 3.7 Actual inversion profile for a von Mises ($k = 2$) profile vis-à-vis RWA case

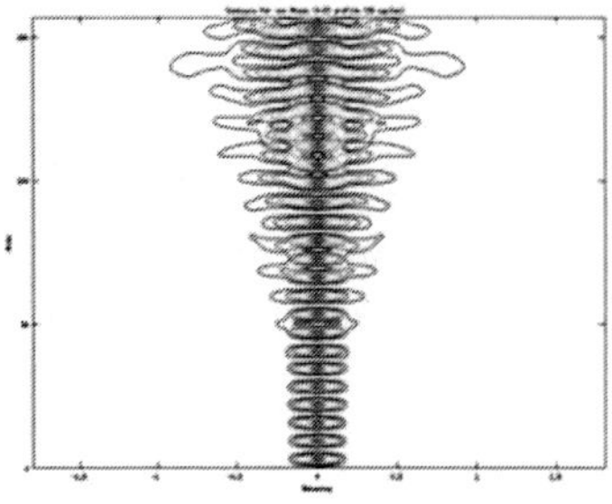

(c) $\chi(16) = 53.4071$

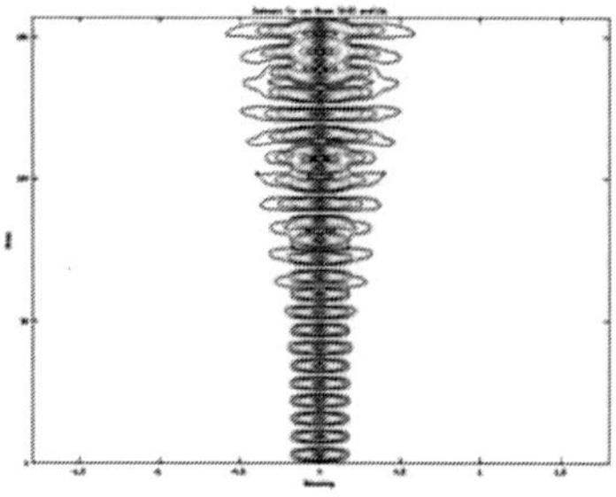

(d) with RWA, 16 cycles

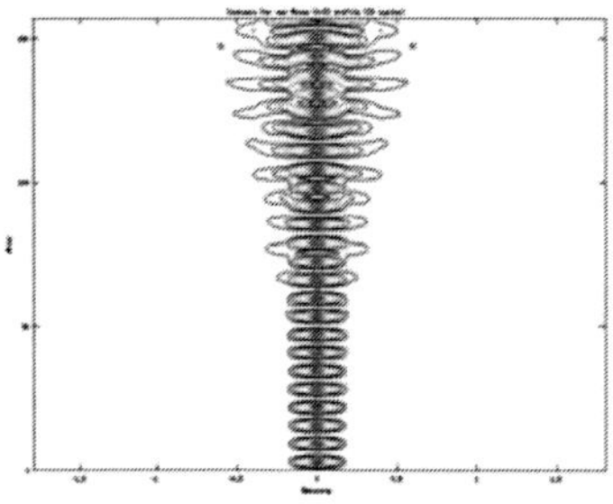

(e) $\chi(21) = 65.9734$

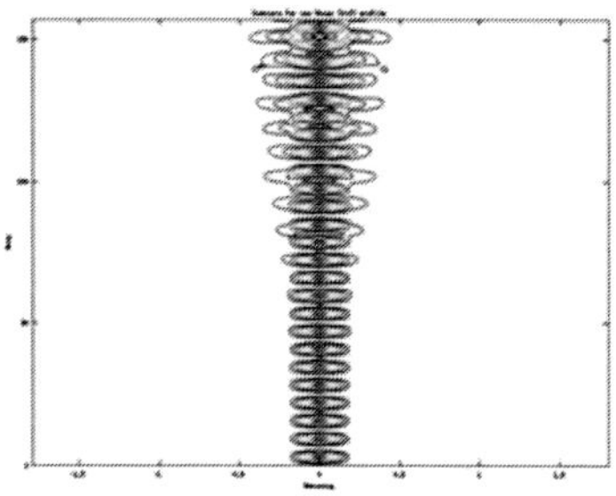

(f) with RWA, 21 cycles

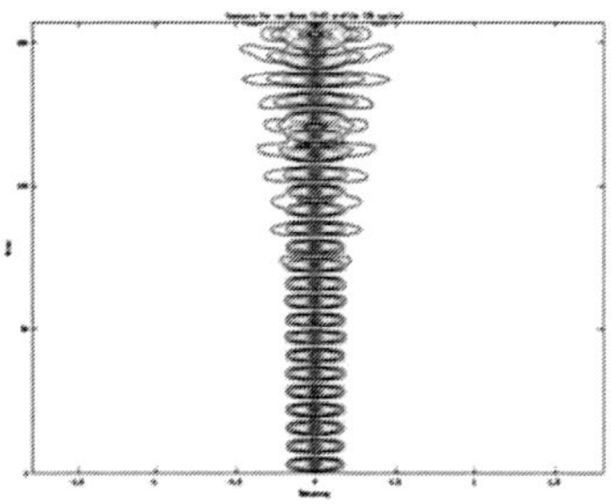

(g) $\chi(26) = 72.2566$

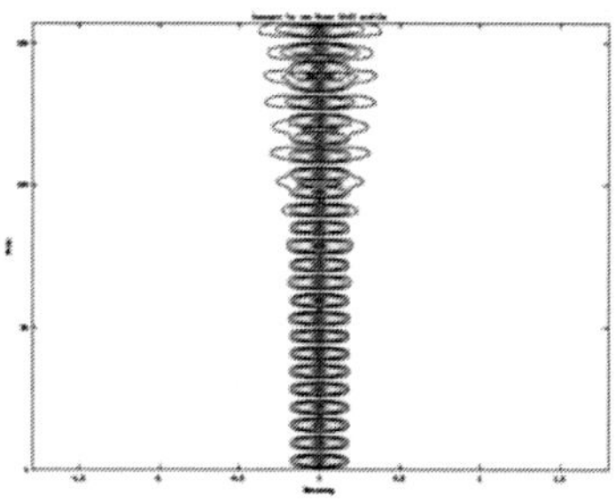

(h) with RWA, 26 cycles

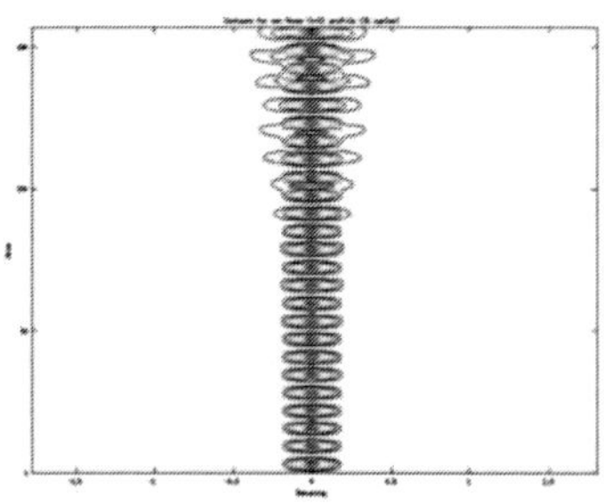

(i) $\chi(31) = 84.8230$

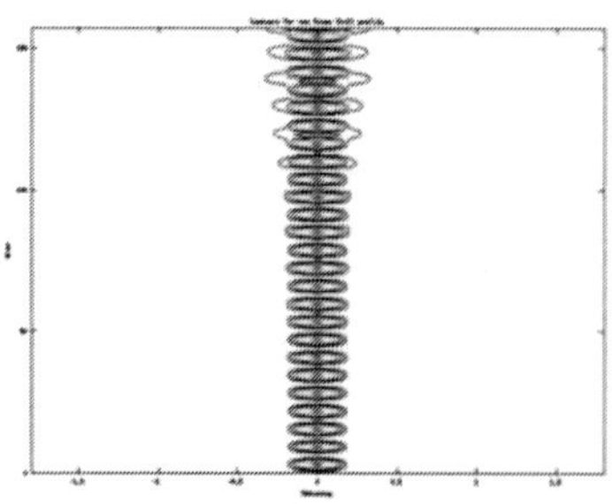

(j) with RWA, 31 cycles

Figure 3.7 contd...

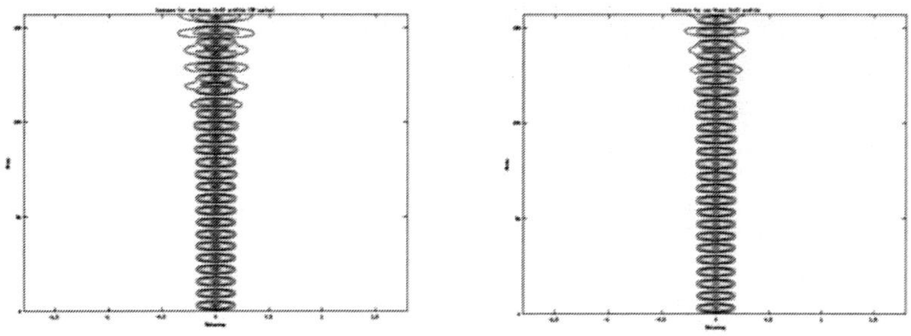

(k) $\chi(36) = 91.1062$ (l) with RWA, 36 cycles

Figure 3.7 contd...

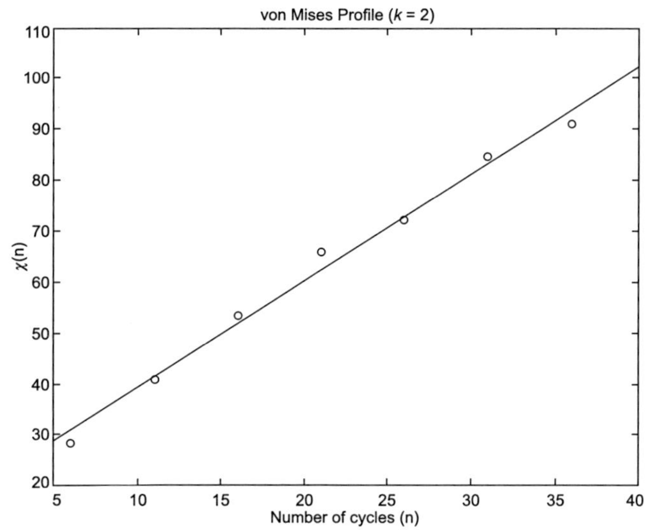

Figure 3.8 Best fit line: $\chi(n) = 2.1094\,n + 18.0866$

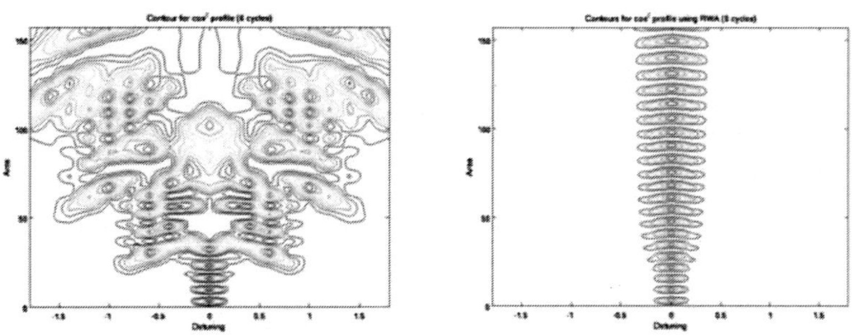

(a) $\chi(6) = 9.4248$ (b) with RWA, 6 cycles

Figure 3.9 Actual inversion profile for a cosine squared profile vis-à-vis RWA case

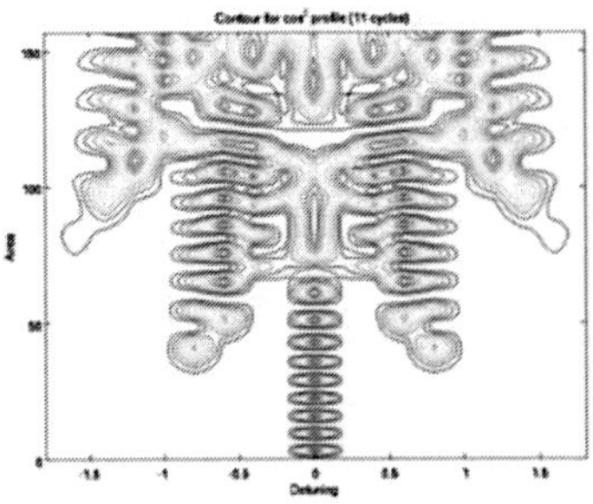

(c) $\chi(11) = 15.7080$

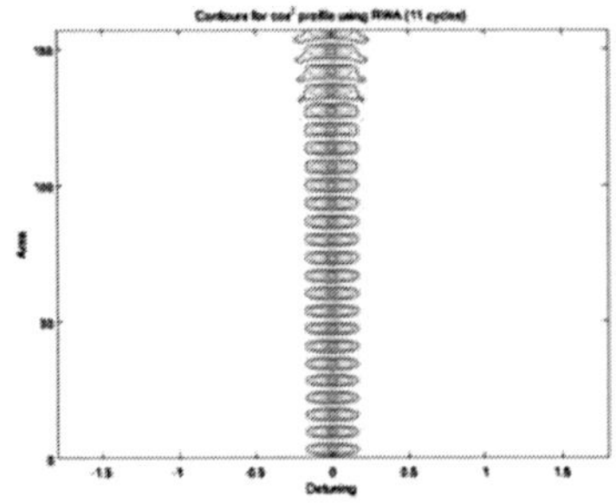

(d) with RWA, 11 cycles

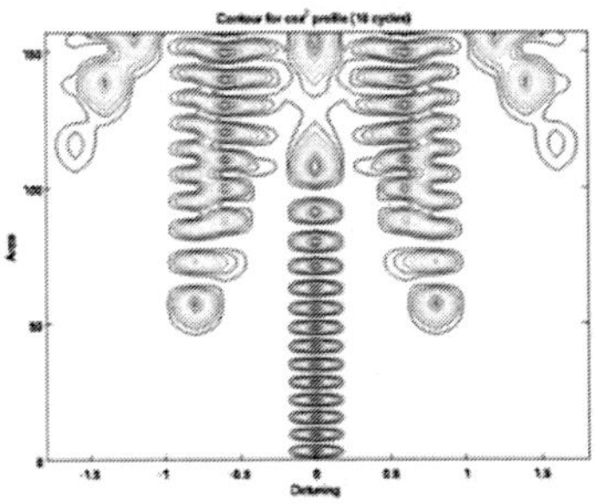

(e) $\chi(16) = 21.9911$

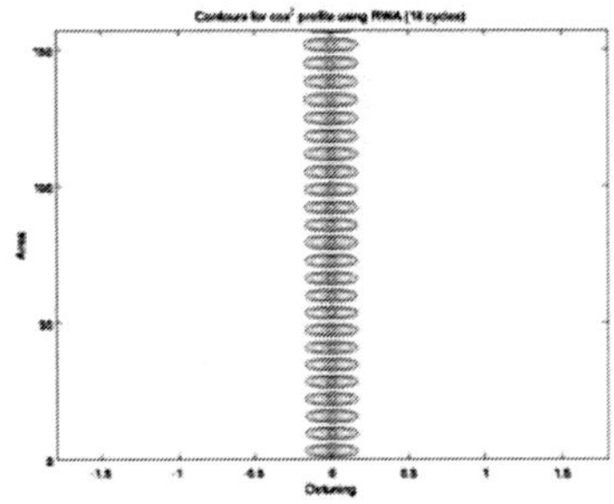

(f) with RWA, 16 cycles

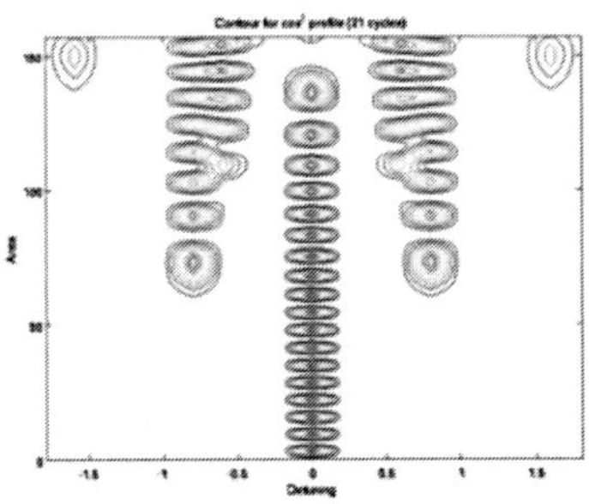

(g) $\chi(21) = 28.2743$

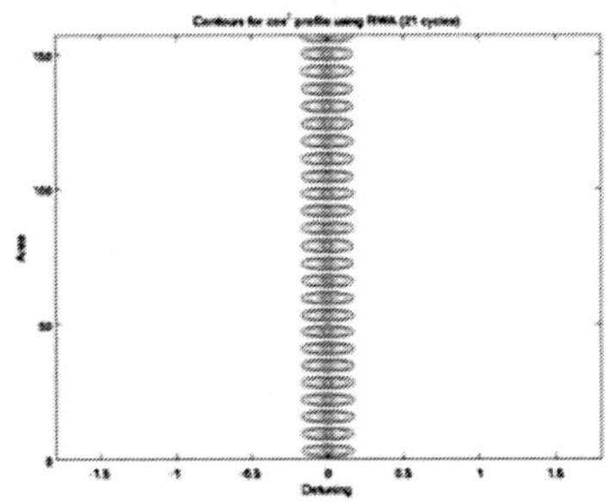

(h) with RWA, 21 cycles

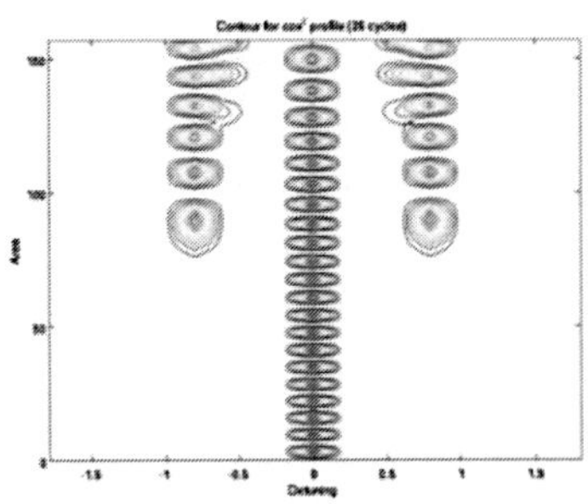

(i) $\chi(26) = 34.5575$

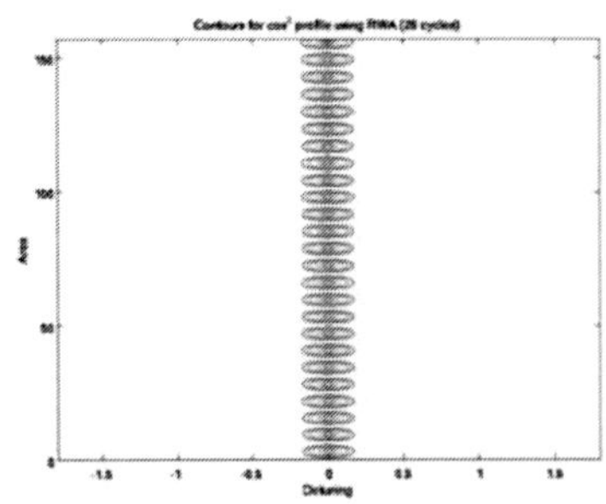

(j) with RWA, 26 cycles

Figure 3.9 contd...

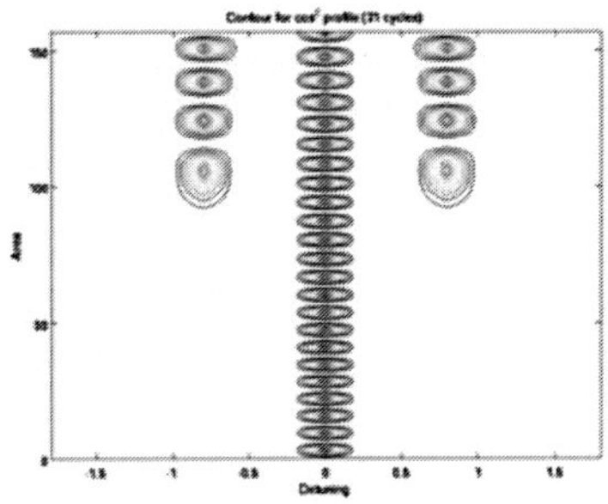

(k) $\chi(31) = 40.8407$

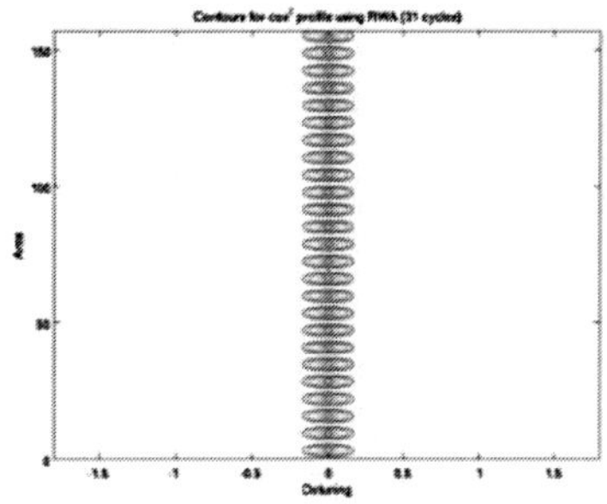

(l) with RWA, 31 cycles

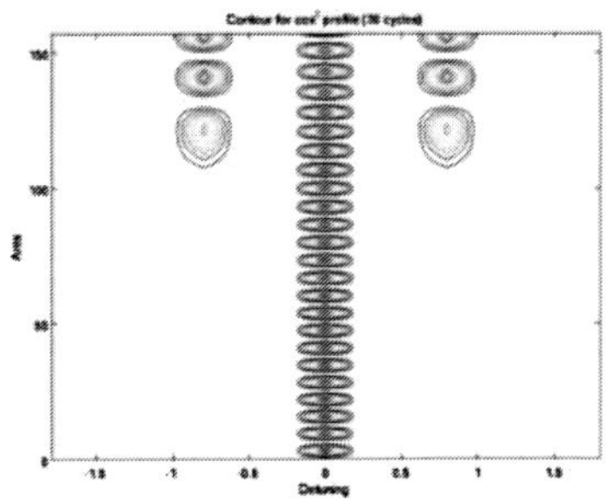

(m) $\chi(36) = 47.1239$

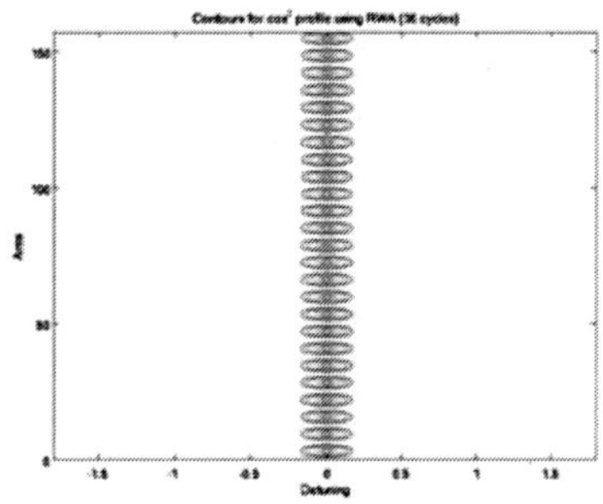

(n) with RWA, 36 cycles

Figure 3.9 contd...

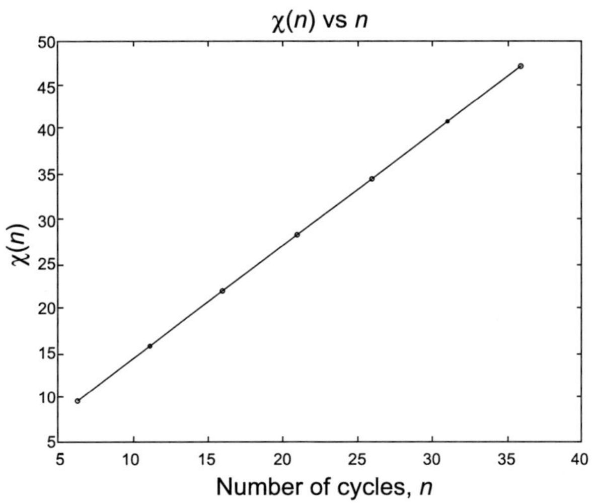

Figure 3.10 Best fit line: $\chi(n) = 1.2566n + 1.8850$

3.3.3 Analysis and observations

The lines of best fit for each of the profile types were plotted and the envelope profile dependence that is suggested by Table 3.1 and Fig. 3.11 was observed.

Table 3.1 *Fit parameters for given profiles*

Profile	$\chi(n)$	$\frac{\partial \chi}{\partial n}$
$\cos^2$	$1.2566n + 1.8850$	1.2566
sech	$0.7630n + 0.5834$	0.7630
Gaussian	$1.0322n + 4.8021$	1.0322
von Mises ($k = 10$)	$0.9874n + 4.8021$	0.9874
von Mises ($k = 2$)	$2.1094n + 18.0866$	2.1094

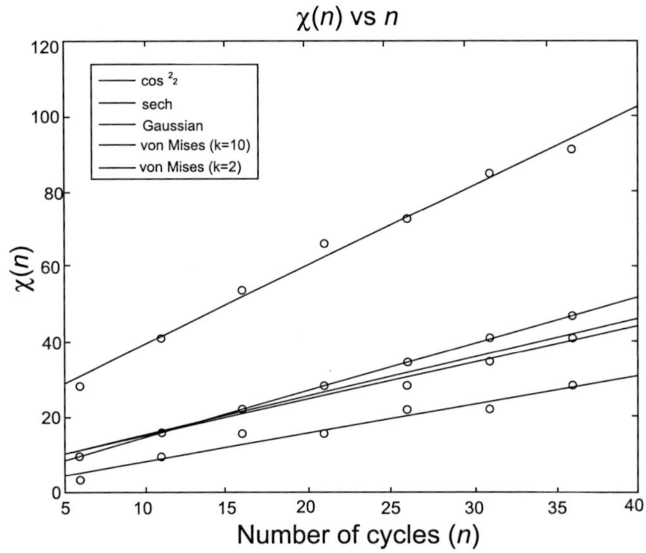

Figure 3.11 χ vs n for various profiles

3.4 Conclusion

$\chi(n)$ characterizes the critical limit of area, after which the cycling effect dominates the envelope profile effect. It was found that this measure is dependent on the envelope profile under question. Thus, despite the overwhelming importance of the cycling nature, the envelope profile still plays an important role, in all cases.

Bibliography

1. A. L. Cavalieri and N. Muller. 2007. 'Attosecond spectroscopy in condensed matter.' *Nature* 449: 1029.
2. G. Farkas and C. Toth. 1992. 'Proposal for attosecond light-pulse generation using laser-induced multiple harmonic conversion processes in rare-gases.' *Physics Letters A*. 168: 447.
3. S. Dasgupta, T. Khushwaha and D. Goswami. 2006. 'Dependence of adiabatic population transfer on pulse profile.' *Pramana* 66: 999.
4. S. Hughes. 1998. 'Breakdown of the area theorem: Carrier-wave rabi flopping of femtosecond optical pulses.' *Physical Review Letters* 81: 3363.
5. R. Parzynski and M. Sobczak. 2004. 'Analytical calculation of atomic response to a few-cycle optical pulse.' *J. Phys. B: At. Mol. Opt. Phys*. 37: 743.
6. Christopher Uiberacker and Werner Jakubetz. 2004. 'Molecular isomerization induced by ultrashort infrared pulses. I. few-cycle to sub-one-cycle Gaussian pulses and the role of the carrier-envelope phase.' *Journal of Chemical Physics* 120: 11532.
7. Yulia V. Bezvershenko and Petro I. Holod. 2011. 'Resonance in a driven two-level system: Analytical results without the rotating wave approximation.' *Physics Letters A*. 375: 3936.
8. Bruce W. Shore. 1990. *The Theory of Coherent Atomic Excitation* 1: chapter 3. New York: JohnWiley and Sons.

II
Laser Spectroscopy

4 Selective Photodissociation of Bonds in $^{18}O^{16}O^{16}O$ Molecule

Bhavesh, K. Shandilya, Manabendra Sarma, Satrajit Adhikari and Manoj K. Mishra

4.1 Introduction

The easy availability of ultrashort high intensity lasers has fuelled a variety of approaches for the selective control of bond dissociation using appropriate laser pulses[1–22, 24, 25, 27–30, 32–68]. Photodissociation is one of the most fundamental chemical processes. The process involves the distribution of products among the various possible quantum channels and provides unique insights into the energetics and the mechanism of bond breaking[8, 11, 43, 56, 67] which is of fundamental importance in chemistry. Considerable experimental and theoretical effort has been directed over the past few years towards the goal of laser assisted control of molecular dynamics[10, 12, 15, 20, 24, 39–42, 49–51, 58]. The important issue is how to control the branching ratio between chemically distinct channels. To change the branching ratio, several ways have been reported[19, 42]. Infrared mutiphoton excitation is one such process that induces selective bond dissociation. However, in this type of process, the intramolecular vibrational relaxation (IVR) creates a road block to selective control. With a properly designed laser pulse, the effect of IVR can be eliminated or minimized. However, the infrared mutiphoton excitation method requires very high intensity to detect dissociation probability and therefore may lead to molecular ionization.

Another way to investigate photo fragmentation is by using the photons in the ultraviolet part of the electro magnetic spectrum. The limitation of this process is that one of the excited states of molecule of interest must be purely repulsive. Also, with two lasers in the Continuous Wave regime, the dynamics can be explored in accordance with the Franck–Condon principle, which is known as passive control[19, 54]. In this approach, the first laser is used to excite the vibrational eigenstate of the molecule and the second laser to dissociate the molecule via excitation to the repulsive electronic state. The above method is suitable only for a molecule with local mode vibrational eigenstates such as the photodissociation of HOD molecule where the vibrational eigenstates are essentially local modes in the O–H and O–D stretch[6, 16, 22, 60, 61]. The method is inapplicable to those molecules where vibrational eigenstates are of normal mode type[2].

To overcome the limitations of passive control, there is another scheme which has been reported, known as the active control scheme[19, 54]. The "pump–dump" scheme given by Tannor and Rice[57] is an example of an active control scheme. In the pump–dump scheme, one requires two lasers, one intense in the infrared region which creates superposition of vibrational eigenstates and another short laser pulse in the ultraviolet region which takes the molecule in the dissociative state with an appropriate time delay between the pulses. Our group has also been involved in photodynamic control of di/tri-atomic molecules utilizing the technique of flux analysis in different channels to choose the form of external field as well as the initial state for selective control of bond cleavage[1, 7, 18, 44–48, 52, 62–66], and we have demonstrated, recently, the effectiveness of our approach in selective dissociation of O–H/O–D bonds in HOD molecule with desired extent of selectivity and sufficient yield[44–48, 52].

In our earlier investigations[44–48, 52] pertaining to HOD molecule, we have already reported a selective photodynamic control of this prototypical molecule. In one of those investigations[52], we used an IR pulse to soften the desired bond (either O–H or O–D bond) followed by an UV pulse to transfer the probability amplitude in the proper region of the upper repulsive surface with a proper time delay to ensure dissociation of the desired bond. The fundamental IR frequency corresponding to O–H bond (3706 cm^{-1}) had been used to get appropriate vibrational churn. This churn prepares a combination of pure O–H modes with a maximum population in the three quanta of excitation ($|3,0>$) in the O–H mode. With proper time delay between IR and UV pulses and a UV carrier frequency of 51,090 cm^{-1} employed to deposit the amplitude to the upper repulsive surface, the O–H bond alone could be broken with full selectivity and sufficient yield. The IR+UV dynamical calculations provided a maximum 82% flux in O–H mode. Likewise for O–D mode, the fundamental frequency of 2727 cm^{-1} had been used to get a linear combination of pure O–D modes with a maximum population in five quanta of excitation ($|0,5>$) in the O–D mode followed by a UV carrier frequency (46,062 cm^{-1}) of sufficient energy to get a maximum 79% flux in the H–O+D channel.

Along this same line, we have started investigating different routes for selective bond dissociation of the isotopic ozone molecule $^{18}O^{16}O^{16}O$ which is another much studied system of current interest[2, 9, 4, 5, 13, 21, 25, 27–30, 32–38, 53, 55, 59, 60].

Freeman et al.[17] have studied experimentally the photodissociation of ozone ($^{16}O^{16}O^{16}O$) in the UV region (240 nm $< \lambda <$ 350 nm) and observed $O_3 \rightarrow O_2(^1\Delta_g) + O(^1D)$ and $O_3 \rightarrow O_2(^3\Sigma g) + O(^1D)$ as products. The absorption spectra obtained in their measurement was for $\lambda >$ 315 nm with small oscillations of about 250 cm^{-1} seen on the top of the UV absorption band. Several theoretical calculations have been performed by using potential energy surfaces (PESs) of Sheppard and Walker[55] where the equation for dynamics were solved using different approximate methods namely semi-classical Gaussian wave packet approach[21] and exact two-dimensional quantum calculations[13] with frozen bending angle, but none of those approaches could produce those oscillations of the experimental absorption spectrum. On the contrary, the Fourier transform of experimental spectrum from frequency domain to the time domain provides the 'experimental' auto correlation function. It shows that before the photodossociation process, the system gets trapped in the Franck–Condon region giving rise to the structures at 250 cm^{-1} in the spectrum. The other experimental observation on the vibrationally and rotationally resolved spectra of the product, $O_2(^1\Delta g)$, brings insight into dissociation dynamics.

As mentioned earlier, selective photodissociation of HOD has been investigated in detail in our group[1, 44–48, 52]. In case of the HOD molecule, deuterium (D) is twice as heavy as the hydrogen (H) atom and is connected via the much heavier oxygen atom which makes it a prototypical bond specific system for numerical investigation. It is our purpose here to present some initial results from our investigation of the photodissociation of $^{18}O^{16}O^{16}O$ molecule using approaches developed in our group for the HOD molcecule. Preliminary results obtained from this investigation provide further insights into selective control of photodissociation of small polyatomics.

4.2 Method

Following the earlier investigations[2, 4, 13, 55, 59, 68], two degrees of freedom model of $^{18}O^{16}O^{16}O$, considering only the ground (X) and the second excited (B) electronic states of the molecule[55] held great scientific fascination. The bending degree of freedom can be neglected for the lowest vibrational energy levels with zero quanta of excitation in the bending mode in the ground electronic state[13]. Though the bending mode is active in the second excited electronic state[4, 29], for the short pulse durations considered here, excitation to and dynamics at the B state could yield qualitatively correct branching ratio with frozen bending mode. The rotational motion can be left out because of the differences in the dynamical time scale. The coordinates describing the $^{18}O-^{16}O$ and $^{16}O-^{16}O$ are denoted by r_1 and r_2 and the associated conjugated momenta by p_1 and p_2, respectively. The internal kinetic energy operator for these two modes, thus, can be expressed as[2]

$$\hat{T} = \frac{\hat{p}_1^2}{2\mu_1} + \frac{\hat{p}_2^2}{2\mu_2} + \frac{\hat{p}_1\hat{p}_2}{2m(^{16}O)}\cos\theta,$$

$$\text{where,} \quad \hat{p}_j = \frac{\hbar}{i}\frac{\partial}{\partial r_j}, \qquad j = 1,2, \tag{4.1}$$

$$\mu_1 = \frac{m(^{18}O)m(^{16}O)}{m(^{18}O)+m(^{16}O)}, \; \mu_2 = \frac{m(^{16}O)m(^{16}O)}{m(^{16}O)+m(^{16}O)}$$

and the bending angle θ is fixed at 116.8°. The subscript '1' refers to $^{18}O-^{16}O$ and that of '2' refers to $^{16}O-^{16}O$ bonds. The ground (X) and excited (B) potential energy surfaces are available from ab initio calculations[55]. The contour plots of these two surfaces are displayed in Fig. 4.1 (left panel for the ground state and right panel for the excited state). As can be seen there is one minimum along both $^{16}O-^{16}O$ and $^{18}O-^{16}O$ channels.

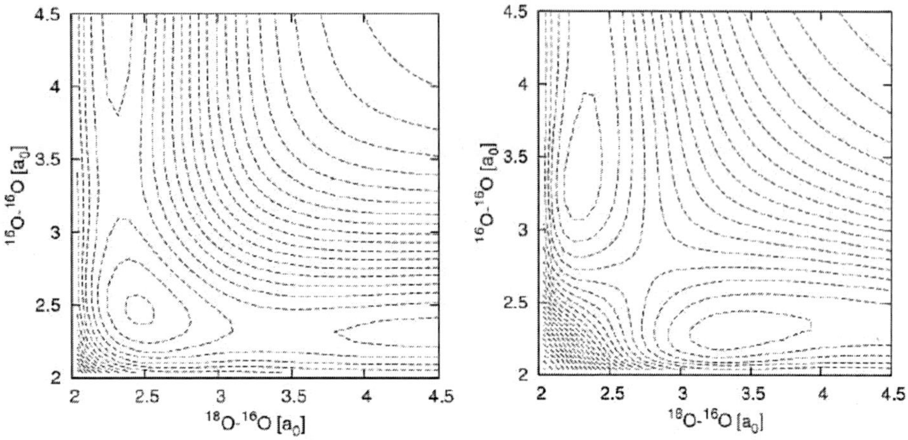

Figure 4.1 Contour plots of ground (X) [left panel] and excited (B) [right panel] potential energy surfaces
Source: [55]

This ground state surface is symmetric along the diagonal passing through the lower bond length. In case of the excited (B) surface, being diagonally symmetric, two shallow minima can be seen, one along the $^{16}O-^{16}O$ bond and the other along $^{18}O-^{16}O$ channel. The dipole moment surface of the electronic ground state is taken from the contribution by Alder-Golden et al.[3] which has two components, μ_x, along the bisector of the two bonds and μ_y, along the perpendicular to the bisector. Both of these are shown in Fig. 4.2 (the left panel corresponds to μ_x and the right panel for μ_y). The interaction of the ground electronic state dipole moment with the electric field vector of the IR laser is the same as that used in the selective photodynamic control of the HOD molecule[52].

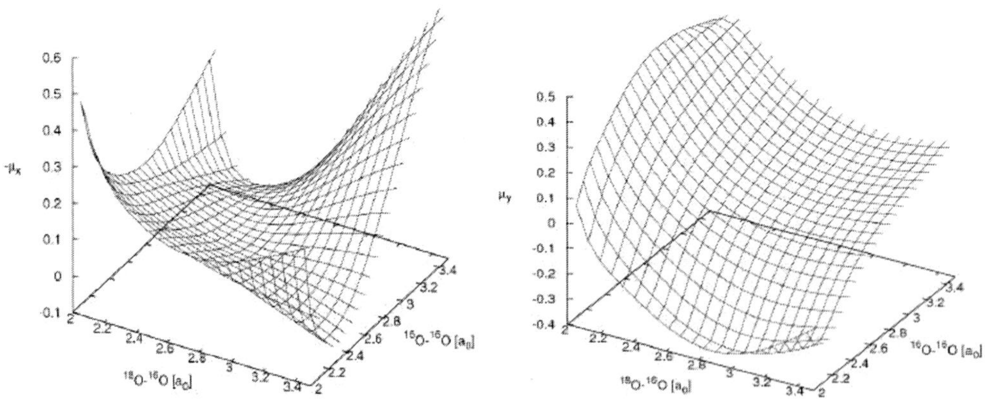

Figure 4.2 *x* component (μ_x) [left panel] and *y* component (μ_y) [right panel] of the ground electronic state dipole moment
Source: [3]

The transition dipole moment (μ_{uv}) (Fig. 4.3) for X → B transitions, which is in the molecular plane and perpendicular to the bisector of the two bonds, is obtained from ab initio calculations[29] for a small range of bond length values. The interactions of this transition dipole moment with the electric field vector of the UV laser are also taken to be similar to that used for HOD[45–47, 52]

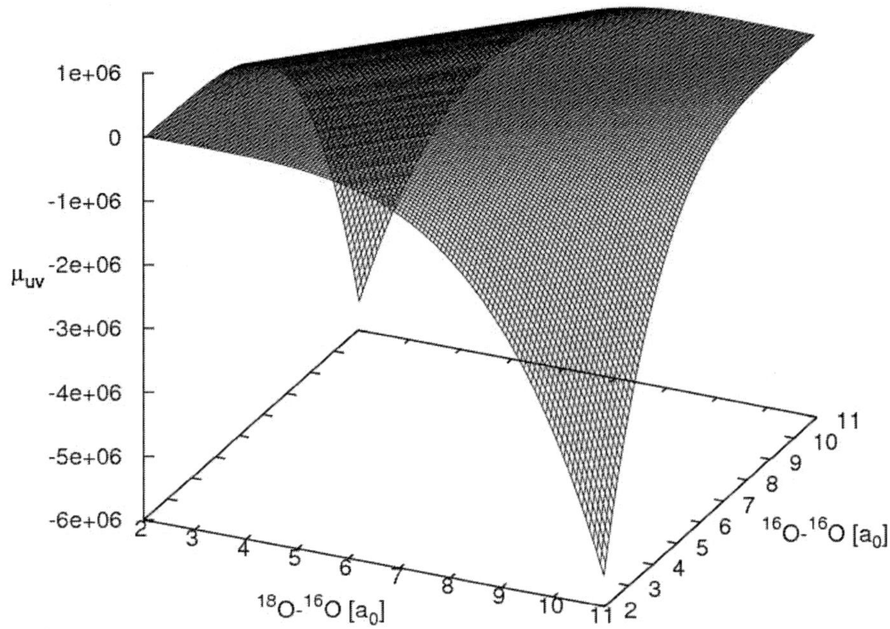

Figure 4.3 Transition dipole moment surface for UV dynamics
Source: [29]

The time evolution of nuclear motion in ground and excited states can be performed using the time dependent Schrödinger equation[2],

$$i\hbar\frac{\partial}{\partial t}\begin{pmatrix}\Psi_g\\ \Psi_e\end{pmatrix}=\begin{pmatrix}\hat{H}_g & \hat{H}_{uv}(t)\\ \hat{H}_{uv}(t) & \hat{H}_e\end{pmatrix}\begin{pmatrix}\Psi_g\\ \Psi_e\end{pmatrix}, \tag{4.2}$$

where $\Psi_g = \Psi_g(r_1, r_2, t)$ and $\Psi_e = \Psi_e(r_1, r_2, t)$ are the wave functions associated with nuclear motion in the ground and second excited electronic states, respectively. $\hat{H}_g = \hat{T} + \hat{V}_g$ and $\hat{H}_e = \hat{T} + \hat{V}_e$ are the nuclear Hamiltonians for the two electronic states, where H_{uv} couples as well as perturbs both the electronic states. We solve Eq. (4.2) with ($t = 0$) initial condition that the ground state wave function Ψ_g is a single, field free, vibrational state or a linear combination of more than one vibrational state(s) of the $^{18}O^{16}O^{16}O$ electronic ground state, and the excited state wave function $\Psi_e = 0$, at $t = 0$.

Vibrational eigenfunctions of the ground electronic state of the $^{18}O^{16}O^{16}O$ molecule were obtained using the Fourier grid Hamiltonian (FGH) method[31] modified for two dimensions[14]. After the preparation of initial wave functions, the propagation of the wave functions $\{\Psi_g(t), \Psi_e(t)\}$ has been performed using Eq. (4.2) where the effect of kinetic energy operators of the Hamiltonian on the wave function is evaluated with a two-dimensional fast Fourier transform (FFT)[23] algorithm and the time propagation is carried out using the Lanczos scheme[26]. The wave function is represented on a spatial grid spanning r_1/r_2 bond lengths between $2a_0$ to $11a_0$ in 128 steps with $\Delta r_1 = \Delta r_2 = 0.0703\, a_0\, \left(0.0371\text{Å}\right)$ and propagation of field effects under the influence of a generic Gaussian UV pulse of the form $E(t) = 0.02a(t)\cos\omega t$, where $a(t) = \exp[-\gamma(t-t_{uv})^2]$ with $FWHM = \sqrt{4\ln 2/\gamma}$, done in the time steps of $\Delta t = 1$ a.u. of time = 0.02418842 fs. Maximum field amplitude for the electric field employed in this investigation is 0.10 GV/cm and maximum field intensity is 22 TW/cm^2.

As we know from our earlier investigations, product yield in the specific channels is related to the time-integrated flux. Therefore, the flux expressions for the total flux J in $^{18}O + ^{16}O - ^{16}O$ and $^{18}O - ^{16}O + ^{16}O$ labeled as subscript 1 and 2 are given by,

$$J_1 = \int_0^{r_{2d}} \int_0^{T} \Psi^*(r_1, r_2, t)\left(\hat{j}_1 + \frac{\mu_2 \cos\theta}{2m(^{16}O)}\hat{j}_2\right)\Psi(r_1, r_2, t)dr_2 dt \tag{4.3}$$

$$J_2 = \int_0^{r_{1d}} \int_0^{T} \Psi^*(r_1, r_2, t)\left(\hat{j}_2 + \frac{\mu_1 \cos\theta}{2m(^{16}O)}\hat{j}_1\right)\Psi(r_1, r_2, t)dr_1 dt, \tag{4.4}$$

where, $\hat{j}_i$ is the flux operator in the ith channel, defined as $\hat{j}_i = \frac{1}{2\mu_i}[\hat{p}_i\delta(r_i - r_i^d) + \delta(r_i - r_i^d)\hat{p}_i] \cdot \mu_i, \hat{p}_i$ and r_i^d are the reduced mass, the momentum operator, and a grid point in the asymptotic region of the ith channel. The flux in $^{18}O + ^{16}O - ^{16}O/^{18}O - ^{16}O + ^{16}O$ channels are evaluated along asymptotic cuts at $r_1 = r_2 = 8.5a_0$. An absorbing optical potential is used at the asymptotic cuts to avoid unphysical reflection from the edges. The second operator in Eqs 4.3 and 4.4 represents the effect of kinetic coupling between the $^{18}O - ^{16}O$ and $^{16}O - ^{16}O$ modes. The field dependence of $H(\vec{r}_1, \vec{r}_2, t)$ manifests itself through $U(t,0) \simeq e^{-iHt/\hbar}$ where $H = H_{\text{molecule}} + H_{uv}(t)$, and we can see that product yield in the desired channel may be altered by altering the field or $\Psi_g(0)$.

4.3 Results and Discusion

In this chapter preliminary results from the calculations of $^{18}O^{16}O^{16}O$ photodissociation where only one initial vibrational state is involved will be discussed. To start with, the potential energy surfaces of Sheppard and Walker[55] (Fig. 4.1) were used for dynamical calculations. There is another set of PESs of Leforestier et al.[29] available in literature, which however, have not been considered in this initial investigation. The vibrational eigenstates of the ground electronic state of $^{18}O^{16}O^{16}O$ molecule were obtained using the Fourier grid Hamiltonian (FGH) method[31] modified for two dimensions[14]. The first six eigen modes are depicted in Fig. 4.4. As can be seen in Fig. 4.4, eigen modes are not along the axes corresponding to either $^{16}O - ^{16}O$ bond or $^{18}O - ^{16}O$ mode. While the ground vibrational state ($|0,0>$) [where the first no. refers to quantum of excitation in $^{18}O - ^{16}O$ mode and the second to the no. of quantum of excitation in $^{16}O - ^{16}O$ mode] is totally symmetric, all other eigen modes are angular and directed in one of the $^{16}O - ^{16}O/^{18}O - ^{16}O$ modes. This angular nature of the eigen modes is a signature of strongly coupled modes which in turns give rise to closely spaced vibrational energy levels as shown in Table 4.1. It is seen from Table 4.1 that the energy difference between ($|1,0>$) and ($|0,1>$) is only ≈ 170 cm^{-1} in the calculation; as the energy difference between these two states calculated by Amstrup and Henriksen[2] was only 53 cm^{-1}. Similar trends are observed in other vibrational states as well. It is, therefore, clear that to prepare pure $^{16}O-^{16}O$ or $^{18}O-^{16}O$ mode is a difficult task unlike in the case of HOD where the difference between two consecutive modes is more than $\approx$ 1000 cm^{-1}. Following the successful investigation of HOD[44–48, 52], here too, a Gaussian UV pulse, being peaked at 100 fs with a pulse length of 250 fs, with the same temporal width (FWHM = 50 fs) was utilized to permit easy separation of frequencies involved in photodynamics. The field employed and its power spectrum is depicted in Fig. 4.5. The single peak shown in the power spectrum confirms the use of single color laser pulse.

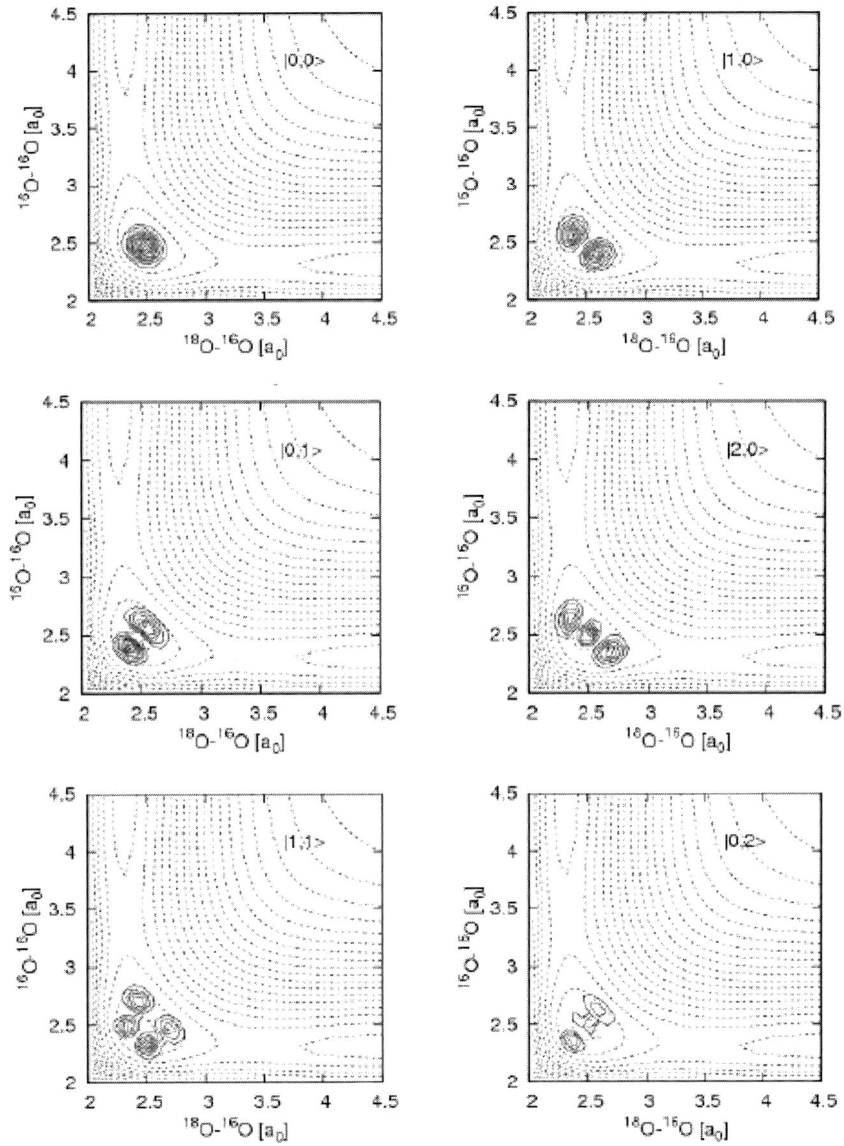

Figure 4.4 First six eigen modes of $^{18}O^{16}O^{16}O$ molecule on contours of ground potential energy surface

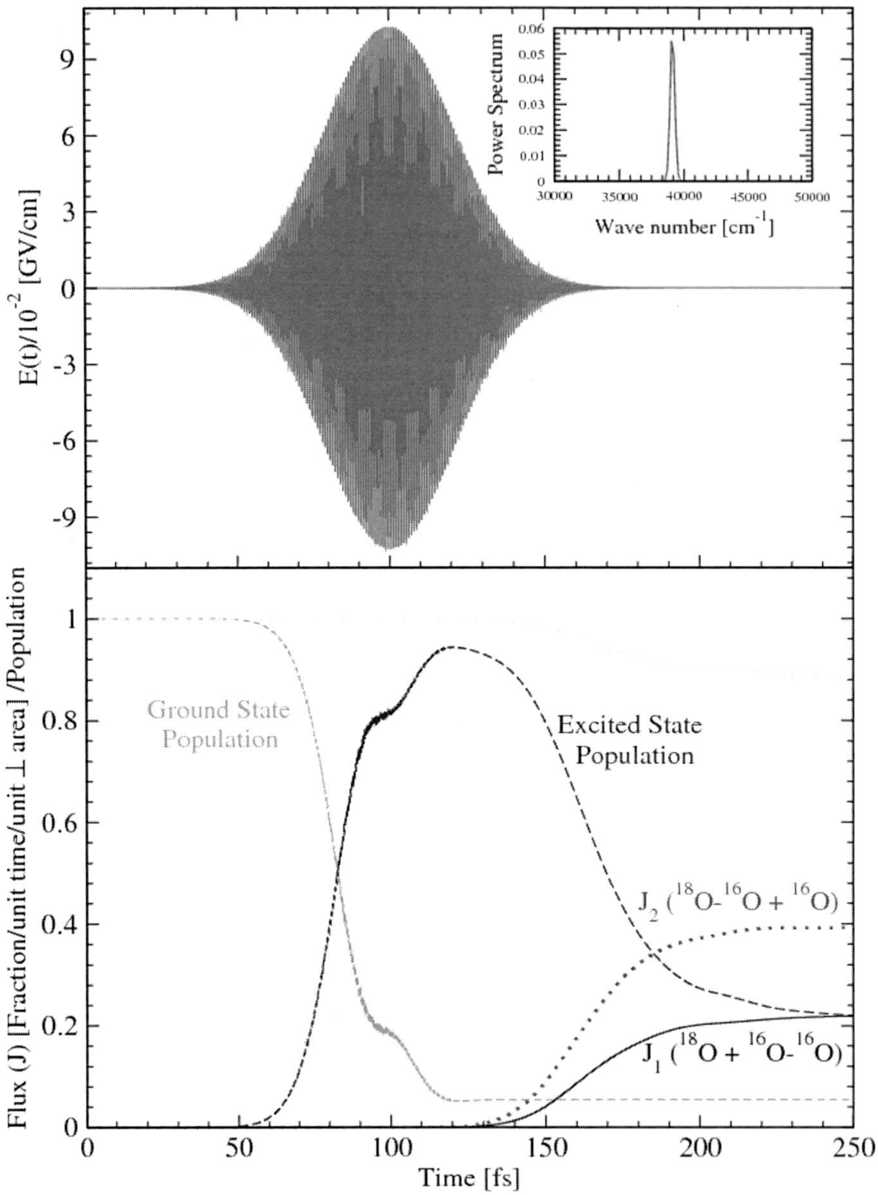

Figure 4.5 UV laser pulse $E(t) = 0.02a(t)\cos\omega t$, where $a(t) = \exp[-\gamma(t-t_{uv})^2$ with FWHM $= \sqrt{4\ln 2/\gamma} = 50$ fs. Maximum field amplitude is 0.10 GV/cm and maximum field intensity is 22 TW/cm^2 . The corresponding power spectrum is shown in the upper panel and in the lower panel is plotted flux/population vs. time with $|0,0>$ as the initial state of the $^{18}O^{16}O^{16}O$ molecule under the influence of the Gaussian UV pulse shown in the top panel

Table 4.1 *Vibrational frequencies of first six modes of* $^{18}O^{16}O^{16}O$ *molecule*

No. of quantum of excitation in $^{18}O-^{16}O$ mode	No. of quantum of excitation in $^{16}O-^{16}O$ mode	Frequency (cm^{-1}) [Present calculation]	Frequency (cm^{-1}) [Earlier calculation] (Amstrup and Henriksen, [2])
0	0	0	0
1	0	957	1002
0	1	1127	1055
2	0	1887	1957
1	1	2024	2012
0	2	2243	2105

Table 4.2 *Mapping of UV frequencies in the Hartley absorption band (32000 – 55000 cm^{-1}). The time integrated flux in* $^{18}O + ^{16}O - ^{16}O(J_1)$ *and* $^{18}O - ^{16}O + ^{16}O(J_2)$ *channels and population distribution in ground and excited states after 250 fs are also provided*

UV frequency ω (cm^{-1})	J_1(^{18}O + $^{16}O^{16}O$ channel) in %	J_2($^{18}O^{16}O$+ ^{16}O channel) in %	Ground state population in %	Excited state population in %	Total population after 250 fs in %
35000	0.28	0.3	98.8	0.6	100
38000	20.9	23.5	28.6	21.6	93.6
38919	22.7	38.4	6.2	21.6	88.8
39000	22.3	39.3	5.9	21.6	88.8
39019	21.9	39.4	5.9	21.7	88.8
39069	21.8	39.5	5.7	21.8	88.7
39119	21.9	39.3	5.4	22.0	88.6
39169	22.2	38.9	5.1	22.3	88.5
39219	22.7	38.2	4.6	22.7	88.3
39319	24.1	36.5	3.5	23.5	87.7
40000	25.3	32.4	1.9	22.5	82.1
41000	19.5	29.8	10.0	20.8	80.0
45000	7.0	9.4	59.4	11.5	87.2
50000	2.6	4.39	73.1	10.7	90.6
55000	1.4	1.6	91.0	2.7	96.6

With the field profile of Fig. 4.5 and the ground vibrational state ($|0,0>$) of the ground electronic state as the initial state, the time propagation has been carried out using Lanczos scheme[29] in time steps of 1 a.u. of time (= 0.024 fs) for 10340 time steps (250 fs) as mentioned earlier.

Employing this Gaussian UV pulse of FWHM 50 fs, UV dynamics was carried out in the Hartley absorption band (32000 cm^{-1}–55000 cm^{-1}). It is shown in Table 4.2 that, at lower UV frequencies, the flux in both channels are negligible due to the insufficient energy for transferring the probability amplitude to the excited surface. It is also remarkable that nil or insignificant population is transferred from ground to excited surface even with a sufficiently energetic pulse which is the signature of the Franck–Condon window. The transfer of population from ground to excited state is feasible for optimal choice of frequencies and thereby flux in both channels can be attained. The maximum flux in the $^{16}O - ^{16}O$ mode is 39.5% at the UV frequency of 39 069 cm^{-1}. However, the sum of ground and excited states populations and flux flowing through both channels deviates from unity for the frequency which is able to transfer sufficient population to the excited state. This may be due to the possibility of opening the totally dissociative $^{16}O + ^{16}O + ^{18}O$ channel to which some of the probability density could be flowing through. In the lower panel of Fig. 4.5, the distribution of flux in both channels and populations in ground and excited states versus time in femtosecond (fs) for 39,069 cm^{-1} is shown as the frequency providing the highest flux in the $^{16}O - ^{16}O$ mode. Since the pulse with FWHM = 50 fs, peaked at 100 fs, the effect of pulse starts significantly from 75 fs and remain powerful up to 125 fs. The population in the excited state starts building even before 75 fs and reaches its maxima at about 125 fs. Finally, it starts diminishing with a kick of flux flowing through both channels, as the effect of pulse is no more significant for pumping the population from ground to excited state. It is seen that after 250 fs, the population remaining in ground and excited states are 5.7% and 21.8% respectively.

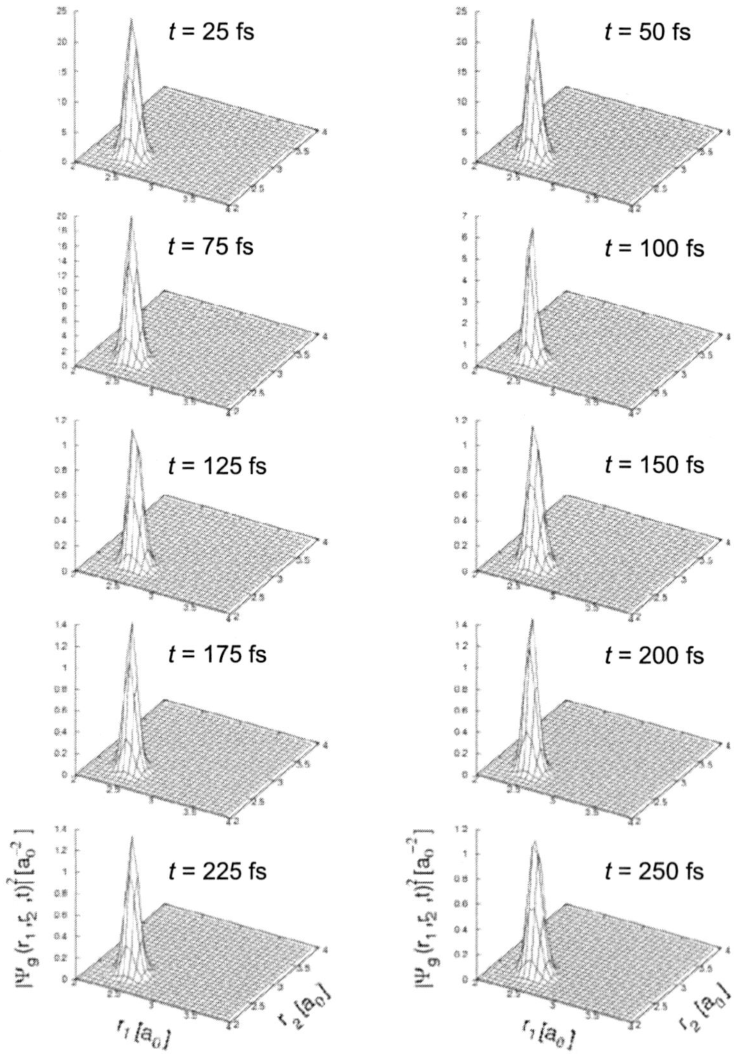

Figure 4.6 Time evolution of (|0, 0 >) state on ground electronic state at t =25, 50, 75, 100, 125, 150, 175, 200, 225, and 250 fs under the influence of Gaussian UV pulse (Fig. 4.5)

The final flux flowing through $^{16}O-^{16}O$ and $^{18}O-^{16}O$ channels are respectively 39.5 % and 21.8%. The total population after 250 fs is 88.8%, and therefore, there is a loss of 11.2% population from the total population. As mentioned earlier, this 11.2% population may end up flowing to possibly the $^{18}O+^{16}O+^{16}O$ channel. The greater flux in the $^{18}O-^{16}O+^{16}O$ channel than that in the $^{18}O+^{16}O-^{16}O$ may be attributed to the lower mass of ^{16}O atom and the consequent ease for larger amplitude vibrations in $^{16}O-^{16}O$ than $^{18}O-^{16}O$, whereby $^{16}O-^{16}O$ bond dissociation dominates $^{18}O-^{16}O$ bond for the field profile chosen in this investigation.

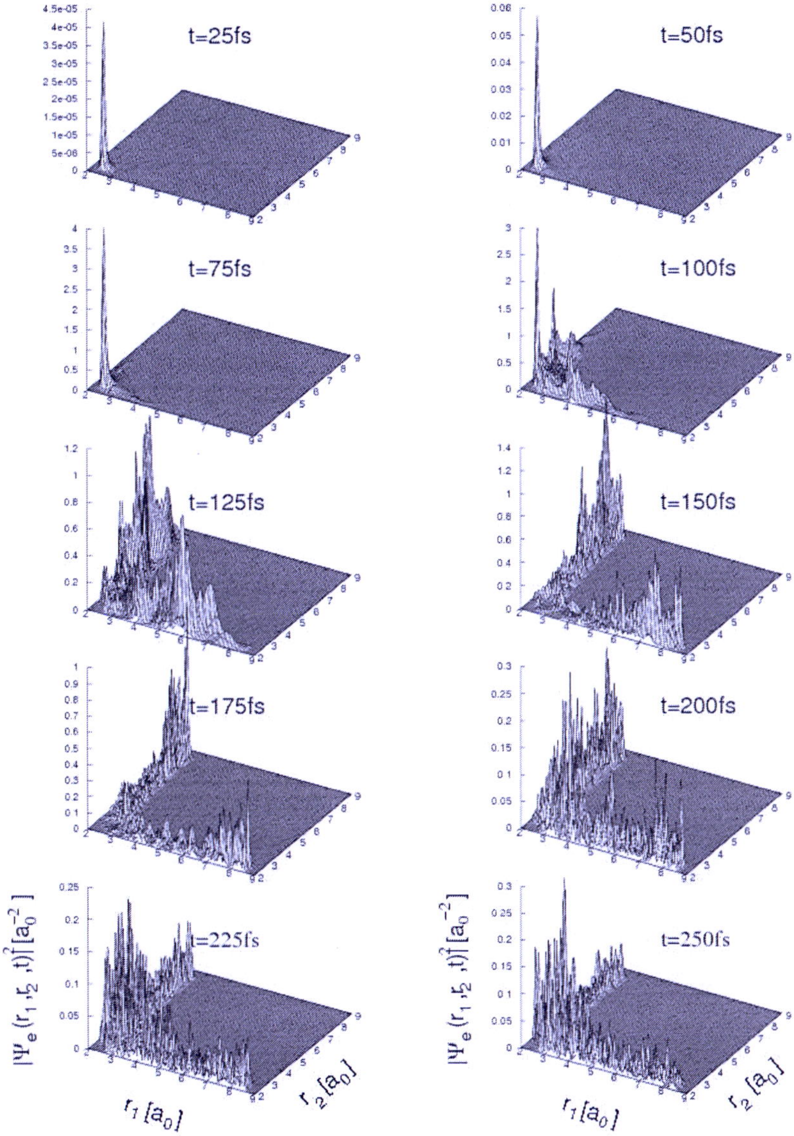

Figure 4.7 Time evolution of (|0, 0 >) state on excited electronic state at t = 25, 50, 75, 100, 125, 150, 175, 200, 225, and 250 fs under the influence of Gaussian UV pulse (Fig. 4.5)

To understand the mechanistic features which may assist in selective control, the time evolution of $|\Psi_g(r_1, r_2, t)|^2$ and $|\Psi_e(r_1, r_2, t)|^2$ on the ground (X) and excited (B) states of the molecule were examined. Results from the time evolution of the ($|0,0>$) state on ground and excited electronic states, are plotted in Figs 4.6 and 4.7 respectively. There is a diminution of probability amplitude from 25 fs itself on the ground electronic state and it is persistent throughout the dynamics with a small change in the amplitude in the ground state after 150 fs. On the excited state, the probability amplitude is accumulated from the very beginning. However, if we compare the plots of $t = 100$ fs and 125 fs on both ground and excited states, we can see that there is a dumping of probability amplitude from the excited electronic state to the ground electronic state. This may be due to the active manipulation by our applied electric field thus causing mixing of other vibrational states. With increasing time, the probability amplitude separates into two modes in which there is more amplitude flow towards the $^{16}O-^{16}O$ bond than the $^{18}O-^{16}O$ bond as can be seen in the time evolution plot of probability density on the excited state surface (Fig. 4.7). This variation of amplitude in both bond modes is active as long as the field is switched on. As soon as the field is switched off, the remaining amplitude flows through the respective channels and there is a kick in flux values which can be seen from the lower panel of flux vs. time plot of Fig. 4.5. The final flux in the $^{18}O-^{16}O+^{16}O$ channel (39.5%) is approximately twice as large as that in the $^{18}O+^{16}O-^{16}O$ channel (21.8%).

4.4 Concluding Remarks

In this chapter, we discussed a fully quantal time dependent investigation for photodynamic control of selective dissociation of $^{18}O-^{16}O$ and $^{16}O-^{16}O$ bonds in $^{18}O^{16}O^{16}O$ molecule using experimentally realizable Gaussian UV pulses and from the ground vibrational state of the ground electronic state ($|0,0>$). The results obtained were analyzed using time evolution of the probability density plots and flux profiles on both ground and excited states. Results from the initial calculations have shown a $^{18}O-^{16}O+^{16}O/^{18}O+^{16}O-^{16}O$ flux branching ratio of $\sim$1.8 and thus demonstrates an effective route to selective photodissociation in the $^{18}O^{16}O^{16}O$ molecule.

We hope the results presented here will provide a good starting point for more extensive future investigations of the $^{18}O^{16}O^{16}O$ molecule. An effort along these lines is under way in our group.

Bibliography

1. S. Adhikari, S. Deshpande, M. Sarma, V. Kurkal and M. K. Mishra. 2006. 'Selective control of photodissociation in deutereted water HOD.' *Radiat. Phys. Chem.* 75: 2106–2118.

2. B. Amstrup and N. E. Henriksen. 1996. 'Two-pulse laser control of bond-selective fragmentation.' *J. Chem. Phys.* 105: 9115–9120.

3. S. M. Adler-Golden, S. R. Langhoff, C. W. Bauschilicher and G. D. Carney. 1985. 'Theoretical calculation of ozone vibrational infrared intensities.' *J. Chem. Phys.* 83: 255–264.

4. N. Balakrishnan and G. D. Billing. 1994. 'Three-dimensional wave packet studies of ozone photodissociation in the Hartley band: Converged autocorrelation functions and absorption spectra.' *J. Chem. Phys.* 101: 2968–2977.

5. E. Baloitcha and G. G. Balint-Kurti. 2005. 'Theory of the photodissociation of ozone in the Hartley continuum: Potential energy surfaces, conical intersections and photodissociation dynamics.' *J. Chem. Phys.* 123: 014306/1–014306/11.
6. I. Bar, Y. Cohen, D. David, S. Rosenwaks and J. J. Valentini. 1990. 'Direct observation of preferential bond fission by excitation of a vibrational fundamental: Photodissociation of HOD (0, 0, 1).' *J. Chem. Phys.* 93: 2146–2148.
7. D. B. Bairagi, P. Gross and M. K. Mishra. 'Field optimized initial state based control of IBr photodissociation.' *J. Phys. Chem. A.* 101: 759–763.
8. A. Ben-Shaul, Y. Haas, K. L. Kompa and R. D. Levine. 1981. *Laser and Chemical Change.* Berlin: Springer Verlag.
9. G. Barinovs, N. Markovic and G. Nyman. 1999. '3D wavepacket calculations of ozone photodissociation in the Hartley band: Convergence of the autocorrelation function.' *Chem. Phys. Lett.* 315: 282–286.
10. I. Bar and S. Rosenwaks. 2001. 'Controlling bond cleavage and probing intramolecular dynamics via photodissociation of rovibrationally excited molecules.' *Int. Rev. Phys. Chem.* 20: 711–749.
11. J. G. Calvert and J. N. Pitts. 1966. *Photochemistry.* New York: Wiley.
12. R. Chakrabarti and H. Rabitz. 2007. 'Quantum control landscapes.' *Int. Rev. Phys. Chem.* 26: 671–735.
13. D. Chasman, D. J. Tannor and D. G. Imre. 1988. 'Photoabsorption and photoemission of ozone in the Hartley band.' *J. Chem. Phys.* 89: 6667–6675.
14. P. Dutta, S. Adhikari and S. P. Bhattacharyya. 1993. 'Fourier grid Hamiltonian method for bound states of multidimensional systems: Formulation and preliminary applications to model systems.' *Chem. Phys. Lett.* 212: 677–684.
15. C. G. Elles and F. F. Crim. 2006. 'Connecting chemical dynamics in gases and liquids.' *Anu. Rev. Phys. Chem.* 57: 273–302.
16. V. Engel and R. Shinke. 1988. 'Isotopic effects in the fragmentation of water: The photodissociation of HOD in the first absorption band.' *J. Chem. Phys.* 88: 6831–6837.
17. D. E. Freeman, K. Yoshino, J. R. Esmond and W. H. Parkinson. 1984. 'High resolution absorption cross-section measurements of ozone at 195K in the wavelength region 240–350 nm.' *Planet Space Sci.* 32: 239–248.
18. P. Gross, A. K. Gupta, D. B. Bairagi, M. K. Mishra. 1996. 'Nitial state laser control of curvecrossing reactions using the Rayleigh-Ritz variational procedure.' *J. Chem. Phys.* 104: 7045–7051.
19. N. E. Henriksen. 1995. 'Theoretical concepts in molecular photodissociation dynamics.' *Adv. Chem. Phys.* 91: 433–509.
20. N. E. Henriksen. 2002. 'Laser control of chemical reactions.' *Chem. Soc. Rev.* 31: 37–42.
21. P. J. Hay, R. T. Pack, R. B. Walker and E. J. Heller. 1982. 'Photodissociation of ozone in the Hartley band. Exploratory potential energy surfaces and molecular dynamics.' *J. Phys. Chem.* 86: 862–865.
22. D. G. Imre and J. Zhang. 1989. 'Dynamics and selective bond breaking in photodissociation.' *Chem. Phys.* 139: 89–121.

23. D. Kosloff and R. Kosloff. 1983. 'A Fourier method solution for the time dependent Schrödinger equation as a tool in molecular dynamics.' *J. Comput. Phys.* 52: 35–53.

24. P. Krl, I. Thanopulos and M. Shapiro. 2007. 'Colloquium: Coherently controlled adiabatic passage.' *Rev. Mod. Phys.* 79: 53–77.

25. L. C. Lee, G. Black, R. L. Sharpless and T. G. Slanger. 1980. 'O(1S) yield from O3 photodissociation at 1700-2400Å.' *J. Chem. Phys.* 73: 256–258.

26. C. Leforetier, R. H. Bisseling, C. Cerjan, M. D. Feit, R. Friesner, A. Guldberg, A. Hammerich, G. Jolicard, W. Karralein, H. D. Meyer, N. Lipkin, O. Roncero and R. Kosloff. 1991. 'A comparison of different propagation schemes for the time dependent schrödinger equation.' *J. Comput. Phys.* 94: 59–80.

27. S. Y. Lin, K. L. Han and G. Z. He. 2001. 'Exact three-dimensional quantum mechanical calculation of ozone photodissociation in the Hartley band.' *J. Chem. Phys.* 114: 10651 –10661.

28. S. Y. Lin, K. L. Han and G. Z. He. 2001. 'Effects of the transition dipole moment function on the dynamics of ozone photodissociation: An exact 3D quantum mechanical study.' *Chem. Phys.* 273: 169–174.

29. C. Leforestier, Le F. Qur, K. Yamashita and K. Morokuma. 1994. 'Theoretical study of the ultraviolet photodissociation of ozone. Comparison with experiments.' *J. Chem. Phys.* 101: 3806–3818.

30. S. K. Lee, D. Townsend, O. S. Vasyutinskii and A. G. Suits. 2005. 'O(1D2) orbital orientation in the ultraviolet photodissociation.' *Phys. Chem. Chem. Phys.* 7: 1650–1656.

31. C. C. Marston and G. G. Balint-Kurti. 1989. 'The Fourier grid Hamiltonian method for bound state eigenvalues and eigenfunctions.' *J. Chem. Phys.* 91: 3571–3576.

32. R. L. Miller, A. G. Suits, P. L. Houston, R. Toumi, J. A. Mack and A. M. Wodtke. 1994. 'The ozone deficit problem: O_2(X, v ge 26) + O(^{3}P) from 226-nm ozone photodissociation.' *Science* 265: 1831–1838.

33. S. A. Ndengu, F. Gatti, R. Schinke, H. D. Meyer and R. Jost. 2010. 'Absorption crosssection of ozone isotopologues calculated with multiconfiguration time-dependent hartree (MCTDH) method: Hartley and Huggins Bands.' *J. Phys. Chem. A.* 114: 9855–9863.

34. G. Parlant. 2000. 'Classical survival probability for ozone photodissociation in the Hartley band.' *J. Chem. Phys.* 112: 6956–6958.

35. Le F. Qur and C. Leforestier. 1990. 'Quantum exact three-dimensional study of the photodissociation of the ozone molecule.' *J. Chem. Phys.* 92: 247.

36. Le F. Qur and C. Leforestier. 1991. 'Hyperspherical formulation of the photodissociation of ozone.' *J. Chem. Phys.* 94: 1118–1126.

37. Z. W. Qu, H. Zhu, S. Yu. Grebenshchikov and R. Shinke. 2005. 'The triplet channel in the photodissociation of ozone in the Hartley band: Classical trajectory surface hopping analysis.' *J. Chem. Phys.* 122: 191102/1–191102/4.

38. Z. W. Qu, H. Zhu, S. Yu. Grebenshchikov and R. Shinke. 2005. 'The photodissociation of ozone in the Hartley band: A theoretical analysis.' *J. Chem. Phys.* 123: 074305/1–074305 /12.

39. S. A. Rice. 2001. 'Interfering for the good of a chemical reaction.' *Nature* (London) 409: 422–426.
40. H. Rabitz. 2006. 'Strong-arming molecular dynamics.' *Science* 314: 264–265.
41. S. A. Rice and M. Zhao. 2000. *Optical control of molecular dynamics*. New York: Wiley.
42. H. Rabitz and W. Zhu. 2000. 'Optimal control of molecular motion: Design, implementation and inversion.' *Acc. Chem. Res.* 33: 572–578.
43. R. Shinke. 1993. *Photodissociation dynamics*. Cambridge: Cambridge University Press.
44. M. Sarma, S. Adhikari and M. K. Mishra. 2006. 'Selective control of HOD photodissociation using low quanta O–D excitation and field optimized initial state (FOIST) based combination of states and colors.' *Chem. Phys. Lett.* 420: 321–329.
45. M. Sarma, S. Adhikari and M. K. Mishra. 2007. 'Selective photodissociation of O–H and O–D bonds from ground vibrational state of HOD using simple UV pulses.' *J. Chem. Phys.* 127: 024305/1–024305/5.
46. M. Sarma, S. Adhikari and M. K. Mishra. 2007. 'Selective control of HOD photodissociation using CW lasers.' *J. Chem. Sci.* 119: 377–384.
47. M. Sarma, S. Adhikari and M. K. Mishra. 2008. 'An examination of the expectation value profiles for average stretch and average momentum in O–H and O–D bonds in the HOD molecule to determine their role in selective photodissociation.' *J. Phys. Chem. A.* 112: 13302–13307.
48. M. Sarma, S. Adhikari and M. K. Mishra. 2009. 'Laser assisted control of selective bond dissociation in HOD-some mechanistic insights.' *Mol. Phys.* 107: 939–961.
49. M. Shapiro and P. Brumer. 2003. *Principles of the quantum control of molecular processes*. New York: Wiley.
50. M. Shapir and P. Brumer. 2003. 'Coherent control of molecular dynamics.' *Rep. Prog. Phys.* 66: 859–942.
51. M. Shapiro and P. Brumer. 2006. 'Quantum control of bound and continuum state dynamics.' *Phys. Rep.* 425: 195–264.
52. M. Sarma and M. K. Mishra. 2008. 'Role of photolysis frequency in enhanced selectivity and yield for controlled bond breaking in HOD.' *J. Phys. Chem. A.* 112: 4895–4905.
53. M. Svanberg, B. C. Jan Petterson and D. Murtagh. 1995. 'Ozone photodissociation in the Hartley band: A statistical description of the ground state decomposition channel $O_2(X^3\sum^- g)$+ O (^{3}P).' *J. Chem. Phys.* 102: 8887–8896.
54. J. Somloi and D. J. Tannor. 1995. 'Controlled dissociation of HgI2 via optical transitions between electronic states.' *J. Phys. Chem.* 99: 2552–2560.
55. M. G. Sheppard and R. W. Walker. 1983. 'Wigner method studies of ozone photodissociation.' *J. Chem. Phys.* 78: 7191–7199.
56. N. J. Turro. 1978. *Modern Molecular Photochemistry*. Menlo Park: Benjamin-Cummings.
57. D. J. Tannor, R. Kosloff and S. A. Rice. 1986. 'Coherent pulse sequence induced control of selectivity of reactions: Exact quantum mechanical calculations.' *J. Chem. Phys.* 85: 5805–5820.

58. A. K. Tiwari, K. B. Moller and N. E. Henriksen. 2008. 'Selective bond breakage within HOD molecule using optimized laser femtosecond ultraviolet laser pulses'. *Phys. Rev. A*. 78: 065402/1–065402/4.

59. M. R. Taherian and T. G. Slanger. 1983. 'Products and yields from O_3 photodissociation at 1576 Å.' *J. Chem. Phys*. 83: 6246–6250.

60. R. L. Vander Wal, J. L. Scott and F. F. Crim. 1990. 'Selectively breaking the O–H bond in HOD.' *J. Chem. Phys*. 92: 803–805.

61. R. L. Vander Wal, J. L. Scott, F. F. Crim, K. Weide and R. Schinke. 1991. 'An experimental and theoretical study of the bond selected photodissociation of HOD.' *J. Chem. Phys*. 94: 3548–3555.

62. K. Vandana, D. B. Bairagi, P. Gross and M. K. Mishra. 1998. 'Field optimized initial state based control of photodissociation.' *Pramana J. Phys*. 50: 521–534.

63. K. Vandana and M. K. Mishra. 1999. 'Photodynamical control using optimized initial state: A mechanistic investigation of selective control with application to IBr and HI photodissociation.' *J. Chem. Phys*. 110: 5140–5148.

64. K. Vandana and M. K. Mishra. 1999. 'Selective photodynamic control of chemical reactions: A Rayleigh-Ritz variational approach.' *Adv. Quantum Chem*. 35: 261–281.

65. K. Vandana and M. K. Mishra. 2000. 'Investigation and control of photodissociation dynamics: An overview.' *Proc. Indian Natl. Sci. Acad. A*. 66: 29–58.

66. K. Vandana and M. K. Mishra. 2000. 'A simplification of selective control using field optimized initial state with application to HI and IBr photodissociation.' *J. Chem. Phys*. 113: 2336–2342.

67. R. P. Wayne. 1988. *Principles and Applications of Photochemistry*. Oxford: Oxford University Press.

68. K. Yamashita, K. Morokuma, F. Le Quéré and C. Leforestier. 1992. 'New ab initio potential surfaces and three-dimensional quantum dynamics for transition state spectroscopy in ozone photodissociation.' *Chem. Phys. Lett*. 191: 515–520.

5 Possibility of Enhancement and Suppression of Chemical Reactions by CW Laser

Yoshiaki Teranishi, Hiroki Nakamura and Sheng H. Lin

5.1 Introduction

Thanks to the recent remarkable progress of laser technology, it has become meaningful and useful to investigate the possibilities of manipulating molecular processes and chemical dynamics by lasers. Various ideas have been proposed so far[1–9]. Typical examples are (i) coherent control in which quantum mechanical interference effect is used to control the branching ratio into the desirable final product, (ii) pump–dump method in which a wave packet is pumped or dumped by laser at appropriate positions to achieve a desirable product, (iii) linear chirping in which the laser frequency is linearly varied as a function of time to attain high efficiency, (iv) various kinds of adiabatic rapid passage that represent applications of linear chirping, and (v) optimal control theory which tries to control wave packet dynamics optimally. Among these, we have especially noticed the significance of non-adiabatic transitions induced by lasers. If the laser frequency ω is not low (< 1000 cm^{-1}), the Floquet (or dressed) state representation holds well and molecular potential energy curves or surfaces are shifted up or down by the amount of laser energy $\hbar\omega$[10]. This means that potential curve (or surface) crossings are created among these states. The interaction between the crossing states is equal to $-\mu\epsilon/2$, where μ is the transition dipole moment and ϵ is the amplitude of the laser field.

The time-dependences of the frequency $\omega(t)$ and the pulse envelope $\epsilon(t)$, which are supposed to be much slower than the time-dependence of the field oscillation $\exp(i\int^t \omega(t)dt)$, cause time-dependent non-adiabatic transitions. The time-dependence of $\omega(t)$ induces the time-dependent Landau–Zener type non-adiabatic transition and the time-dependence of $\epsilon(t)$ induces the time-dependent Rosen–Zener type non-adiabatic transition[9]. In the case of CW (continuous wave) laser, on the other hand, both ω and ϵ are constant and potential curve crossings are created along the molecular spatial coordinate, i.e., the internuclear distance. Non-adiabatic transitions are generally well localized at avoided crossings between adiabatic states regardless of whether they are time-dependent or time-independent, and molecular processes can be nicely described as a sequence of adiabatic wave propagations along adiabatic states and non-adiabatic transitions at avoided crossings. Since the laser–molecule interaction at the potential curve crossing is determined by the laser parameters as mentioned above, these non-adiabatic transitions can be controlled by appropriately designing the laser parameters. In the case of time-dependent transition, the parameters should be appropriately and quickly manipulated before the molecular wave packet moves away from the crossing region. This method has been demonstrated to work well for controlling wave packet motion in molecular systems[9, 11]. Molecular processes such as photo-excitation and photo-dissociation can be made highly efficient by properly controlling the corresponding non-adiabatic transitions. In the case of time-independent non-adiabatic transition induced by CW lasers, the non-adiabatic tunneling type of transition, in which two diabatic potential curves cross with opposite signs of slopes, appears and the intriguing phenomenon of complete reflection occurs[9, 12]. Selective photo-dissociation of molecules and molecular switching have been proposed by using this phenomenon[9, 13–17]. If the excited state is repulsive, we can create two non-adiabatic tunneling type potential surface crossings by dressing up the attractive ground state. Selective photo-dissociation can be achieved by appropriately choosing the dressing laser frequency. This idea was applied to selective photo-dissociation of HOD, HI, and CH_3SH. Not relying on this phenomenon, we can still use CW laser to control photo-dissociation branching[18].

In this chapter, the possibility of enhancement and suppression of chemical reactions is discussed by using a CW laser. If complete reflection occurs at relatively high energies where chemical reaction proceeds with high probability, we can suppress the reaction. If complete transmission occurs, on the other hand, at low energies where chemical reaction cannot proceed efficiently, we can enhance the reaction. Even if these phenomena are not complete, we may be able to enhance or suppress the reaction. More generally, laser induced interferences can be manipulated by appropriately choosing the laser frequency and intensity; we may be able to enhance or suppress reactions. These possibilities are demonstrated by using one-dimensional model potentials.

This chapter is organized as follows. In the next section, basic theories and ideas about dressed state representation, non-adiabatic transition, complete reflection and complete transmission phenomena are explained. In Section 5.3, the above-mentioned numerical demonstrations are presented by employing two types of one-dimensional potentials. Section 5.4 is devoted to concluding remarks.

5.2 Preliminaries

5.2.1 Dressed state representation and non-adiabatic transition

The time-dependent Schrödinger equation for a molecule in a laser field in the dipole approximation is given by

$$i\hbar\frac{\partial\Psi}{\partial t} = \left[H_0 - \mu\epsilon(t)\cos\left(\int^t \omega(t)dt\right)\right]\Psi, \tag{5.1}$$

where H_0 is the unperturbed Hamiltonian of the molecule and μ is the transition dipole moment. Assuming that the time-dependences of $\epsilon(t)$ and $\omega(t)$ are slow compared to that of field oscillation $\exp[i\int\omega(t)dt]$ (rotating wave approximation) and taking a time average over the laser period, we obtain

$$i\hbar\frac{\partial\Phi}{\partial t} = H^F\Phi, \tag{5.2}$$

where H^F is called a Floquet Hamiltonian and is given as follows in the two-state approximation[9, 10]

$$H^F = \begin{pmatrix} V_1(x) & -\mu\epsilon(t)/2 \\ -\mu\epsilon(t)/2 & V_2(x) - \hbar\omega(t) \end{pmatrix}. \tag{5.3}$$

In general, the Floquet Hamiltonian H^F is given by an infinite size matrix whose diagonal elements are the photon-dressed states $V_j(x) \pm n\hbar\omega$, where V_j represents the jth electronic state and n represents the photon number [+(−) means absorption(emission)]. As is well known, extracting the physically essential portion [like Eq. (5.3)] from this infinite matrix works well. The representation with the use of $V_1(x) + \hbar\omega(t)$ and $V_2(x)$ is the same as that of Eq. (5.3). If the original two states, $V_1(x)$ and $V_2(x)$, are coupled, then we have the following 4×4 matrix under the laser field:

$$i\hbar\frac{\partial\Phi}{\partial t} = \begin{pmatrix} V_1(x) & V_{12}(x) & 0 & V_{14} \\ V_{12}(x) & V_2(x) & V_{23} & 0 \\ 0 & V_{23} & V_3(x) & V_{34}(x) \\ V_{14} & 0 & V_{34}(x) & V_4(x) \end{pmatrix}\Phi, \tag{5.4}$$

where $V_{12}(x) = V_{34}(x)$ is the diabatic coupling between the original two states, $V_1(x)$ and $V_2(x)$, $V_3(x) = V_1(x) - \hbar\omega(t)$, $V_4(x) = V_2(x) - \hbar\omega(t)$, and $V_{14} = V_{23} = -\mu\epsilon(t)/2$. This representation will actually be used in Section 5.3.2.

In the case of CW laser, ϵ and ω are both constant, and thus Eqs (5.2) and (5.4) describe the time-independent non-adiabatic transition phenomena, since the two states $V_1(x)$ and $V_2(x) - \hbar\omega$ can have an additional curve crossing due to laser–molecule interaction (see Fig. 5.8 in Section 5.3.2).

5.2.2 Complete reflection and transmission phenomena

As was mentioned in Section 5.1, complete reflection phenomenon was found to occur in the non-adiabatic tunneling type curve crossing (see Fig. 5.1). The Zhu–Nakamura semiclassical theory gives the overall transmission probability at energies higher than the bottom of the upper adiabatic potential $E_2(x)$ as

$$P = \frac{4\cos^2(\Psi)}{4\cos^2(\Psi) + p^2/(1-p)} \tag{5.5}$$

with

$$\Psi = \int_{T_L}^{T_R} K_2(x)dx - \phi_S \tag{5.6}$$

and

$$K_2(x) = \sqrt{\frac{2m}{\hbar^2}[E - E_2(x)]}, \tag{5.7}$$

where m is the mass, and ϕ_S is the additional phase due to the non-adiabatic coupling between the two adiabatic states $E_1(x)$ and $E_2(x)$. The explicit expression of ϕ_s is not given here, but is given in literature[9]. Equation (5.5) clearly tells us that the complete reflection ($P = 0$) occurs when $\cos(\Psi) = 0$ which is satisfied at a series of discrete energies. This condition is similar to the Bohr–Sommerfeld quantization condition, but the modification ϕ_S due to the non-adiabatic coupling between the two states is necessary. This phenomenon was also found to occur more generally in other potential systems[9, 19]. Complete transmission phenomenon is rather well known to occur in any periodic potential system and leads to the formation of conduction band in the solid phase[9, 20]. This phenomenon has also been found to occur in other potential systems, if some conditions are satisfied[9, 19].

Thus, due to the appropriate phase interferences, enhancement and suppression of transmission, if not complete, can occur in non-adiabatically coupled potential systems.

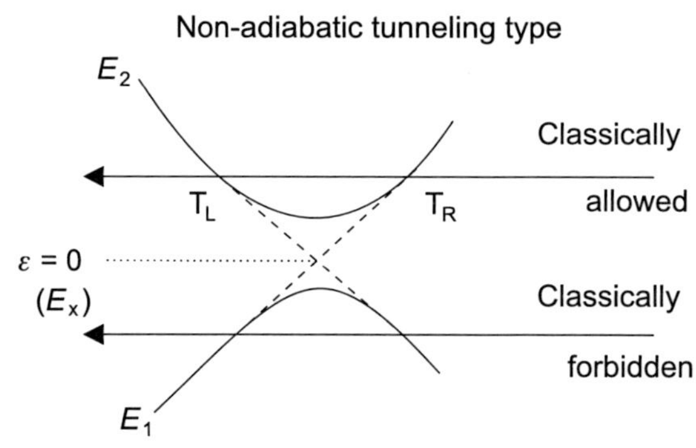

Figure 5.1 Non-adiabatic tunneling type of potential curve crossing. $T_{L(R)}$ represents the turning point on the upper adiabatic potential

5.3 Enhancement and Suppression of Chemical Reactions: Model Calculations

5.3.1 Barrier penetration reaction

The first case we consider is a reaction which goes through the ground state potential energy surface that requires thermal activation because of the potential barrier. The first excited state is assumed to be attractive, located a bit away from the ground state; it is assumed not to affect the reaction along the ground state. The one-dimensional model potentials used are as follows:

$$V_1(x) = V_0 \text{sech}^2(ax) \tag{5.8}$$

and

$$V_2(x) - \hbar\omega = b(x - x_0)^2 + c, \tag{5.9}$$

where $V_1(x)$ is chosen to mimic approximately the $H + H_2$ reaction with the parameters $V_0 = 0.457/27.21$ and $a = 1.36$ in atomic units[21]. The excited state $V_2(x)$ is chosen to be a shifted harmonic oscillator and the parameters used are $b = 0.01$, $x_0 = 0.3$, and $c = -0.003$ in atomic units. The original excited state potential $V_2(x)$ is assumed to be higher than the ground state by 0.07 a.u. This is actually not important, unless the excited state is too high and very high laser frequency is required to shift it down. Figure 5.2 shows the two adiabatic potentials obtained from Eqs (5.8) and (5.9) with the coupling strength $V = 1.0 \times 10^{-3}$ a.u. which corresponds to the laser intensity $I \simeq 140$ GW/cm^2 in the case of transition dipole moment $\mu = 1.0$ a.u.

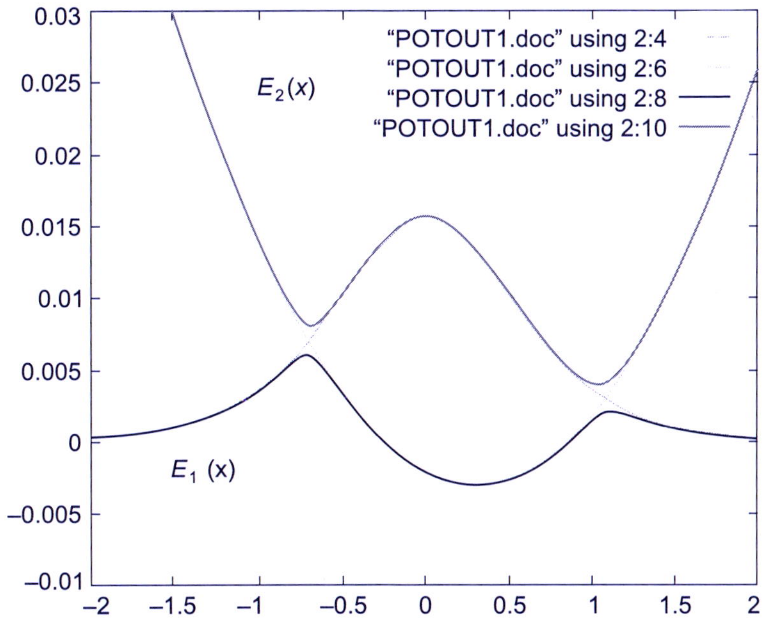

Figure 5.2 Adiabatic potentials $E_1(x)$ and $E_2(x)$ obtained from the diabatic potentials given by Eqs (5.8) and (5.9) with the laser–molecule interaction $V = 1.0 \times 10^{-3}$ a.u.

Numerically calculated transmission (reaction) probabilities are shown in Fig. 5.3 as a function of translational energy. The reduced mass of the system is taken to be 1060 a.u. which corresponds to the reduced mass of the $H + H_2$ system. The thin grey curve is the reaction probability in the original one potential case. The potential barrier height is around 0.015 a.u., as is seen from Fig. 5.2. The black curve with peaks and dips is the result under the CW laser. It is clear that we can enhance the reaction quite a bit at low energies when the original reaction probability is very small. On the other hand, the reaction is suppressed at around $E \sim 0.021$ a.u. when the original reaction probability is almost unity.

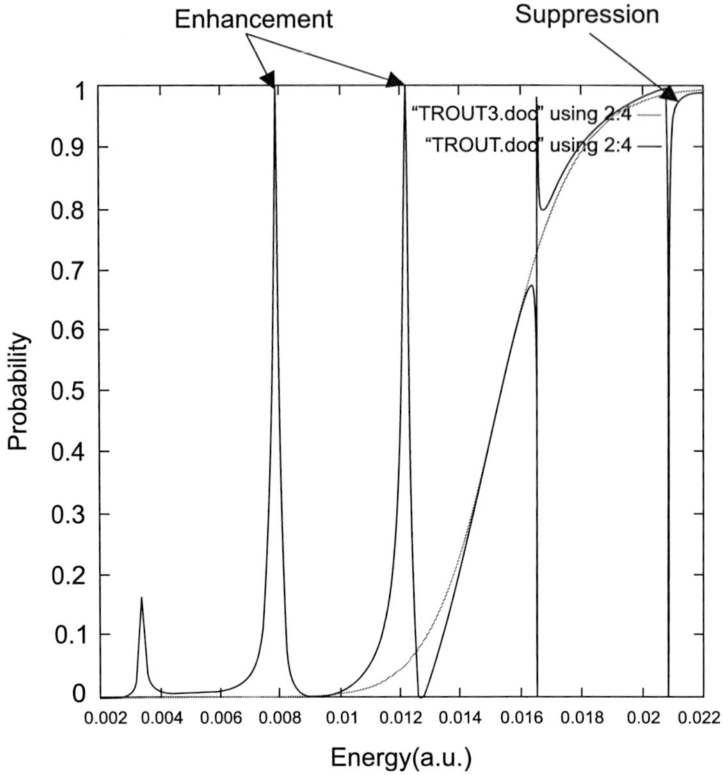

Figure 5.3 Reaction probability as a function of energy. The thin grey curve represents the result of the original single potential case. The black curve indicates the effect of laser dressing. The three peaks at low energies represent the enhancement of the reaction. The dips at energy around $E \simeq 0.0125, 0.0165$, and 0.021 a.u. suppress the reaction

Figure 5.4 shows the reaction rate constant against temperature in the Arrhenius plot in the temperature range 300 ~ 1000 K. The enhancement is clearly seen. The widths of the probability enhancement peaks and suppression dips are dependent on the coupling strength V or the laser intensity and the potential shapes. Figure 5.5 shows the result for $V = 2.0 \times 10^{-3}$ a.u. The widths become slightly broader. In the case of a simple one crossing non-adiabatic tunneling type potentials (see Fig. 5.1), the complete reflection dip becomes wider when the diabatic coupling is weaker[9]. When the system has more than one crossing like in the present case, the dependence on the diabatic coupling is not monotonous. The peaks and dips also depend on the relative position of the potentials. Figure 5.6 shows the case of further shifting down the excited state $V_2(x)$ by 0.002 a.u. The Fano-type resonance discussed by Vardi and Shapiro[22] under the name of laser catalysis can be reproduced by the present scheme which gives a general mechanism. The peak and dip at $E \simeq 0.0165$ a.u. in Fig. 5.3 is this type of resonance. At the other resonances, the peak or dip portion appears strongly.

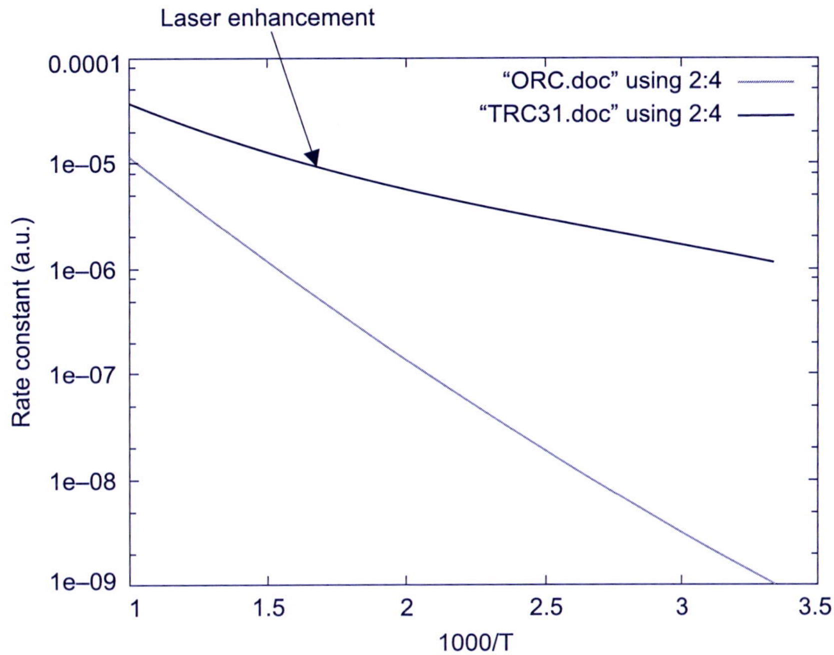

Figure 5.4 Reaction rate constant corresponding to Fig. 5.3 in the Arrhenius plot

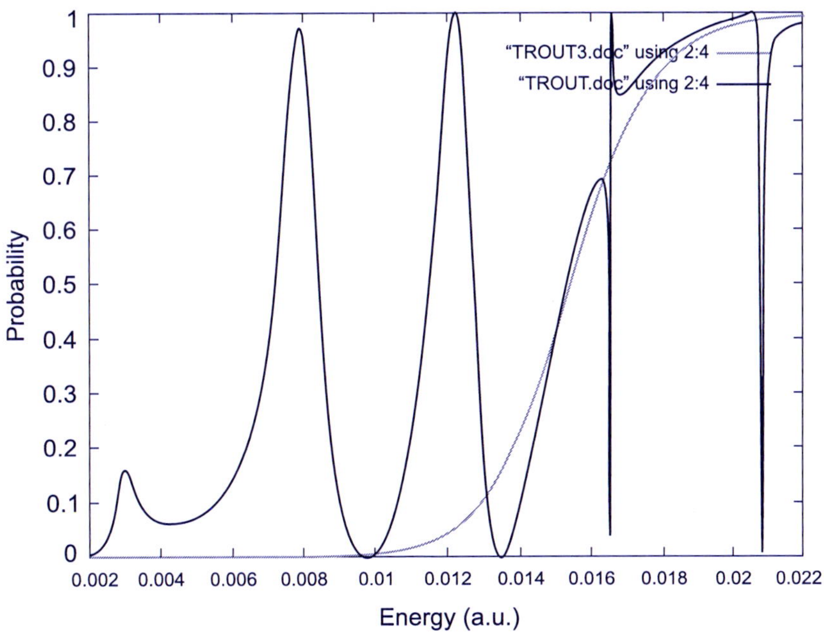

Figure 5.5 The same as Fig. 5.3 for the laser–molecule coupling strength $V = 2.0 \times 10^{-3}$ a.u. The peaks and dip are broader. The suppression occurs at $E \simeq 0.0135$, 0.0165, and 0.021 a.u.

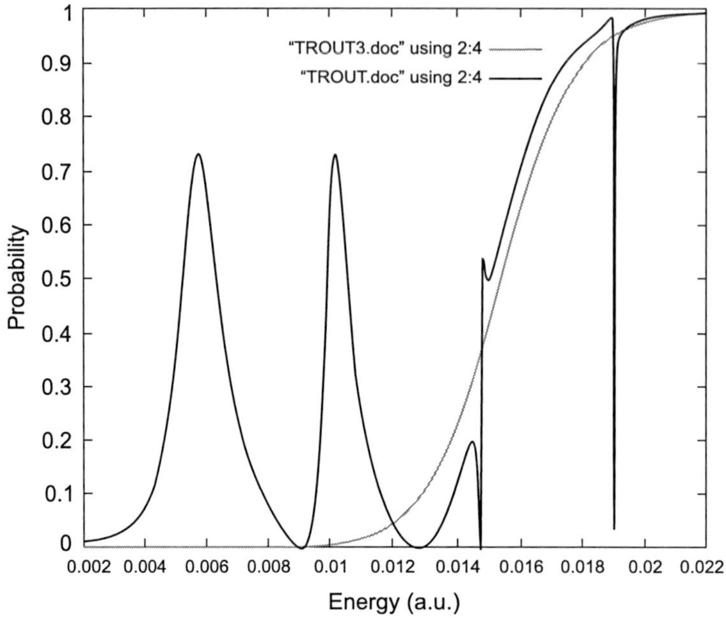

Figure 5.6 The same as Fig. 5.3 for a further down shift of the excited state potential from Fig. 5.2 by 0.002 a.u. The laser–molecule coupling strength is the same as in Fig. 5.3 ($V = 1.0 \times 10^{-3}$ a.u.)

The semiclassical analysis using the Zhu–Nakamura theory of non-adiabatic transition[9] can be done in order to have a deeper understanding of the phenomenon. First of all, the corresponding diagram can be drawn as in Fig. 5.7. The arrows indicate the wave propagation from a certain position a to another b along the adiabatic potential n ($\gamma_n(a,b)$), the squares represent non-adiabatic transitions at avoided crossings and tunneling, and the circles with t indicate the turning points. $I^\alpha(\alpha = R, L)$ is a matrix, representing the non-adiabatic transition at the right (R) or left (L) side avoided crossing. O^α is a transposed matrix of I^α. The element I_{21}, for instance, gives the transition amplitude from state 1 on the right side to state 2 on the left side. These are given in analytical forms as

$$\gamma_{(n)}(a,b) = \exp\left[i \int_a^b K_n(x)dx\right], \tag{5.10}$$

$$I^\alpha = \begin{pmatrix} \sqrt{1-p_\alpha}e^{i\phi_S^\alpha} & \sqrt{p_\alpha}e^{i\sigma_0^\alpha} \\ \sqrt{p_\alpha}e^{-i\sigma_0^\alpha} & \sqrt{1-p_\alpha}e^{-i\phi_S^\alpha} \end{pmatrix}, \tag{5.11}$$

and

$$O^\alpha = \text{Transpose of } I^\alpha, \tag{5.12}$$

where

$$K_n(x) = \sqrt{\frac{2m}{\hbar^2}[E - E_n(x)]}. \tag{5.13}$$

p_α in Eq. (5.11) is the non-adiabatic transition probability for one passage of the avoided crossing point α. The various quantities including this p_α in Eq. (5.11) are provided by the Zhu–Nakamura theory [9].

S^T represents the transition matrix of barrier penetration. If we use the comparison equation method based on the Weber differential equation[23], we have

$$S^T = \begin{pmatrix} i\frac{e^{-\pi\epsilon}}{\sqrt{1+e^{-2\pi\epsilon}}}e^{-i\phi} & \frac{e^{-i\phi}}{\sqrt{1+e^{-2\pi\epsilon}}} \\ \frac{e^{-i\phi}}{\sqrt{1+e^{-2\pi\epsilon}}} & i\frac{e^{-\pi\epsilon}}{\sqrt{1+e^{-2\pi\epsilon}}}e^{-i\phi} \end{pmatrix}, \tag{5.14}$$

where

$$\epsilon = \begin{cases} \frac{-1}{\pi}\int_{T_L}^{T_R} |K_2(x)|dx & \text{for } E \le E_t \\ \mathrm{Re}\left[\frac{1}{\pi i}\int_{T_L^*}^{T_R^*} K_2(x)dx\right] & \text{for } E \ge E_t \end{cases}, \tag{5.15}$$

and

$$\phi = \arg\Gamma\left(\frac{1}{2}+i\epsilon\right) - \epsilon\ln|\epsilon| + \epsilon. \tag{5.16}$$

T_L^* and $T_R^*(\mathrm{Im}T_L^*\langle 0, \mathrm{Im}T_R^*\rangle 0)$ are the complex turning points, and E_t represents the energy of the barrier top. Accurate quantitative reproduction of the numerical results cannot be obtained because of the quadratic potential barrier approximation to S^T. The qualitative features of the whole process can, however, be analyzed as described below.

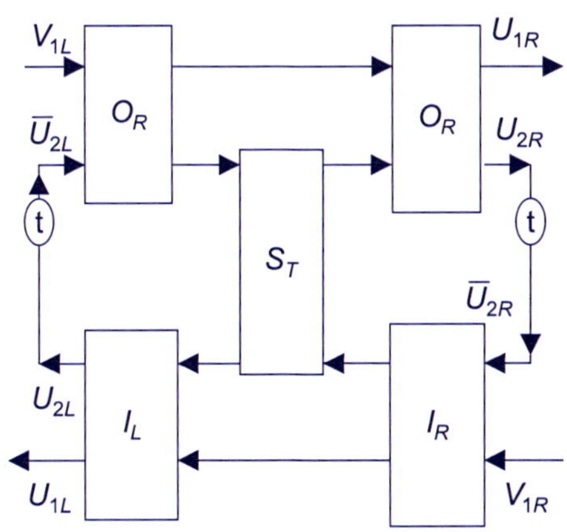

Figure 5.7 The semiclassical diagram. Arrows denote the adiabatic wave propagation. The rectangles with matrices I and O represent non-adiabatic transitions at the right (R) and left (L) side crossings and the rectangle with matrix S^T represents the potential barrier penetration or reflection

The labels attached to the arrows in Fig. 5.7 represent the coefficients of wave functions propagating in the direction of the arrow. The overall scattering matrix S can now be defined as

$$\begin{pmatrix} U_{1R} \\ U_{1L} \end{pmatrix} = S \begin{pmatrix} V_{1R} \\ V_{1L} \end{pmatrix}. \tag{5.17}$$

The adiabatic state $E_2(x)$ is an energetically closed channel as depicted by the closed loop in the middle of Fig. 5.7. The whole dynamic process inside the loop is described by the following χ matrix:

$$\begin{pmatrix} U_{1R} \\ U_{1L} \\ U_{2R} \\ U_{2L} \end{pmatrix} = \chi \begin{pmatrix} V_{1R} \\ V_{1L} \\ \bar{U}_{2R} \\ \bar{U}_{2L} \end{pmatrix}. \tag{5.18}$$

The coefficients $U_{2\alpha}$ and $\bar{U}_{2\alpha}(\alpha = R, L)$ are connected by the adiabatic propagation through the turning point and are related as

$$\bar{U}_{2\alpha} = \gamma^2_{(2)}(x_s, x_l) \exp\left[i\frac{\pi}{2}\right] U_{2\alpha}, \tag{5.19}$$

where $x_{s(l)}$ is the smaller (larger) of the corresponding crossing point and turning point. The overall S matrix is now expressed in terms of the χ matrix as

$$S = \chi_{OO} + \chi_{OC}[\Theta - \chi_{CC}]^{-1}\chi_{CO}, \tag{5.20}$$

where

$$\Theta^{-1}_{nm} = \delta_{nm}\gamma^2_{(2)}(x_s, x_l) \exp\left[i\frac{\pi}{2}\right] \tag{5.21}$$

and

$$\chi = \begin{pmatrix} \chi_{OO} & \chi_{OC} \\ \chi_{CO} & \chi_{CC} \end{pmatrix}. \tag{5.22}$$

χ_{OC}, for instance, means the 2×2 submatrix representing the open–closed portion of the 4×4 χ matrix. All the χ matrix elements are expressed in terms of the elements of the matrices I, O, and S^T. The phase $\exp[i\pi/2]$ in Eqs (5.19) and (5.21) represents the effect of turning point. All the elements of the χ matrix are explicitly given in terms of I, O and S^T matrix elements as given in the Appendix. The diagonal (off diagonal) elements of the S matrix given by Eq. (5.20) represent the overall reflection (transmission) amplitude. The first term in this equation represents the direct process without trapping by the upper closed state and the second term represents the process trapped by this upper state. The destructive interference between these two waves at the exit to the left or at the exit to the right leads to complete reflection or complete transmission. Even if this interference is not complete, the reaction can be enhanced or suppressed by it as demonstrated above.

5.3.2 Non-adiabatic tunneling type reaction

Let us next consider a non-adiabatic tunneling type system of two coupled potentials. If we apply a CW laser and shift these potentials downward, we have four coupled potential systems as shown in Fig. 5.8. The solid lines represent the original coupled diabatic potentials. These are defined as

$$V_1(x) = V_0[1+\tanh(\alpha x)], \tag{5.23}$$

$$V_2(x) = V_0[1+\tanh(\beta x)], \tag{5.24}$$

with the diabatic coupling

$$V_{12}(x) = V\exp[-\gamma x^2]. \tag{5.25}$$

They correspond approximately to the potentials along the reaction coordinate used in[24]. The parameters used are as follows in atomic units,

$$V_0 = 0.1,\ \ \alpha = 0.8,\ \ \beta = 0.9, V = 0.035,\ \text{ and }\ \gamma = 0.28. \tag{5.26}$$

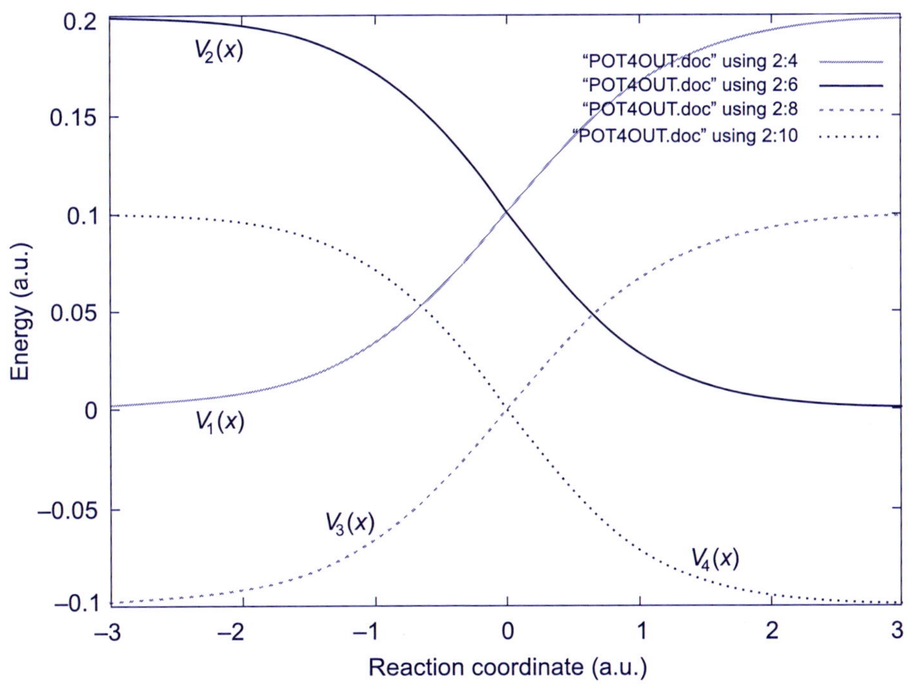

Figure 5.8 Four coupled diabatic potential curves. $V_1(x)$ and $V_2(x)$ are the original coupled diabatic states with the coupling V_{12}. $V_3(x) = V_1(x) - \hbar\omega$ and $V_4(x) = V_2(x) - \hbar\omega$ with $\hbar\omega = 0.1$ a.u. are the laser dressed diabatic states with the same coupling V_{12}. $V_3(x)[V_4(x)]$ couples with $V_2(x)[V_1(x)]$ by the laser–molecule dipole interaction $V_{23} = V_{14} = 4.0 \times 10^{-3}$ a.u.

The potentials $V_3(x)$ and $V_4(x)$ are the laser dressed states by the photon energy $\hbar\omega = 0.1$ a.u. and coupled by $V_{12}(x)$. The potentials $V_3(x)$ and $V_4(x)$ are coupled with $V_2(x)$ and $V_1(x)$, respectively, by the dipole interaction $V_{23} = V_{14} = 4.0 \times 10^{-3}$ a.u. which corresponds to the laser intensity $I =$ 2.2 TW/cm^2 in the case of the transition dipole moment $\mu = 1.0$ a.u.

The numerical results of the transition probabilities as a function of energy are shown in Fig. 5.9. The thin grey curve is the probability in the original coupled two state case. The black curve indicates the enhancement by the laser. In the energy range shown here, the upper adiabatic states, $E_3(x)$ and $E_4(x)$ (lower adiabatic states $E_1(x)$ and $E_2(x)$) $[E_1(x) < E_2(x) < E_3(x) < E_4(x)]$, are closed (open), and the reaction probability is the sum of the probabilities coming out along these open channels. The corresponding rate constants are shown in Fig. 5.10. The laser enhancement is clearly seen. The coupled two state system like $V_1(x)$ and $V_2(x)$ considered here represents the basic one for electron transfer, namely the Marcus' normal case[9, 25] and the present result suggests a possibility of its enhancement at low temperatures. The complete reflection dips appear at higher energies at $E \sim 0.14$ a.u. and can be used to suppress the reaction, if necessary. The enhancement of the reaction probability at low energies is, however, more interesting.

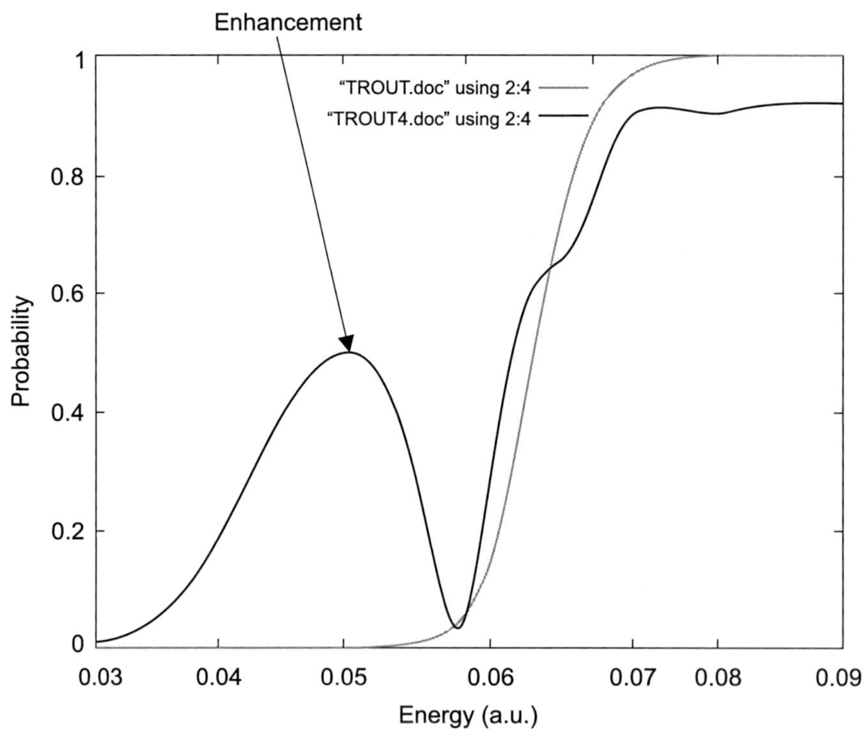

Figure 5.9 Reaction probabilities as a function of energy in the case of the potential system of Fig. 5.8. The thin grey curve is the result in the case of the original coupled two states and the black curve shows the result of the four state calculations, indicating the laser enhancement

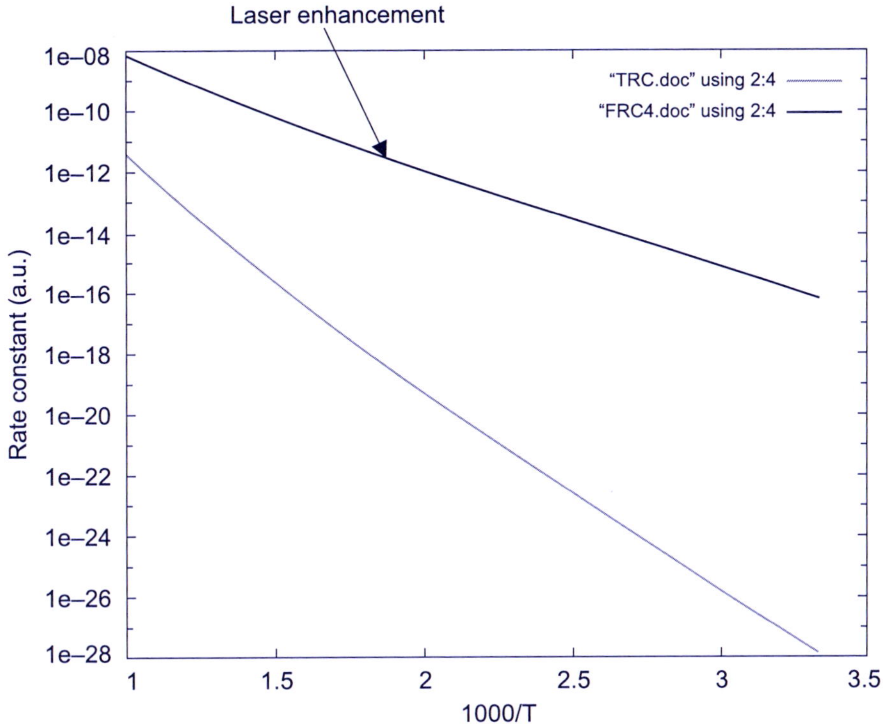

Figure 5.10 Reaction rate constants corresponding to the results in Fig. 5.9

The semiclassical analysis can be performed in the same way as before. The diagram corresponding to the potential system of Fig. 5.8 is depicted in Fig. 5.11. The matrices I_C and O_C (= transpose of I_C) represent the non-adiabatic transition at the original crossing. The overall S matrix and the χ matrix are defined as follows:

$$\begin{pmatrix} \tilde{U}_{1R} \\ U_{1R} \\ \tilde{U}_{1L} \\ U_{1L} \end{pmatrix} = S \begin{pmatrix} \tilde{V}_{1R} \\ V_{1R} \\ \tilde{V}_{1L} \\ V_{1L} \end{pmatrix} \tag{5.27}$$

and

$$\begin{pmatrix} \tilde{U}_{1R} \\ U_{1R} \\ \tilde{U}_{1L} \\ U_{1L} \\ U_{2R} \\ U_{2L} \end{pmatrix} = \chi \begin{pmatrix} \tilde{V}_{1R} \\ V_{1R} \\ \tilde{V}_{1L} \\ V_{1L} \\ \bar{U}_{2R} \\ \bar{U}_{2L} \end{pmatrix}. \tag{5.28}$$

The expression of the S matrix in terms of the χ matrix is the same as Eq. (5.20). The matrix elements of χ can be obtained in the same way as before in terms of the elements of I, O and S^T.

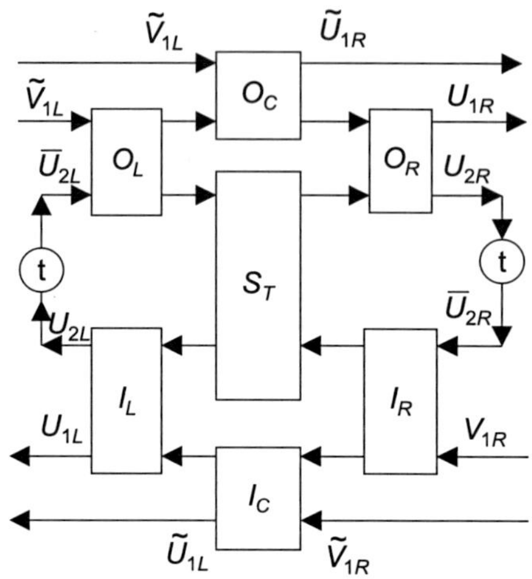

Figure 5.11 The semiclassical diagram corresponding to Fig. 5.8. The meaning of each element is the same as that of Fig. 5.7

5.4 Concluding Remarks

Using the one-dimensional model potentials, the possibility of enhancing and suppressing of chemical reactions by a CW laser is discussed. Reaction probabilities at low energies where the original reaction does not proceed efficiently can be enhanced and those at high energies where the original reaction proceeds with high probability can be suppressed. Two cases are considered: (i) barrier penetration type reaction and (ii) non-adiabatically coupled type of reaction, an example of which is the normal case of electron transfer. The effectiveness of the laser induced potential energy surface crossings and non-adiabatic transitions there is clearly demonstrated. This is true even in the simplest case of a CW laser with constant frequency and intensity. The frequency and intensity can be appropriately chosen to realize a desirable control. A laser pulse can also be used, if its width is wide enough to cover the reaction zone. The laser intensity, required at least in the present model calculations, ranges from 100 $\mathrm{GW/cm^2}$ to a few $\mathrm{TW/cm^2}$ and is weaker than a sort of threshold ($\sim 10\ \mathrm{TW/cm^2}$) which causes various undesirable multi-photon processes[3, 8].

The control based on the present idea would be efficient, when the complete reflection and transmission phenomena occur. More generally, however, these phenomena are not necessarily complete and the quantum mechanical interferences induced by laser can lead to desirable enhancement or suppression, as demonstrated by the actual numerical computations. The semiclassical analysis based on the Zhu–Nakamura theory was carried out and found to be useful to understand the mechanisms. Applications of the present idea to multi-dimensional realistic systems are naturally very important and present a good subject for future study.

Appendix

The matrix elements of χ defined by Eq. (5.22) is a symmetric matrix and can be explicitly expressed as follows:

$$\chi_{11} = O^R_{12} S^T_{11} I^R_{21} \gamma^2_{(2)}(r_{tn}, r_x), \tag{5.29}$$

$$\chi_{12} = O^R_{11} O^L_{11} \gamma_{(1)}(l_x, r_x) + O^R_{12} S^T_{12} O^L_{21} \gamma_{(2)}(r_{tn}, r_x)\gamma_{(2)}(l_x, l_{tn}), \tag{5.30}$$

$$\chi_{13} = O^R_{12} S^T_{11} I^R_{22} \gamma^2_{(2)}(r_{tn}, r_x), \tag{5.31}$$

$$\chi_{14} = O^R_{11} O^L_{12} \gamma_{(1)}(l_x, r_x) + O^R_{12} S^T_{12} O^L_{22} \gamma_{(2)}(r_{tn}, r_x)\gamma_{(2)}(l_x, l_{tn}), \tag{5.32}$$

$$\chi_{22} = I^L_{12} S^T_{22} O^L_{21} \gamma^2_{(2)}(l_x, l_{tn}), \tag{5.33}$$

$$\chi_{23} = I^L_{11} I^R_{12} \gamma_{(1)}(l_x, r_x) + I^L_{12} S^T_{21} I^R_{22} \gamma_{(2)}(r_{tn}, r_x)\gamma_{(2)}(l_x, l_{tn}), \tag{5.34}$$

$$\chi_{24} = I^L_{12} S^T_{22} O^L_{22} \gamma^2_{(2)}(l_x, l_{tn}), \tag{5.35}$$

$$\chi_{33} = O^R_{22} S^T_{11} I^R_{22} \gamma^2_{(2)}(r_{tn}, r_x), \tag{5.36}$$

$$\chi_{34} = O^R_{21} O^L_{12} \gamma_{(1)}(l_x, r_x) + O^R_{22} S^T_{12} O^L_{22} \gamma_{(2)}(r_{tn}, r_x)\gamma_{(2)}(l_x, l_{tn}), \tag{5.37}$$

and

$$\chi_{44} = I^L_{22} S^T_{22} O^L_{22} \gamma^2_{(2)}(l_x, l_{tn}), \tag{5.38}$$

where $l_x(r_x)$ and $l_{tn}(r_{tn})$ are the left (right) side crossing point and the left (right) side turning point of the central potential barrier.

Bibliography

1. P. Brumer and M. Shapiro. 1997. 'Principles of the quantum control of molecular processes.' *Annu. Rev. Phys. Chem.* 48: 601. (John Wiley & Sons, 2003).
2. S. A. Rice and M. Zhao. 2000. *Optical Control of Molecular Dynamics*. John Wiley & Sons, 2000.
3. A. D. Bandrauk, Y. Fujimura and R. J. Gordon. 2002. 'Laser control and manipulation of molecules'. *ACS Symposium series* 821. Washington DC: American Chemical Society.
4. V. Bonacic-Koutecky and R. Mitric. 2005. *Chemical Reviews* 105: 11.
5. K. Yamanouchi, S. Chin, P. Agostini and G. Ferrante Eds. 2006. *Progress in Ultrafast Intense Laser Science Vol. I and II*. Berlin: Springer.
6. H. Nakamura. 2008. *Adv. Chem. Phys.* 138: 95.
7. L. G. G. Rego, L. Santos and V. S. Batista. 2009. *Annu. Rev. Chem. Phys.* 60: 293.

8. A. Kondorskiy, S. Nanbu, Y. Teranishi and H. Nakamura. 2010. *J. Phys. Chem. A*. 114: 6171.
9. H. Nakamura. December, 2012. *Non-adiabatic Transition: Concepts, Basic Theories and Applications*, 2nd edition. Singapore: World Scientific.
10. S. I. Chu. 1986. *Advances in Multiphoton Processes and Spectroscopy*. 2, Singapore: World Scientific; 1989. *Adv. Chem. Phys.* 73: 739.
11. S. Zou, A. Kondorskiy, G. V. Mil'nikov and H. Nakamura. 2005. *J. Chem. Phys.* 122: 95–117.
12. C. Zhu, H. Nakamura, N. Re and V. Aquilanti. 1992. *J. Chem. Phys.* 97: 1892.
13. K. Nagaya, Y. Teranishi and H. Nakamura. 2000. *J. Chem. Phys.* 113: 6197.
14. H. Fujisaki, Y. Teranishi and H. Nakamura. 2002. *J. Theo. Comp. Chem.* 1: 245.
15. S. Nanbu, H. Nakamura and F. O. Goodman. 1997. *J. Chem. Phys.* 107: 5445.
16. H. Nakamura. 1999. *J. Chem. Phys.* 110: 10253.
17. S. Nanbu, T. Ishida and H. Nakamura. 2006. *Chem. Phys.* 324: 721.
18. K. Nagaya, S. H. Lin and H. Nakamura. 2006. *J. Chem. Phys.* 125: 214311.
19. L. Pichl, H. Nakamura and J. Horacek. 2000. *J. Chem. Phys.* 113: 906.
20. R. de L. Kronig and W. G. Penney. 1931. *Proc. Roy. Soc. London A*. 130: 499.
21. W. H. Miller, Y. Zhao, M. Ceotto and S. Yang. 2003. *J. Chem. Phys.* 119: 1329.
22. A. Vardi and M. Shapiro. 1998. *Phys. Rev. A*. 58: 1352.
23. M. S. Child. 1991. *Semiclassical Mechanics with Molecular Applications*, Clarendon: Oxford.
24. C. Shin and S. Shin. 2000. *J. Chem. Phys.* 113: 6528.
25. R. A. Marcus and N. Sutin. 1985. *Biochim. Biophys. Acta*. 811: 265.

6 Doubly Charged Ions Formation from the Multiphoton Atomic Ionization of Alkaline Earth Metals

V. V. Suran and I. I. Bondar

6.1 Introduction

The creation of doubly charged ions upon the multiphoton ionization of atoms and molecules was observed for the first time in[1]. Discovery of the effect of the creation of doubly charged ions upon multiphoton atomic ionization laid the foundation of a new branch in the physics of nonlinear ionization processes that involved the creation of multicharged ions due to the multiphoton or tunneling ionization of atoms and molecules.

In later experiments, the creation of doubly charged ions was revealed in experiments with atomic barium, samarium, europium, and ytterbium[2–4]. The creation of doubly charged ions was also observed upon multiphoton ionization of inert gas atoms and other atoms and molecules[5]. Experimental results have demonstrated the general character of the creation of doubly charged ions upon multiphoton ionization of various groups of atoms.

In this chapter, we will discuss the creation of singly and doubly charged ions upon multiphoton atomic ionization of alkaline-earth elements using laser radiation with nanosecond pulse duration at laser field strengths $\varepsilon = 5\times10^6 - 10^7$ V/cm, which are significantly less than the intra-atomic field $\varepsilon_a = 5\times10^9$ V/cm. The results of using this field strength significantly differ from the recent results obtained at the laser field strength $\varepsilon \leq 5\times10^9$ V/cm or $\varepsilon > 5\times10^9$ V/cm in experiments with femtoseconds and atto second laser pulses[6]. The differences include the domination of the tunneling ionization of atoms and molecules under such conditions and the creation of doubly charged ions. The main processes arc electron rescattering, shaking, etc.[6].

The processes that lead to the creation of doubly charged ions upon multiphoton ionization of the alkaline-earth elements under the given conditions differ from the above processes. Two effects (cascade and two-electron) were proposed in[7, 8] to account for the creation of doubly charged ions upon the multiphoton ionization of alkaline-earth elements. In the cascade effect, doubly charged ions (A^{2+}) are created owing to the multiphoton ionization of singly charged ions (A^+) that appear in the presence of the same laser pulse due to the multiphoton ionization of atoms $A : A + K_0\, \hbar\omega - A^+ + e$ and $A^+ + K_1\, \hbar\omega - A^{2+} + e$. The two-electron effect leading to the creation of doubly charged ions involves the simultaneous detachment of two outer electrons from the atom ($A + K_2\hbar\omega - A^{2+} + 2_e$).

The creation of doubly charged ions upon the multiphoton atomic ionization of alkaline-earth elements was recently studied in several spectral ranges and at various durations of laser pulses[5]. In this work, we review study of the creation performed at the Department of Quantum Electronics, Uzhgorod National University.

6.2 Experimental Details

The experiments conducted employed the method of crossed atomic and laser beams and resonance ionization spectroscopy. Laser radiation is focused to the center of the atomic beam. Ions are pulled from the beam interaction region by a dc electric field, are separated with respect to charge and mass in the time-of-flight mass spectrometer, and are detected by an electron multiplier.

Several lasers were used in the experiments: dye lasers (DLs) with the total frequency range $\omega = 14800 - 18700\ \text{cm}^{-1}$, color-center laser (CCL) with the tuning range $\omega = 8200 - 9100\ \text{cm}^{-1}$, neodymium laser with the frequency range $\omega = 9395 - 9500\ \text{cm}^{-1}$ and laser on yttrium–aluminum garnet (LYAG) with the frequency $\omega = 9395\ \text{cm}^{-1}$. We also employed the second harmonic radiation of IR lasers with $\omega = 18780 - 19000\ \text{cm}^{-1}$ and $\omega = 18780\ \text{cm}^{-1}$. The radiation line width ranged from 1 to 10 cm^{-1} and the pulse durations were $\tau \approx 40$ ns. The linear polarization of the laser beams was provided by the Glan–Foucault polarizer. A Fresnel rhomb was used to transform the linear polarization into the circular polarization.

6.3 Experimental Results and Discussion

As was mentioned, the first observation of the creation of doubly charged ions can be found in[1], where the multiphoton ionization of atomic strontium was studied. Figure 6.1 demonstrates the oscillogram of ion signals from this work. The first and second pulses correspond to the creation of the Sr^{2+} and Sr^+ ions, respectively. It is seen that the ion yields are comparable, so that an anomalously high yield of the doubly charged ions is observed. Further experiments were aimed at the interpretation of the creation of doubly charged ions upon multiphoton atomic ionization.

Several experiments were performed, which involved the study of the creation of the A^+ and A^{2+} ions at various laser frequencies and intensities in several spectral ranges (in the visible and IR ranges) for linear and circular polarizations; the investigation of the effect of variations in the concentrations of neutral atoms A and singly charged ions A^+ on the creation of the A^{2+} ions; the study of the creation of the A^+ and A^{2+} ions in the presence of two laser fields with different frequencies; and the analysis of the effect of the dynamic Stark shift on the creation of the A^{2+} ions.

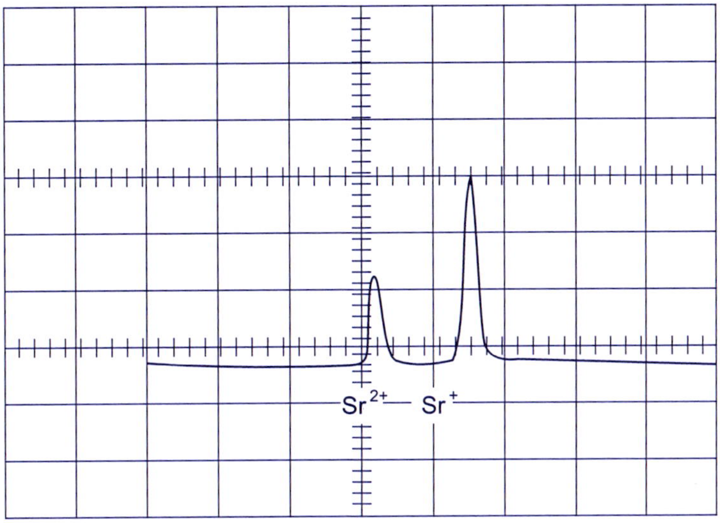

Figure 6.1 Oscillogram of the ion signals corresponding to the multiphoton ionization of the strontium atoms by the neodymium laser radiation with frequency $\omega \approx 9440\ cm^{-1}$ at the laser field strength $\varepsilon = 5 \times 10^6$ V/cm on the scale 2.5 μ s/div.
Source: [1]

6.3.1 Dependence of the creation of doubly charged ions on the laser frequency

The first experiments yielded a resonance structure on the curve of the $A^{2+}(N^{2+})$ ion yield versus laser frequency ω[9]. This fact indicates that the creation of doubly charged ions can be related to the spectrum of the atom or the singly charged ions. Therefore, the identification of the peaks can assist in the characterization of the creation of the A^{2+} ions. The resonance structure in the creation of singly and doubly charged calcium, strontium, and barium ions was extensively studied in the visible ($\omega = 14800 - 19000\ cm^{-1}$)[10–17] and IR($\omega = 8200 - 9100$ and $9395 - 9500\ cm^{-1}$)[9, 18–23] spectral ranges. Note that significantly different dependencies $N^{2+}\omega$ were obtained for the visible and IR spectral ranges.

Dependence of the creation of doubly charged ions on the laser frequency for the visible spectral range.

We consider the doubly charged ions resulting from the multiphoton ionization of calcium, strontium, and barium atoms by the visible laser radiation. Figures 6.2 and 6.3 show typical dependencies of the yields of singly (N^+) and doubly (N^{2+}) charged barium ions on laser frequency ω for the linear and circular polarizations. The dependences $N^+(\omega)$ and $N^{2+}(\omega)$ were studied for these polarizations at identical laser field strengths ε. It is seen that a variation in the laser frequency leads to variation in the ratio of the Ba^{2+} and Ba^+ ion yields from $N^{2+}/N^+ \approx 10^{-4}$ to $N^{2+}/N^+ \approx 10^{-1}$. Note that there are significantly different shapes, widths, and amplitudes of the resonance peaks on the curves $N^{2+}(\omega)$ in Figs 6.2 and 6.3.

The character of the resonance peaks on the curves $N^{2+}(\omega)$ for the Sr^{2+} ions[11, 12] almost coincides with the character of the resonance peaks for the Ba^{2+} ions (Figs 6.2 and 6.3). However, the resonance peaks for the Ca^{2+} ions[12, 14] are significantly different. The difference lies in greater widths and smaller amplitudes of the peaks corresponding to Ca^{2+} ions.

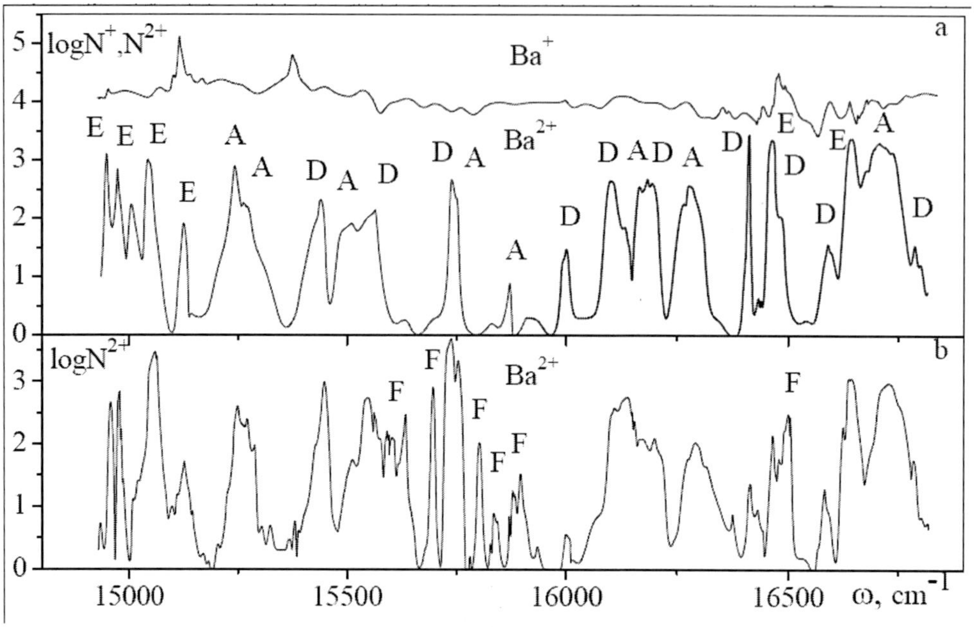

Figure 6.2 Plots of the Ba^+ and Ba^{2+} ion yields vs. the frequency of the dye (phenalemine 160) laser for (a) linear and (b) circular polarizations[17]. The ratio of the Ba^+ and Ba^{2+} ion yields for the linear polarization is absolutized. The laser field strength is $\varepsilon = 2 \times 10^6$ V/cm. The notation for the resonance peaks corresponds to the classification in the text

A detailed analysis of the resonance peaks on the dependencies $N^+(\omega)$ for the Ba^+ ions at relatively low laser field strengths ($\varepsilon \approx 4 \times 10^4$ V/cm) and their identification can be found in[24]. Note that these peaks are related to the two-photon excitation of various bound states of the Ba atom (one- and two-electron, singlet and triplet). Relatively small amplitudes of the resonance peaks on the curves $N^+(\omega)$ (Figs 6.2 and 6.3) are due to the saturation of the three-photon ionization of the Ba atom.

Several conclusions can be drawn from the analysis of the curves $N^+(\omega)$ and $N^{2+}(\omega)$ (Figs 6.2 and 6.3) and the results obtained for singly and doubly charged ions resulting from the multiphoton ionization of calcium, strontium, and barium atoms by the laser radiation in the spectral range $\omega = 14800 - 19000\ \text{cm}^{-1}$. In most cases, the resonance peaks in the yields of the A^+ and A^{2+} ions emerge at different laser frequencies. The peak widths for the yields of the A^{2+} ions range from 5 to 40 cm^{-1} and the amplitudes are varied by one to four orders of magnitude. Several resonance peaks on the curves $N^{2+}(\omega)$ are observed at both linear and circular polarizations whereas a few peaks can be obtained at either linear or circular polarization. The resonance peaks on the curves $N^{2+}(\omega)$ for the Ca^{2+}, Sr^{2+} and B^{2+} ions were identified with the aid of the known data on the spectra of the states of atoms A and ions A^+[25]. The comparison of the frequencies that correspond to the resonant transitions in the spectra of the A^+ ions (ω_t) and the laser frequencies (ω_m) that correspond to the resonance maxima on the curves $N^{2+}(\omega)$ and the analysis of the maxima at various laser polarizations yield two groups. The resonance peaks of the first group can be identified using the resonant transitions in the spectrum of the A^+ ion. The resonance maxima of the second group cannot be identified using the resonant transitions in the spectra of the A^+ ions and A atoms.

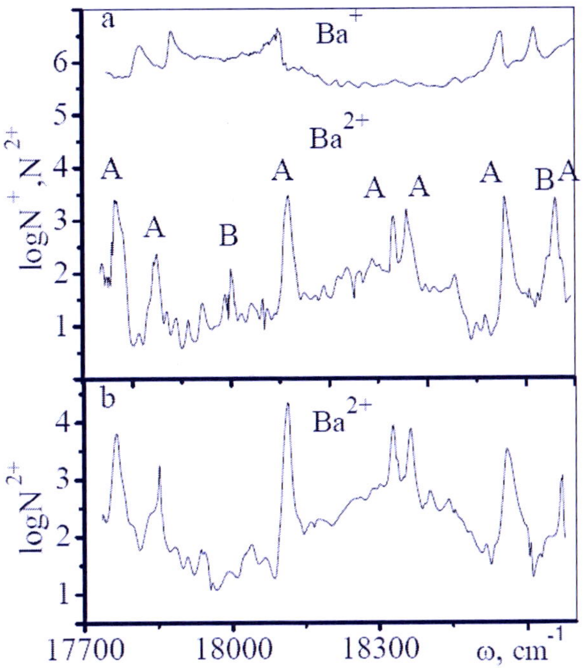

Figure 6.3 Plots of the Ba^+ and Ba^{2+} ion yields vs. the frequency of the dye (Rhodamine 110) laser for (a) linear and (b) circular polarizations[16]. The ratio of the Ba^+ and Ba^{2+} ion yields for the linear polarization is absolutized. The laser field strength is $\varepsilon = 2 \times 10^6$ V/cm. The notation for the resonance peaks corresponds to the classification in the text

The following resonance peaks belong to the first group: A maxima, which are manifested at linear and circular polarizations (the resonant transitions in the spectrum of the A^+ ion at the corresponding frequencies ω_t are allowed for both polarizations); *B* maxima, which are observed only at linear polarization (the resonant transitions in the spectrum of the A^+ ion at the corresponding frequencies ω_t are allowed only for linear polarization); and C maxima, which are observed only at linear polarization (the resonant transitions in the spectrum of the A^+ ion at the corresponding frequencies ω_t are allowed for both polarizations). The presence of such peaks indicates the cascade creation of the A^{2+} ions in the vicinity of frequencies ω_m, corresponding to the resonance maxima on the curves $N^{2+}(\omega)$.

In the case of the multiphoton ionization of the alkaline-earth atoms, the singly charged ions are effectively created in both ground ($ns^2S_{1/2}$) states and in the first excited ($nd^2D_{3/2,\,5/2}$ and $np^2\ P^0_{1/2,\,3/2}$) states. The supporting evidence can be found in[26–28]. The creation of the A^+ ions in these excited states can be owing to the electron decay of the auto-ionization states (AISs) whose excitation results from the multiphoton absorption including the above-threshold absorption of one photon[26–28]. Therefore, the measured dependencies $N^{2+}(\omega)$ exhibit maxima related to the resonant transitions from the three initial states. Figure 6.4 demonstrates the general scheme of the cascade creation of the A^{2+} ions.

The ratio of the yields of the A^+ and A^{2+} ions was calculated in[12] for the frequencies that correspond to the inter resonance intervals of the dependencies $N^+(\omega)$ and $N^{2+}(\omega)$ on the assumption of the cascade creation of the A^{2+} ions. The calculated ratios are in good agreement with the experimental data. Such a coincidence of the ion yields indicates that the cascade creation of the A^{2+} ions is implemented in the inter resonance intervals of the dependencies $N^{2+}(\omega)$.

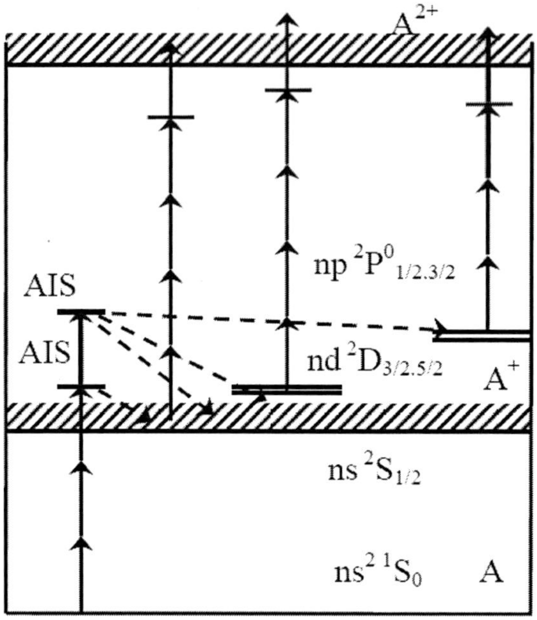

Figure 6.4 The scheme of the cascade process of the creation of doubly charged ions upon multiphoton atomic ionization: (horizontal lines) resonance states of singly charged ions and (dashed lines) possible decay channels of the auto-ionization states

As was mentioned, the dependencies $N^{2+}(\omega)$ also exhibit resonance peaks that cannot be identified. Such peaks are marked with the symbol D, E and F in Fig. 6.2. These peaks belong to the second group: D maxima, which are manifested at linear and circular polarizations, or only at linear polarization and cannot be identified with the resonance transitions in the spectrum of Ba atoms and Ba^+ ions; E maxima, which are manifested at linear and circular polarizations and can be identified with the resonance transitions in the spectrum of Ba^+ ions only at linear polarizations, because this transition is forbidden at circular polarization.

Other unusual maxima (type F) are observed only for circular polarizations of radiation and cannot be identified with the resonance transitions in the spectrum of Ba atoms and Ba^+ ions. The presence D, E and F peaks indicate the non-cascade creation of the A^{2+} ions in the vicinity of the corresponding frequencies ω_m

Note that the creation of the Ca^{2+}, Sr^{2+} and Ba^{2+} ions for the visible spectral range was also studied in[29–34]. The investigations were performed for relatively narrow spectral intervals and, mostly, for linear polarization. The identification of the resonance maxima in the yield of the A^{2+} ions is in general agreement with our identification.

The experimental results for the visible spectral range show that the creation of the Ca^{2+}, Sr^{2+} and Ba^{2+} ions upon multiphoton ionization of atoms by nanosecond pulses (see, for example,[12, 14, 16, 17]) is similar to the creation of ions by picoseconds pulses[30–32].

Thus, the analysis of the dependencies $N^+(\omega)$ and $N^{2+}(\omega)$ indicates the dominant contribution of the cascade creation of the A^{2+} ions upon multiphoton ionization of alkaline-earth atoms in the visible spectral range ($\omega = 14800 - 19000\,\text{cm}^{-1}$). In several small spectral fragments of this range, the effect that leads to the creation of ions cannot be classified as cascade.

Dependence of the creation of doubly charged ions on the laser frequency for the IR spectral range

We consider the dependence of the creation of doubly charged ions on the laser frequency for the multiphoton ionization of strontium[20, 22, 23] and barium [9, 18, 19, 21, 22] atoms by IR radiation with frequencies $\omega = 8200 - 9100$ and $9395 - 9500\,\text{cm}^{-1}$. Note the absence of other works devoted to the creation of the A^{2+} ions in the given spectral ranges. Figure 6.5 shows the dependences $N^2 + (\omega)$ measured for the Ba^{2+} ions at several field strengths ε. The dependences $N^{+}(\omega)$ and $N^{2+}(\omega)$ measured for the IR spectral range substantially differ from the corresponding dependencies obtained for the visible spectral range. In particular, the peak amplitudes for the IR spectral range are significantly less than the amplitudes for the visible spectral range, the widths for the IR spectral range are greater and the peaks are mostly asymmetric. The ratio of the yields of the A^{2+} and A^{+} ions is varied from $N^{2+}/N^{+} \approx 10^{-3}$ to $N^{2+}/N^{+} \approx 10^{-1}$ with a variation in the frequency. In most cases, the resonance peaks on the curves $N^{+}(\omega)$ and $N^{2+}(\omega)$ are observed at the same frequencies. In addition, the dependences $N^{2+}(\omega)$ for the A^{2+} ions exhibit peaks that are manifested only at certain field strengths ε.

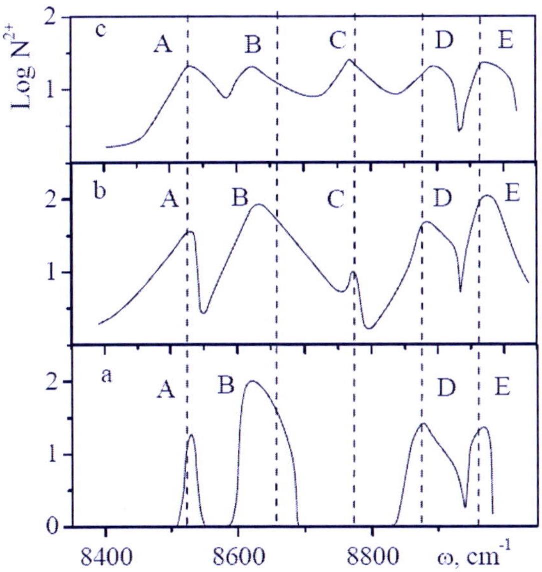

Figure 6.5 Plots of the Ba^{2+} ion yield vs. the frequency of the linearly polarized radiation of the CCL laser for the laser field strengths ε = (a) 4×10^{6}, (b) 5.5×10^{6}, and (c) 8×10^{6} V/cm[22]. The dashed lines show frequencies ω_{nm} (see Table). The notation for the resonance maxima corresponds to Table 6.1

The analysis shows that the peaks observed on the $N^{2+}(\omega)$ curves for Sr^{2+} and Ba^{2+} ions cannot be identified using the resonant transitions in the spectra of the singly charged ions as in the case of the visible spectral range. This fact indicates the non-cascade creation of Sr^{2+} and Ba^{2+} ions for the IR spectral range. The peaks are related to the resonances that involve the bound states of neutral atoms and are induced owing to the dynamic Stark effect. The first observation of such peaks can be found in[35], where the creation of Ba^+ ions was studied in the range $\omega = 8200 - 9100\ cm^{-1}$. Such peaks on the curves $N^{2+}(\omega)$ are manifested in the vicinity of frequencies ω_{nm} that correspond to the one-photon transitions between the n and m states (none of these states is the ground state). Under such conditions, the dynamic polarizabilities of the states strongly depend on the laser frequency and the absolute values of the polarizabilities are relatively large. States n and m exhibit large shifts here. Therefore, frequencies ω_{mn} corresponding to the resonance peaks can substantially differ from the frequencies that correspond to the normal multiphoton excitation of these unperturbed states from the ground state.

By way of example, Fig. 6.6 demonstrates the scheme of implementation of such induced resonance peaks in the vicinity of frequency ω_{mn}. Table 6.1 presents frequencies ω_{mn} and frequencies ω_m that correspond to resonance peaks on the curve $N^{2+}(\omega)$ upon ionization of the Ba atoms in the frequency range $\omega = 8200 - 9100\ cm^{-1}$ (Fig. 6.5). The values of frequencies ω_m and ω_{mn} are close to each other. This fact indicates that the resonance maxima on the curve $N^{2+}(\omega)$ are due to the multiphoton excitation of strongly perturbed *n* or *m* states of the Ba atom. Note that such peaks are also observed on the curve $N^{2+}(\omega)$ for Sr^{2+} ions upon the ionization of Sr atoms in the frequency range $\omega = 8200 - 9100\ cm^{-1}$[22, 23]. A variation in the energy of the *n* and *m* states by 100–2000 cm^{-1} is needed for the excitation of these states in the vicinity of frequency ω_{nm}. Such a variation in the energy is possible by the dynamic Stark effect when the dynamic polarizability of atomic levels is $\alpha \approx 10^4$a.u. This level is real for the dynamic polarizabilities of the Ba atomic levels[35].

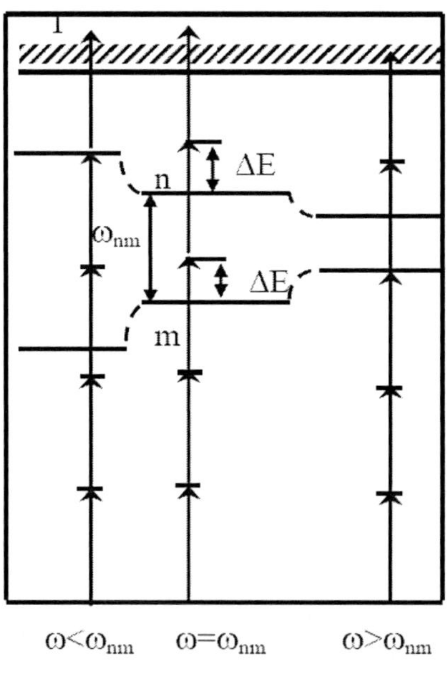

Figure 6.6 The scheme of the induced resonances in the vicinity of frequencies ω_{nm}

Thus, the results show that the resonance peaks in the yield of A^{2+} ions upon the ionization of alkaline-earth atoms in the IR spectral range are related to the excitation of strongly perturbed states of neutral atoms. The resonance maxima on the curves $N^{2+}(\omega)$ for the Sr^{2+} and Ba^{2+} ions are caused by the multiphoton excitation of the bound atomic states. This fact indicates the two-electron creation of doubly charged ions.

Table 6.1 *Identification of the resonance maxima on the curves $N^{2+}(\omega)$ for the Ba^{2+} ions (Fig. 6.5). The types of maxima correspond to the notation in Fig. 6.5, and the remaining notation can be found in the text*

Maximum	ω_m (cm^{-1})	Transition $n - m$	ω_{nm}, (cm^{-1})	ΔE, (cm^{-1})
A	8530	5d6p $^1F_3^0 - 6p^2\ ^1D_2$	8528	1224
B	8665	5d6p $^3P_1^0 - 6p^2\ ^1S_0$	8630	294
C	8780	5d6p $^3P_1^0 - 6p^2\ ^3P_0$	8788	634
D	8880	5d6p $^1P_1^0 - 6s7d\ ^1D_2$	8880	1905
E	8070	5d6p $^1F_3^0 - 6s7d\ ^3D_3$	8968	93

6.3.2 Effect of the concentration of neutral atoms on the creation of doubly charged ions in the IR spectral range

The next experimental series was aimed at the analysis of the effect of atomic concentration on the creation of A^{2+} ions in the IR spectral range. Such experiments were performed for the creation of doubly charged barium ions. In the experiments, the concentration of neutral atoms was varied both over the duration of the laser pulse needed for the creation of Ba^{2+} ions and prior to the pulse.

In the first case, a variation in the atomic concentration results from an increase in the atomic ionization probability due to the additional excitation[36–38]. The Ba^{2+} ions are created by the IR irradiation with the fixed frequency $\omega = 9395$ cm^{-1} (the LYAG radiation)[36, 37] or $\omega = 8620$ cm^{-1} (the CCL radiation)[38]. The DL radiation with the frequency tuning in the range $\omega = 17860 - 18160$ cm^{-1} is employed for the additional excitation of the Ba atoms. The action of the DL radiation on the Ba atoms in the ground ($6s^2\ ^1S_0$) state leads to either one-photon excitation of the $6s6p^1P_1^0$ state ($\omega = 18060$ cm^{-1}) or two-photon excitation of the $6s7d^3D_2$ state ($\omega = 17881$ cm^{-1}) and $5d6d^3D_2$ state ($\omega = 18100$ cm^{-1}). We compared the yields of the Ba^+ and Ba^{2+} ions created under the simultaneous and separate action of the IR and DL beams on the Ba atomic beam. Figure 6.7 demonstrates the results obtained in the case when the creation of the doubly-charged ions (Ba^{2+}) is induced by the LYAG radiation[36, 37].

It is seen that the yield of the Ba^+ ions is higher for the simultaneous action of the LYAG and DL radiations. The greatest increase in the Ba^+ ion yield is observed at the DL radiation frequency $\omega = 18060$ cm^{-1}, which corresponds to the one-photon excitation of the 6s6p $^1P_1^0$ state. The reason for an increase in the Ba^+ ion yield is as follows. The action of the DL radiation provides the transition of almost one half of the atoms from the ground state to the first resonance (6s6p $^1P_1^0$) state. The creation of the Ba^+ ions under the action of the LYAG radiation is transformed from the five-photon process (the ionization from the ground ($6s^2\ ^1S_0$) state) to the three-photon process (the ionization from the ($6s^2\ ^1S_0$) state). Thus, the ionization probability of the Ba atoms sharply increases.

In contrast to the Ba^+ ion yield, the Ba^{2+} ion yield is lower in the case of the simultaneous action of the two laser beams. Note that the minimum Ba^{2+} ion yields are observed in the vicinity of the DL frequency, which corresponds to the excitation of neutral atoms. The strongest decrease in the Ba^{2+} ion yield is observed in the vicinity of the DL frequency $\omega = 18060$ cm^{-1}. Recall that the maximum increase in the Ba^+ ion yield is also observed in the vicinity of this frequency. Such simultaneous decrease in the Ba^{2+} ion yield and increase in the Ba^{+} ion yield cannot be interpreted in the framework of the cascade model of the creation of doubly charged ions.

The above behavior of the ion yields caused by the additional excitation of the Ba atoms indicates that neutral Ba atoms serve as targets upon the creation of the Ba^{2+} ions in the presence of LYAG irradiation. In particular, an increase in the ionization probability of the Ba atoms under the saturation conditions for this process can lead to a significant decrease in the concentration of the Ba atoms in the interaction region. A decrease in the Ba atomic concentration during the laser pulse prior to the moment at which the effective creation of the Ba^{2+} ions is started must result in a decrease in the ion yield. Such a scenario is in agreement with the experimental data.

Thus, we have described the effect of the additional excitation of the Ba atoms on the creation of the Ba^{2+} ions under the action of the LYAG radiation. Note that the same results were obtained for the creation of the Ba^{2+} ions under the action of the CCL radiation with the frequency $\omega = 8620$ cm^{-1}[38]. Thus, the results show that neutral Ba atoms serve as targets upon the creation of the Ba^{2+} ions induced by the IR radiation.

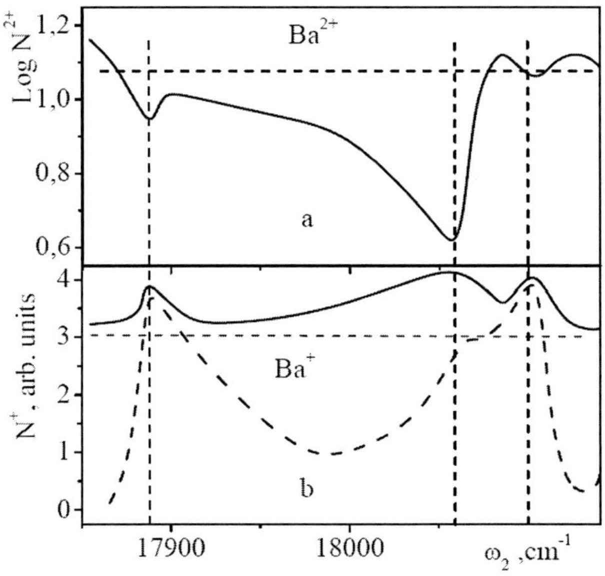

Figure 6.7 Plots of (a) Ba^{2+} and (b) Ba^{+} ion yields vs. the DL frequency for (solid lines) the ionization process involving the simultaneous action of the LYAG and DL radiations, (dashed lines) the ionization induced by the DL radiation, (dashed horizontal lines) the ionization induced by the LYAG radiation and (vertical dashed lines) the DL radiation frequencies that correspond to the resonant transitions in the spectrum of the Ba atom (see text for details).
Source: [37]

Experiments were also performed in which a variation in the concentration of neutral atoms in the beam preceded the arrival of the laser pulse creating the doubly charged ions[39]. Under such conditions, the creation of the Ba^{2+} ions under the action of the linearly polarized LYAG radiation was studied. A variation in the atomic concentration is implemented with the aid of an additional laser beam. The LYAG radiation serves as the additional radiation whose parameters are chosen in such a way that the ionization of the Ba atoms in the beams occurs under saturation conditions or close to saturation. For the correctness of the experiments, the circularly polarized additional radiation were employed, since the Ba^{2+} ion yield is strongly suppressed in the presence of the circularly polarized LYAG radiation (in comparison with the case of the linearly polarized radiation)[18]. The additional radiation pulse acts upon the Ba atomic beam earlier (by $t = 4 \times 10^{-8}$ s) than the main LYAG pulse that induces the creation of the Ba^{2+} ions.

The action of the additional radiation prior to the action of the main radiation pulse makes it possible to substantially increase the Ba^+ ion concentration and decrease the concentration of the neutral Ba atoms. In the experiments, the intensity of the additional radiation is varied and the intensity of the main radiation remains constant. The yields of ions created due to the action of the additional and main beams were determined. Figure 6.8 shows the results. It is seen that an increase in the intensity of the additional radiation leads to an increase in the amount of the Ba^+ ions in the interaction region. Note the decrease in the yields of the Ba^+ and Ba^{2+} ions created by the main radiation pulse delayed relative to the additional pulse.

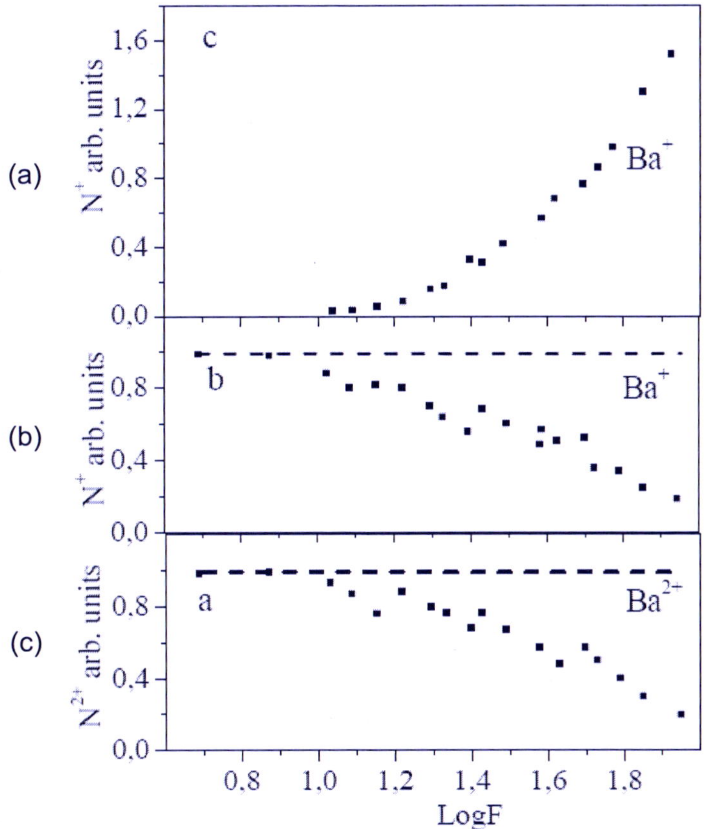

Figure 6.8 Plots of (a) the Ba^{2+} ion yield resulting from the action of the fundamental LYAG radiation in the case when the LYAG radiation pulse is delayed relative to the pulse of additional radiation, (b) the Ba^+ ion yield resulting from the action of the fundamental LYAG radiation in the case when the LYAG radiation pulse is delayed relative to the pulse of additional radiation, and (c) the Ba^+ ion yield resulting from the action of the additional LYAG radiation. The dashed horizontal lines show the mean Ba^+ and Ba^{2+} ion yields resulting from the action of the fundamental LYAG radiation in the absence of the preliminary irradiation of the Ba atoms using the additional radiation.

Source: [39]

Such a behavior of the Ba^{2+} ion yield cannot be interpreted in the model of the cascade creation of the doubly charged ions but can easily be explained using the two-electron model. Under the conditions for the cascade process, the preliminary increase in the Ba^{+} ion concentration in the interaction region due to the action of the additional radiation must lead to an increase in the amount of the Ba^{2+} ions created by the main radiation. At least, the amount of the Ba^{2+} ions must not decrease.

In the case of the two-electron process (when neutral atoms serve as targets upon the creation of the doubly charged ions), a decrease in the concentration of the neutral Ba atoms in the interaction region must result in a decrease in the Ba^{2+} ion yield. This scenario is in agreement with the experimental data.

Thus, the results show that neutral atoms serve as targets upon the creation of the doubly charged Ba atoms in the IR spectral range, and the two-electron process of the creation of the Ba^{2+} ions is implemented upon the ionization of the Ba atoms by the IR radiation.

6.3.3 Effect of the spectral perturbation of neutral atoms on the creation of ions

As was demonstrated, the resonance structure in the yield of the A^{2+} ions resulting from the multiphoton ionization of alkaline-earth atoms by IR radiation is caused by the excitation and ionization of the bound atomic states that are strongly perturbed owing to the dynamic Stark effect. Based on this fact, the effect of the perturbation of atomic states on the creation of the doubly charged ions of the Ba atoms were experimentally studied. The first laser beam was used for the ionization of the Ba atoms, while the second laser beam was employed for the perturbation of the bound states.

In the experiment from[40], the Ba atoms were ionized by the second-harmonic radiation of the LYAG (ω_1 = 18790 cm^{-1}). The atomic states were perturbed by the CCL radiation with the frequency tuning in the range $\omega_2 = 8500 - 8950$ cm^{-1}. The frequency ω_{nm} = 8880 cm^{-1}, which corresponds to the one-photon transition between the $6s7d^1D_2$ and $5d6p^1P_1^0$ states of the Ba atom, belongs to this range. In the vicinity of this frequency, the given states exhibit relatively high dynamic polarizabilities. Thus, the states are strongly perturbed in the presence of the CCL radiation. The perturbed states can be excited by the absorption of two photons of the second-harmonic radiation of the CCL (excitation of the $6s7d^1D_2$ state and the Raman process involving the absorption of two photons of the LYAG second-harmonic radiation and the emission of one photon with the frequency of the CCL radiation (excitation of the $5d6p^1P^01$ state). For the excitation of the $6s7d^1D_2$ and $5d6p^1P_1^0$ states, the minimum shifts of the states must be Δ_1 = 146 cm^{-1} and Δ_2 = 76 cm^{-2}, respectively. Figure 6.9 demonstrates the scheme of the perturbation and excitation of the Ba atomic states corresponding to the experimental conditions.

In the experiments, the yield of the Ba^{2+} ions created due to the simultaneous action of the LYAG second-harmonic radiation and the CCL radiation were measured. The results are presented in Fig. 6.10. It is seen that the Ba^{2+} ion yield significantly increases owing to the simultaneous action of the two beams when the CCL radiation frequency is close to ω_{nm} = 8880 cm^{-1}. Note that the Ba^{2+} ion yield is significantly higher than the ion yield in the case when the Ba atoms are ionized by either the CCL radiation or the LYAG second-harmonic radiation.

The analysis shows that the sharp increase in the Ba^{2+} ion yield is related to the excitation and subsequent ionization of the perturbed Ba atomic states $6s7d^1D_2$ and $5d6p^1P_1^0$. This conclusion is proven by the presence of the dip on the curve of the Ba^{2+} ion yield versus frequency at $\omega_2 = \omega_{mn}$ = 8880 cm^{-1}. The dip emerges due to the fact that the dynamic polarizabilities of the $6s7d^1D_2$ and

$5d6p^1P_1^0$ levels are zeros. Thus, the $6s7d^1D_2$ and $5d6p^1P_1^0$ levels are neither shifted nor excited. The dip is relatively small owing to a large spectral width of the CCL radiation.

Note an important experimental result: The frequency $\omega_2 = 8551\ cm^{-1}$ belongs to the above range of the CCL frequency. The summation of two photons with this frequency and one photon of the LYAG second-harmonic radiation must lead to the excitation of the unperturbed state $6s7d^1D_2$. However, Fig. 6.10 shows that this process does not result in the creation of Ba^{2+} ions.

Thus, the experimental results indicate that the excitation of the strongly perturbed states of the neutral Ba atoms causes the creation of the doubly charged ions. At the same time, the excitation of unperturbed states does not lead to the creation of ions. creation of ions.

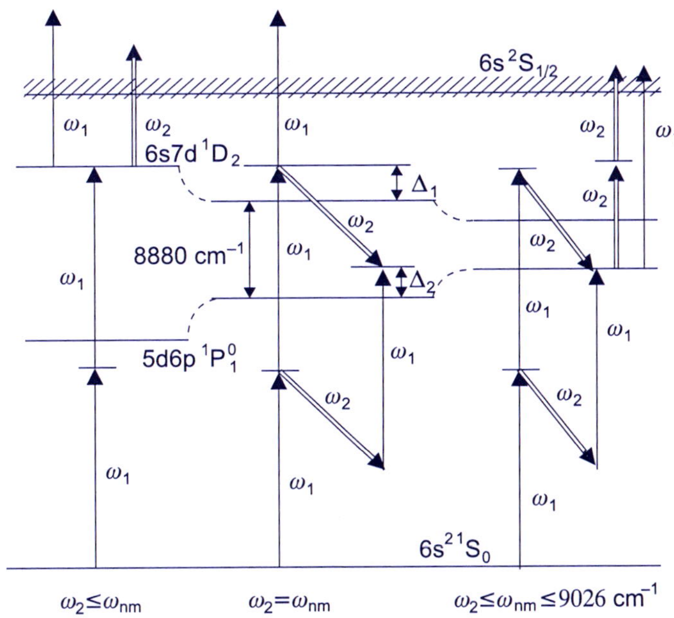

Figure 6.9 The schemes for the perturbation and excitation of the $5d6p^1P_2^0$ and $6s7d^1D_2$ states of the Ba atom under the simultaneous action of the LYAG second-harmonic radiation (ω_1) and the CCL radiation (ω_2). The dashed lines show variations in the energies of the states in the presence of the CCL radiation field

In the experiment from[41] on the effect of the excitation of perturbed atomic states on the creation of doubly charged ions, we employed CCL radiation with the frequency tuning in the range $2\omega_1 = 8800 - 8880\,cm^{-1}$ and the second-harmonic radiation with the frequency $\omega_1 = 17600 - 17760\,cm^{-1}$. We studied the Ba atoms. The radiation of the given frequency range strongly perturbs the spectrum of the Ba atom and, hence, provides the conditions for the creation of the Ba^{2+} ions owing to the two-electron process. Note that several frequencies ω_{nm} that correspond to the one-photon transitions between the bound states n and m belong to the ω_1 frequency interval: $\omega_{nm} = 8845\ cm^{-1}$ (transition between the $6s5d^3D_2$ and $6s6p^1P_1^0$ states), $\omega_{nm} = 8867\ cm^{-1}$ (transition between the states), and $\omega_{nm} = 8790\ cm^{-1}$ (transition between the $5d6p^3P_2^0$ and $6p^{23}P_1$ states). In accordance with the above analysis, these states in the vicinity of the frequency ω_{nm} must be strongly perturbed. In the case of the simultaneous action of the two laser beams, the perturbed states can be excited by the conventional multiphoton excitation of the CCL second-harmonic radiation and the Raman excitation processes involving the CCL radiation photons and the second-harmonic photons. The excitation schemes of the perturbed states are similar to the schemes in Fig. 6.9. In addition, the frequency $2\omega_1 = 17672\ cm^{-1}$, which corresponds to the two-photon excitation of the unperturbed state $6p^{2D}_{2}$ also belongs to the frequency interval of the CCL second-harmonic radiation.

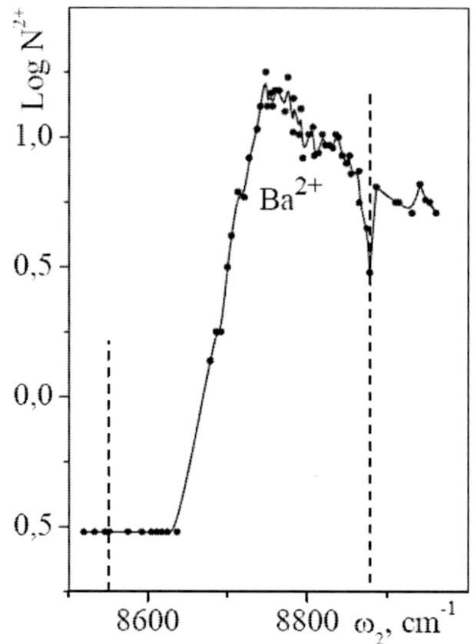

Figure 6.10 Plot of the Ba^{2+} ion yield vs. the CCL radiation frequency for the ionization of the Ba atoms under the simultaneous action the CCL radiation and the LYAG second-harmonic radiation[40]. The vertical dashed lines show frequencies of 8551 and 8880 cm^{-1}

In the experiment, the yield of the doubly charged Ba^{2+} ions created owing to the simultaneous action of the CCL radiation and the second-harmonic radiation were measured. Figure 6.11 demonstrates the corresponding results.

The analysis shows that the simultaneous action of the CCL radiation and the second-harmonic radiation upon the Ba atoms leads to the creation of Ba^{2+} ions due to the excitation and the subsequent ionization of the perturbed Ba atomic states. The supporting evidence is as follows. First, the simultaneous action of the two laser beams in the entire frequency interval leads to a relatively high excitation probability of the strongly perturbed Ba atomic states, so that the Ba^{2+} ion yield appears to be substantially greater than the ion yields corresponding to the separate action of the laser beams. Second, the dip on the curve of the Ba^{2+} ion yield in the vicinity of the frequency $2\omega_1 = 17672\ cm^{-1}$ (Fig. 6.11) and the resonance maximum in the Ba^+ ion yield have the same positions on the frequency scale.

The maximum in the Ba^+ ion yield emerges owing to the two-photon excitation of the unperturbed state $6p^{21}D_2$ by the CCL second-harmonic radiation. Thus, a relatively high excitation probability of unperturbed state $6p^{21}D_2$ does not lead to an increase in the Ba^{2+} ion yield. The reason for this is as follows. At a relatively high excitation, probability of the state $6p^{21}D_2$ lead to an excited state, the saturation of the three-photon ionization of the Ba atoms (the total atomic ionization in the beam interaction region) takes place over the laser pulse duration during the time when the CCL radiation field strength does not reach the level that is sufficient for a significant perturbation of the n and m states. This leads to the dip on the curve of the Ba^{2+} ion yield.

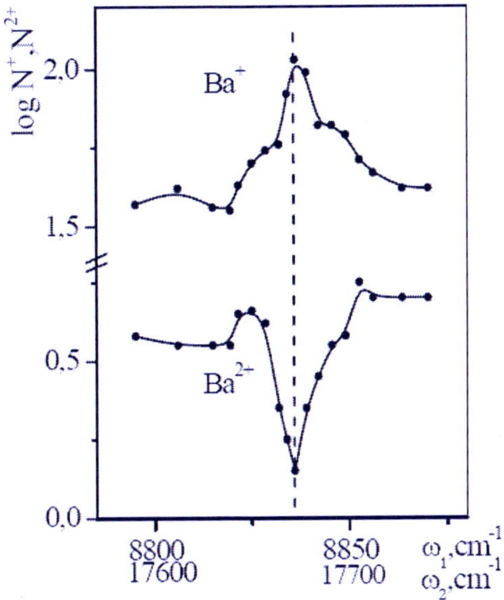

Figure 6.11 Plots of the Ba^+ and Ba^{2+} ion yields vs. frequency for the ionization of the Ba atoms under the simultaneous action of the fundamental CCL radiation (ω_1) and the second-harmonic radiation ($\omega_2 = 2\omega_1$)[41]. The vertical dashed line shows the excitation frequencies of the unperturbed $6p^{2\,1}D_2$ state ($\omega_1 = 8836\ cm^{-1}$) and $2\omega_1 = 17672\ cm^{-1}$

Thus, the creation of the doubly charged Ba^{2+} ions under the simultaneous action of the CCL radiation and the second-harmonic radiation (as well as in the case of the simultaneous action of the CCL radiation and the LYAG second-harmonic radiation) is related to the excitation and ionization of the strongly perturbed states of neutral Ba atoms. Hence, the two-electron creation of the Ba^{2+} ions occurs under the above conditions of the two experiments.

6.4 Conclusions

The existing data on the creation of doubly charged ions upon the multiphoton ionization of alkaline-earth atoms by laser radiation with nanosecond pulse duration show that the process of the creation of A^{2+} ions depends on the spectral range of the ionizing laser radiation. In particular, the cascade creation of the doubly charged ions dominates when the visible radiation is used the ions result from the multiphoton ionization of singly charged ions that are created during the same laser pulse. In the case of multiphoton ionization of the alkaline-earth atoms by IR laser radiation, the two-electron process occurs: the doubly charged ions are directly created from neutral atoms by the ionization via the induced resonances with strongly perturbed states of these atoms.

The fact that the two-electron process of the A^{2+} ion creation is implemented in the IR spectral range and not implemented in the visible spectral range is interpreted in the following way[42]. A significantly greater number of absorbed photons (five or six) are needed for the multiphoton ionization of the alkaline-earth atoms in the IR spectral range (in comparison with three absorbed photons needed for the ionization in the visible spectral range). Therefore, the probability of multiphoton ionization of atoms in the IR spectral range is substantially less than the probability corresponding to the visible spectral range. In this regard, the atomic ionization in the IR spectral range is saturated at substantially higher intensities (in comparison with the ionization in the visible spectral range). In experiments with IR radiation, the irradiation of neutral atoms occurs at significantly higher intensities. In addition, note the specific features of the spectra of the alkaline-earth atoms: the frequencies that correspond to the one-photon transitions between two states (ω_{mn}) (in the vicinity of such frequencies, the states can be strongly perturbed) often belong to the IR spectral range of the laser radiation employed in the experiments. Thus, the IR laser irradiation provides more favorable conditions for significant perturbation of the atoms (in comparison with the visible irradiation).

The probability of atomic ionization by visible radiation is higher, and the saturation of this process over the laser pulse precedes a significant perturbation of the atomic spectrum. Thus, neutral atoms are absent in the interaction region at the moment when the conditions for the two-electron process are satisfied during the laser pulse. On the other hand, a relatively high concentration of the singly charged ions is created in the interaction region prior to the moment corresponding to the maximum laser intensity when visible radiation is employed. Thus, more favorable conditions are provided for the cascade creation of the doubly charged ions in the case of visible laser irradiation.

The results show that an additional strong perturbation of spectrum of alkaline-earth atoms in the course of ionization by visible laser radiation also results in the two-electron creation of doubly charged ions.

Bibliography

1. V. V. Suran and I. P. Zapesochnyi. 1975. *Sov. Tech. Phys. Lett.* 1: 420.
2. I. S. Aleksakhin, I. P. Zapesochnyi and V. V. Suran. 1976. *Ukr. Fiz. Zh.* 21: 1383.
3. V. V. Suran. 1976. *Opt. Spektrosk.* 41: 901. *J. Opt. Spectrosc.* 41: 532.
4. I. I. Bondar', M. I. Dudich, V. V. Suran and L. L. Shimon. 1988. *Opt. Spektrosk.* 64: 476. *J. Opt. Spectrosc.* 64: 284.
5. N. A. Delone, V. P. Krainov and V. V. Suran. 1992. *Laser Phys.* 2: 815.
6. F. Krausz and M. Ivanov. 2009. *Rev. Mod. Phys.* 81: 1163.
7. I. S. Aleksakhin, I. P. Zapesochnyi and V. V. Suran. 1977. *JETP Lett.* 26: 11.
8. I. S. Aleksakhin, N. B. Delone, I. P. Zapesochnyi and V. V. Suran. 1978. *JETP* 49: 447.
9. I. I. Bondar', I. P. Zapesochnyi, N. B. Delone and V. V. Suran. 1981. *Pis'ma Zh. Tekh. Fiz.* 7: 243. *Sov. Tech. Phys. Lett.* 7: 104.
10. V. V. Bondar' and V. V. Suran. 1983. *Opt. Spektrosk.* 55: 784. *J. Opt. Spectrosc.* 55: 471.
11. I. I. Bondar', I. I. Dudich and V. V. Suran. 1986. *JETP* 63: 1142.
12. I. I. Bondar', N. B. Delone, M. I. Dudich and V. V. Suran. 1988. *J. Phys. B: At. Mol. Opt. Phys.* 21: 2763.
13. V. V. Suran. 1997. *Laser Phys.* 7: 660.
14. V. V. Suran. 1997. *Laser Phys.* 7: 665.
15. V. V. Suran. 1997. *Laser Phys.* 7: 1220.
16. V. V. Suran and I. I. Bondar'. 2005. *Opt. Spectrosc.* 98: 175.

17. V. V. Suran and I. I. Bondar. 2008. *in Proc. of the 4th Conf. of Elementary Processes in Atomic Systems* 113, Romania: Cluj Napoca.
18. N. B. Delone, I. I. Bondar', V. V. Suran and B. A. Zon. 1982. *Opt. Commun.* 40: 268.
19. N. B. Delone, V. V. Suran and B. A. Zon. 1984. *Multiphoton Ionization of Atoms*, 235. Canada: Academic Ontario.
20. T. T. Bernat, I. I. Bondar' and V. V. Suran. 1991. *Opt. Spectrosc.* 71: 22.
21. I. I. Bondar' and V. V. Suran. 1998. *JETP Lett.* 68: 837.
22. I. I. Bondar', V. V. Suran and M. I. Dudich. 2000. *J. Phys. B: At. Mol. Opt. Phys.* 33: 4243.
23. I. I. Bondar' and V. V. Suran. 2000. *Opt. Spectrosc.* 88: 830.
24. I. I. Bondar' and V. V. Suran. 1990. *Quantum Electron.* 20: 954.
25. Ch. E. Moore. 1949, 1952, 1958. *Atomic Energy Levels* 1–3. Washington DC: CNBS.
26. G. Petite and P. Agostini. 1986. *J. Physique* 47: 795.
27. S. G. Nakhate, S. A. Apmad, M. A. N. Razvi and G. D. Saksena. 1991. *J. Phys. B: At. Mol. Phys.* 24: 4973.
28. H. K. Haugen and H. Stapelfeldt. 1992. *Phys. Rev. A.* 45: 1847.
29. D. Feldman, H. J Krautwald and H. J. Welge. 1982. *J. Phys. B: At. Mol. Phys.* 15: L529.
30. P. Agostini and G. Petite. 1985. *J. Phys. B: At. Mol. Phys.* 18: L281.
31. P. Agostini and G. Petite. 1984. *J. Phys. B: At. Mol. Phys.* 17: L811.
32. D. A. Tate, D. G. Papaioannou and T. F. Gallagher. 1991. *J. Phys. B: At. Mol. Opt. Phys.* 24: 1953.
33. I. Liontos, S. Cohen and A. Bolovinos. 2008. *J. Phys. B: At. Mol. Opt. Phys.* 41: 045601.
34. I. Liontos, S. Cohen and A. Luras. 2010. *J. Phys. B: At. Mol. Opt. Phys.* 43: 095602.
35. I. I. Bondar' and V. V. Suran. 1998. *JETP* 86: 276.
36. I. I. Bondar' and V. V. Suran. 1992. *JETP Lett.* 56: 78.
37. I. I. Bondar' and V. V. Suran. 1993. *Zh. Eksp. Teor. Fiz.* 103: 774 [*JETP* 76: 381].
38. I. I. Bondar' and V. V. Suran. 1993. *Laser Phys.* 3: 863.
39. I. I. Bondar' and V. V. Suran. 1994. *Laser Phys.* 4: 1146.
40. I. I. Bondar' and V. V. Suran. 1999. *Kvant. Elektron.* 28: 162 [*Quantum Electron.* 28: 719].
41. I. I. Bondar' and V. V. Suran. 2002. *J. Phys. B: At. Mol. Opt. Phys.* 35: 3391.
42. I. I. Bondar' and V. V. Suran. 2003. *Opt. Spectrosc.* 94: 483.

7 Laser Induced Non-linear Multiphoton Processes

Siddhartha Lahon, Manoj Kumar, Rinku Sharma, P. K. Jha and Man Mohan

7.1 Introduction

The electromagnetic field–matter interactions are studied using both perturbative as well as non-perturbative methods. Both these methodologies have their advantages. Perturbative methods are suitable and are very efficient in electromagnetic field–matter interaction in the regime where the interaction term is negligibly small as compared to the unperturbed Hamiltonian of the system. Prturbative techniques fail when the interaction term (perturbation) becomes comparable to the Hamiltonian of the system[1] like in the case of laser–matter interaction. Further, Cohen-Tannoudji et al.[2] have shown that perturbative methods are valid only when the energy difference between initial and final states is very large as compared to the corresponding transition matrix element. In the present day scenario, very high intensity lasers are available, made possible by the advances in short pulse laser technology. We have the prerogative to study high intensity laser interactions with matter, bulk and nanostructures as well as atoms and molecules. In this high intensity regime of laser, the interactions are best studied using non-perturbative methods. Floquet theory is one of the most accurate non-perturbative methods. It relates the solution of the Schrödinger equation involving a periodic Hamiltonian to the solution of another equation with a time-independent Hamiltonian represented by an infinite dimensional matrix (called the Floquet matrix). It also has the advantage of reducing the problem into a time-independent one. This approach has earlier been applied in various atomic systems[3 – 6], and has the distinct advantage of handling both discrete and continuum states in a single treatment.

The Floquet theorem, as proposed by Floquet[7] describes the solution of a second order partial differential equation, with one independent variable time 't', and containing a time-periodic coefficients of periodicity 'T' as a combination of a periodic function $\Phi(\mathbf{r},t)$, with the same periodicity T, and another appropriate function. The Floquet method then was used extensively by Poincare in the 1890s[8]. Later, Autler and Townes[9] used it for obtaining the wave functions of a two level system in terms of infinite continued functions. However, it was only in the 1960s, when the Floquet method found a way into quantum mechanical approaches. Shirley[10] contributed significantly to the further development of the technique by reformulating the time-dependent problem of interaction of a two level quantum system with an oscillating classical field into an equivalent time-independent problem involving an infinite-dimensional matrix, the Floquet matrix. It reduces the time-dependent Schrödinger equation into finding the eigenvalues and the eigenvectors of the equivalent infinite-dimensional matrix. Here we study the multiphoton processes in quantum dots (QD) using the Floquet technique. Quantum dots (QD) are nanostructures where the charge carrier's movements are confined in all three spatial directions. QDs are sometimes also referred to as 'artificial atoms' with typical dimensions of a few nanometers to tens of nanometers, containing about $1-200$ charge carriers[$11-13$]. The quantum size effect and quantum confinement together results in the formation of discrete energy levels with the conduction and valence band themselves, giving rise to interesting linear as well as non-linear properties.[$14-17$]. Recent advances in nanofabrication techniques have enabled us to manipulate these energy levels and their wave functions by tailoring the QDs into desired shape and size. This controllability of energy levels and wave functions generates a considerable attraction towards these quantum confined structures, named QDs, as this adaptability extends to a fair amount of optical and electronic properties of the QDs. It also propels its study for understanding the quantum behavior as well as for its applications in semi conductor lasers, microelectronics, non-linear optics and many other fields.

The properties associated with the intersubband transitions of QDs show many interesting behaviors. The generally larger value of intersubband dipole matrix elements gives rise to significant non-linear behavior. These behaviors are manifested in the form of non-linear absorption coefficient, susceptibilities refractive index change etc. The non-linear properties are thus found to be greatly enhanced in semiconductor QDs as compared to bulk semiconductors[$18-19$]. In these semiconductor based QDs, the inter-level transitions are accomplished by optical excitations. This makes the optical response of the QD very appealing from the point of view of applications as well as from the basic scientific prospect.

Further, there has been a great advancement in terahertz (THz) laser technology which makes the electromagnetic spectrum, of frequency range 300 GHz to a few THz, a very rich range for application in chemical sensing, astronomy, condense matter spectroscopy, bioimaging and last but not the least, communications. Co-incidentally, the energy separation between the discrete intersubband levels of QDs and other nanostructures happens to be of the order of a few meV, which is the energy level corresponding to the THz regime of the electromagnetic field. Therefore, QDs and other nanostructures couple very strongly with THz frequency lasers. Hence, these days, quantum nanostructures are studied experimentally as well as theoretically extensively for its applications in the above-mentioned fields. Many interesting properties are revealed by these kind of studies using THz intense lasers, to name a few, resonant absorption[20], THz photon assisted tunneling[21], optical stark Shifts[22], excitonic excitations[23] etc. Nowadays, QDs are also used for multiphoton related studies and non-linear processes for potential applications in different areas, for example, THz frequency up-conversion[$24-26$], multiphoton bioimaging[27, 28], THz multiphoton photo detector[29, 30] and optical switching[31]. Bioimaging of tissue using quantum cascade laser and THz technology has been performed by Kim. et al.[32] and Darmo et al.[33].

In this chapter, the intraband and interband multiphoton transitions in a stable disc shaped harmonic potential QD, placed in a static magnetic field is presented. The results are correlated to the application in making ultrasensitive THz bioimaging and photodetecting devices. For finding the solution of the dynamic problem, Floquet theory has been used. Floquet theory relates the solution of the time-dependent Schrödinger equation involving a periodic Hamiltonian to the solution of a time-independent matrix equation involving an infinite-dimensional matrix called Floquet matrix. This method, which was preciously applied to various atomic systems[3, 4, 34, 35], has the distinct advantage of handling both discrete and continuum states in a single treatment. It is used here to explore the interesting cases of near resonance frequency taking into account the effects of laser intensity and magnetic field as the tuning parameter for the intraband transitions. Further, the study is extended into a case involving both the intraband and the interband transitions. In this extended study, the transition probability as well as susceptibility of the QD to laser is obtained. The results thus obtained is related to the efficiency of a hypothetical photodetecting device.

7.2 Theoretical Framework

7.2.1 The Floquet treatment

We consider a system whose unperturbed Hamiltonian is $\mathrm{H}_0(\mathbf{r})$ such that

$$\mathrm{H}_0(r)|\alpha(r)\rangle = E_\alpha^{(0)}|\alpha(r)\rangle, \tag{7.1}$$

where eigenfunctions $|\alpha(r) > s$ form a complete set of orthnormal functions, i.e.,

$$\langle\beta(r)|\alpha(r)\rangle = \int \beta^*(r)\alpha(r)d\mathbf{r} = \delta_{\beta\alpha} \tag{7.2}$$

and

$$\sum_\alpha |\alpha(r)\rangle\langle\alpha(r)| = 1. \tag{7.3}$$

As this system is driven by a time periodic potential (electromagnetic field) with angular frequency $'\omega'$ and time period $T(=2\pi/\omega)$, the process is defined by the Schrödinger equation

$$\mathbf{H}_T(r,t)\psi(r,t) = 0. \tag{7.4}$$

$\mathbf{H}_T$ is defined as (taking $\hbar = 1$),

$$\mathbf{H}_T(r,t) = H(r,t) - i\frac{\partial}{\partial t}, \tag{7.5}$$

where the Hamiltonian H_T is the total Hamiltonain of the interacting system described by

$$H(r,t) = H_0(r) + H_I(r,t). \tag{7.6}$$

The interaction Hamiltonian H_I is time periodic with period T. The wavefunction $\psi(r,t)$ can be written in accordance with the Floquet theorem[7] in terms of a time periodic function $\Phi(r,t) = \Phi(r,t+T)$ as

$$\Psi(r,t) = \exp(-i\lambda t)\Phi(r,t). \tag{7.7}$$

$\Psi(r,t)$ is called the quasienergy state and λ is the Floquet characteristic value or the quasienergy[36, 37]. The name 'quasienergy state' is drawn from the analogy with the Bloch eigenfunctions in solids with space periodic potential. Using the peridicity of $\Phi(r,t)$ and Eqs (7.5) and (7.7) in (7.4) we obtain the eigenvalue equation

$$\mathbf{H}_T(r,t)\Phi_i(r,t) = E_i\Phi_i(r,t). \tag{7.8}$$

A composite Hilbert space in $R(+)T$ can be introduced for the eigenfunctions of the Hermitian operator $H_T(r,t)$[38], where the spatial part is composed of orthonormal square integrable functions and the temporal part is spanned by the complete orthonormal set of functions, $\mathbf{e}^{(im\omega t)}$, where m is the Fourier index which take any integral value. Thus, the eigenvectors of H_T will be orthonormal, i.e.,

$$\langle\langle\Phi_i(r,t)|\Phi_j(r,t)\rangle\rangle = \frac{1}{T}\int_0^T \langle\Phi_i(r,t)|\Phi_j(r,t)\rangle dt = \delta_{ij} \tag{7.9}$$

and form a complete set

$$\sum_i |\Phi_i(r,t)\rangle\langle\Phi_i(r,t)| = 1. \tag{7.10}$$

In the next section, the quasienergy states thus obtained are briefly described.

7.2.2 Quasienergy states

A sufficiently strong perturbation of the electromagnetic field gives rise to a condition whence the concept of quasienergy state becomes relevant. A quasienergy state can be thought of as a state with characteristic energy (time averaged) of the combined system of the atom/molecule/nanostructure and that of the electromagnetic field together. The use of the Floquet theorem[7] to this kind of system of a combination of monochromatic laser and the atom/molecule/nanostructure gives a wholesome description of the system dynamics due to the interactions. Floquet quasienergy states of a system can be considered as identical to the dressed atomic picture introduced by Cohen Tannoudji and Haroche[39]. The quasienergy states are mutually orthogonal at any instant of time and they form a complete set vide Eqs (7.9) and (7.10).

7.2.3 Quantum dots

QDs are theoretically described using many confinement potential approaches like finite square well potential[40], harmonic potential[41 – 43] etc. Further, harmonic potential confinement approach has been considered to be nearest to real world QDs. We here consider an isolated two dimension harmonic potential QD such that

$$H_0^{(0)}(\mathbf{r})\phi^{(0)}(\mathbf{r}) = E^{(0)}\phi^{(0)}(\mathbf{r}), \tag{7.11}$$

where $H_0(\mathbf{r})$ is the unperturbed Hamiltonian defined as

$$H_0^{(0)}(\mathbf{r}) = \frac{p^2}{2m^*} + \frac{1}{2}m^*\omega_0^2 r^2. \tag{7.12}$$

Here m^* is the effective mass of the charge carrier, p is the momentum, ω_0 is the harmonic potential frequency, and $r = x^2 + y^2$. Using Eq. (7.12) in (7.11) and writing p in the derivative form, the Schrödinger equation transforms into

$$\frac{1}{r}\frac{\partial}{\partial r}\left(r\frac{\partial}{\partial r}\phi^{(0)}\right) + \frac{1}{r^2}\frac{\partial^2}{\partial\theta^2}\phi^{(0)} - \left(m^*\omega_0^2 r^2 - \frac{2m^* E^{(0)}}{2}\right)\phi^{(0)} = 0. \tag{7.13}$$

Using separation of variables and taking $\phi^{(0)}(\mathbf{r}) = R(r)U(\theta)$, the solution of the angular part becomes out to be

$$U(\theta) = e^{-il\theta}, \tag{7.14}$$

where 'l' is an integer. Using this, the radial part becomes

$$\frac{d^2R}{dr^2} + \frac{1}{r}\frac{dR}{dr} - \frac{l^2}{r^2}R + (k^2 - \lambda^2 r^2)R = 0, \tag{7.15}$$

where

$$\lambda^2 = \frac{m^{*2}\omega_0^2}{\hbar^2} \tag{7.16}$$

and

$$k^2 = \frac{2m^* E^{(0)}}{\hbar^2} \tag{7.17}$$

Making a transformation of variables from r to $t = \lambda r^2$ and taking

$$R(r) = r^{|l|} e^{-\lambda r^2} F(\lambda r^2). \tag{7.18}$$

Equation (7.15) becomes the associated Laguerre equation

$$t\frac{d^2F}{dt^2} + (|l| + 1 - t)\frac{dF}{dt} - \frac{1}{2}\left(|l| + 1 - \frac{k^2}{2\lambda}\right)F = 0, \tag{7.19}$$

whose solution is the associated Laguerre polynomial such that

$$L_n^{|l|}(t) = L_n^{|l|}(\lambda r^2) \tag{7.20}$$

is normalize-able only if

$$\frac{1}{2}\left(|l| + 1 - \frac{k^2}{2\lambda}\right) = -n \tag{7.21}$$

which results in the energy eigenvalues of the Schrödinger equation as

$$E_{nl}^{(0)} = (2n + |l| + 1)\hbar\omega_0 \tag{7.22}$$

and eigenvectors

$$\phi_{nl}^{(0)} = \sqrt{\frac{n!}{(n+|l|)!}}\frac{1}{\sqrt{2\pi a_{(0)}}}\frac{1}{\left(\sqrt{2a_{(0)}}\right)^{|l|}}r^{|l|}e^{-\frac{r^2}{4a_{(0)}^2}}L_n^{|l|}\left(\frac{r^2}{2a_{(0)}^2}\right)e^{-il\theta}, \tag{7.23}$$

where $a_{(0)}$ is the effective radius of the QD given as

$$a_{(0)}^2 = \frac{\hbar}{2m^*}\frac{1}{\omega_0}. \tag{7.24}$$

When a magnetic field B is introduced perpendicular to the plane of the QD, the time-independent Schrödinger becomes

$$H_0(\mathbf{r})\phi(\mathbf{r}) = E\phi(\mathbf{r}), \tag{7.25}$$

where

$$H_0(\mathbf{r}) = \frac{p^2}{2m^*} + \frac{1}{2}m^*\omega_0^2r^2 - \frac{1}{2}\omega_c\mathbf{l}.\mathbf{B}. \tag{7.26}$$

Here ω_c is the cyclotron frequency and l is the orbital angular momentum and other quantities are as defined earlier. Equation (7.25) is exactly solved with the procedure used in the case of absence of magnetic field. The solutions are obtained as

$$\phi_{nl} = \frac{1}{\sqrt{2\pi a}}\sqrt{\frac{n!}{(n+|l|)!}}\left(\frac{r}{\sqrt{2}a}\right)^{|l|}e^{-\frac{r^2}{4a^2}}L_n^{|l|}\left(\frac{r^2}{2a^2}\right)e^{-il\theta} \tag{7.27}$$

with

$$a^2 = \frac{\hbar}{2m^*}\frac{1}{\omega}, \tag{7.28}$$

where

$$\omega = \sqrt{\omega_0^2 + \frac{1}{4}\omega_c^2} \tag{7.29}$$

and

$$E_{nl}^{(0)} = (2n + |l| + 1)/\omega - \frac{1}{2}l\hbar\omega_c. \tag{7.30}$$

7.2.4 Quantum dot in laser field and its description in Floquet theory

We consider the two-dimensional harmonic potential QD in a dc magnetic field interacting with a periodic laser field. The semiclassical time-dependent Hamiltonian in the presence of the laser field is given as

$$H = H_0(r) + H_I(r,t), \tag{7.31}$$

where $H_0(r)$ is the unperturbed Hamiltonian of the QD such that

$$H_0(r)\varphi_q(r) = E_q\varphi_q(r). \tag{7.32}$$

The wavefunction φ and eigen values E are defined in Eqs (7.27) and (7.30) and $H_I(\mathrm{r},t)$ is the interaction Hamiltonian, given as

$$H_I(r,t) = -\mu(r)E_0\cos(\omega t)\cos(\theta). \tag{7.33}$$

Here $\mu(r)$ is the QD dipole moment operator and θ is the angle between the polarization vector and the fixed body dot axis, E_0 is the amplitude of the laser field. The time evolution of the system wave function is determined by the time-dependent Schrödinger equation in atomic unit

$$H_T\psi_p(r,t) = 0, \tag{7.34}$$

where

$$H_T = H(r,t) - i\frac{\partial}{\partial t}. \tag{7.35}$$

The system wavefunction can be expanded in terms of the eigenfunctions of the isolated dot Hamiltonian H_0 as

$$\psi_p(r,t) = \sum_q a_{qp}(t)\varphi_q(r). \tag{7.36}$$

Here q increments with increasing eingenvalue E_q. Using this expansion of the wavefunction $\psi_p(r,t)$ and orthogonality of the eigenstates in (7.33) a set of coupled equations is obtained as

$$i\frac{\partial}{\partial t}a_{qp}(t) = E_q a_{qp}(t) + E_0\cos(\omega t)\sum_s V_{qs}a_{sp}(t) \tag{7.37}$$

where

$$V_{qs} = \langle\varphi_q|\mu(r)\cos(\theta)|\varphi_s\rangle \tag{7.38}$$

denotes the dipole matrix elements for the QD interacting with the laser. Written in matrix notation, Eq. (7.37) transforms into an equation of the form

$$i\frac{\partial}{\partial t}\,\mathbf{A}(t) = \mathbf{H}_c(t)\,\mathbf{A}(t), \tag{7.39}$$

where $H_C(t)$ is a time periodic matrix with periodicity $T = 2\pi/\omega$ whose diagonal terms are the eigenenergies and the off-diagonal terms are the time periodic interaction terms of the Hamiltonian.

The general solution of (7.39) as asserted by the Floquet theorem[36, 39] in matrix notation is of the form

$$A\chi(t) = \chi(t)e^{-i\lambda_s t}, \tag{7.40}$$

where $\chi(t) = \chi(t+T)$ and λs are the quasienergies or the Floquet characteristics exponent. The Fourier expansions of $A(t)$ and $Hc(t)$ are given as

$$A(t) = \sum_{m=-\infty}^{\infty} f_{vs}^{m} e^{im\omega t} e^{-i\lambda_s t} \tag{7.41}$$

and

$$\mathbf{H}_c(t) = \sum_{m=-\infty}^{\infty} H_{vs}^{m} e^{im\omega t}. \tag{7.42}$$

Here the indices v and s range over the dimension N_q, where N_q, is the total number of QD states included in the study. These expansions, Eqs (7.41) and (7.42) reduce Eq. (7.39) into the eigenvalue equation[44] of Fourier components fvs^m and characteristic values λ_s,

$$\sum_{jk} [H_{vj}^{m-k} + m\omega\delta_{v,j}\delta_{k,m}] f_{js}^{k} = \lambda_s f_{vs}^{m}. \tag{7.43}$$

The first multiplier (within the bracket) on the left-hand side represents the Floquet Hamiltonian matrix H_F.

7.2.5 Computation of Floquet matrix and its application to find the transition probabilities

The diagonal terms of H_F, i.e., $E'_{nl} = (E_{nl} - E_{00} - m\omega)$ refer to the mth detuning and off diagonal terms are the interaction terms which are given as

$$V_{q,q'} = \frac{1}{2} E_0 \langle \varphi_q | \hat{\varepsilon} \cdot r | \varphi_{q'} \rangle, \tag{7.44}$$

where $\varepsilon\wedge$ is the unit direction vector along E_0.

The Floquet Hamiltonian matrix is diagonalized to obtain the eigenvalue of the Floquet matrix, and corresponding eigenvector. The transition probability from the initial state $\varphi_p(r)$ to the final state $\varphi_q(r)$, when the QD is immersed in a laser field for a time period t, can be determined by using the eigenvalues (λ_s) and the corresponding eigenvectors f_{us}^{m} in Eq. (7.44) and is given by

$$P_{p\to q}(t) = \sum_s \sum_v [f_{ps}^{m} f_{pv}^{m} f_{qs}^{m'} f_{qv}^{m'} e^{-i(\lambda_v - \lambda_s)t}]. \tag{7.45}$$

For continuous coherent operation of the laser for an infinitely long duration, the time averaged transition probability from $\varphi_p(\mathbf{r})$ to the final state $\varphi_q(\mathbf{r})$ is

$$P_{q\to q'} = \lim_{t_0\to\infty} \frac{1}{t_0} \int_0^{t_0} dt[P_{q\to q'}(t)] = \sum_s \left| f_{ps}^{m} f_{qs}^{m'} \right|^2, \tag{7.46}$$

where the summation is over all the eigenvalues of the concerned Floquet matrix. The time averaged transition probability characterizes the optical transition.

7.2.6 Non-linear susceptibilities

The polarization of the QD, induced by the laser field, is obtained using the quasienergy states as

$$\begin{aligned} P_p(t) &= \langle \psi_p(\mathbf{r},t) | \mu(r) | \psi_q(\mathbf{r},t) \rangle \\ &= \sum_{n,m,q,q'} [(f_{q',p}^{m})^* \langle \varphi_{q'} | \hat{\varepsilon}.r | \varphi_q \rangle f_{q,p}^{m+n}] e^{-in\omega t} \\ &= \sum_n P_p(n\omega) e^{-in\omega t}, \end{aligned} \tag{7.47}$$

where $P_p(n\omega)$ is the nth Fourier component of the polarization such that

$$P_1 = \sum_{m,q,q'} [(f_{q',1}^{m})^* \langle \varphi_{q'} | \hat{\varepsilon}.r | \varphi_q \rangle f_{q,1}^{m+1}] e^{in\omega t}. \tag{7.48}$$

The total susceptibility of the system is thus obtained as

$$\chi(\omega, E_0) = P_1(1\omega)/E_0. \tag{7.49}$$

The imaginary part of the linear susceptibility gives rise to absorption, an interlevel transition accompanied by the absorption of one photon. Similarly, imaginary parts of non-linear susceptibility imply multiphoton absorptions.

7.3 Results and Discussion

In this chapter, the dynamics of GaAs QD placed in a dc magnetic field and subjected to THz laser fields was studied. The magnetic field is assumed to be applied perpendicular to the plane of the QD prior to the application of the laser fields as is feasible in experimental setups. One can obtain the eigenvalues and eigenfunctions of the QD for a Hamiltonian in the presence of the magnetic field as shown in Eqs (7.26) and (7.27). $\hbar\omega_{0c}$ is taken as 3 meV, where ω_{0c} is the harmonic oscillator frequency in the conduction band and the effective mass of the charge carriers in the conduction band and valence band are $0.067\mathrm{m}_0$ and $0.45\mathrm{m}_0$, respectively.

Figure 7.1 demonstrates the labeling and energy spectrum of the QD conduction band eigenstates prior to degeneracy breaking by the magnetic field. As the magnetic field is increased, the degeneracy for quantum numbers n, l is broken for different levels. When the laser field is applied, this spectrum is modified by the involvement of photons. The Floquet theory was applied for finding the photon-dressed states.

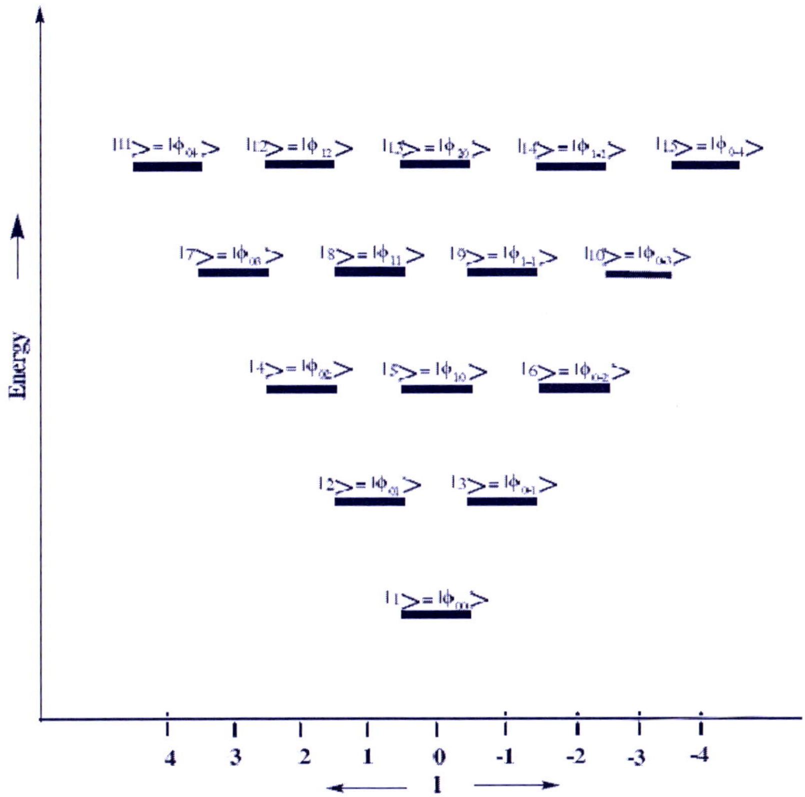

Figure 7.1 The energy eigenstates φ_{nl} of QD in the conduction band with increasing energy prior to the application of magnetic field B

In Fig. 7.2, the variation of time averaged transition probability from ground state $|1>$, with laser frequency in THz at laser intensity $I = 1\,\mathrm{W/cm^2}$, for $B = 1.0$ T was shown. As can be seen, two distinct patterns is obtained, one around $\omega = 3.43$ THz and the second around $\omega = 6.05$ THz. The first pattern around $\omega = 3.43$ THz corresponds to transitions from ground state $|1>$ to the left ladder states, i.e., $|2>, |4>, |7>$, etc., while the second pattern around $\omega = 6.05$ THz corresponds to transition from the ground state $|1>$ to the right ladder states due to selection rules $\Delta l = (+1, -1)$, irrespective of n. Both the patterns show the hole-burning non-linear effect due to multiphoton transitions climbing higher levels (e.g., to $|4>$ transitions involving two photons). Further, the doublet in $|1>$ to $|2>$ transition is quite visible around the position $\omega = 3.43$ THz. Similar type of doublet is seen in the second pattern for $|1>$ to $|3>$ transition. Thus, both laser and magnetic fields can be used externally for tuning the QD in a biological system for use as a multiphoton bioimaging switch.

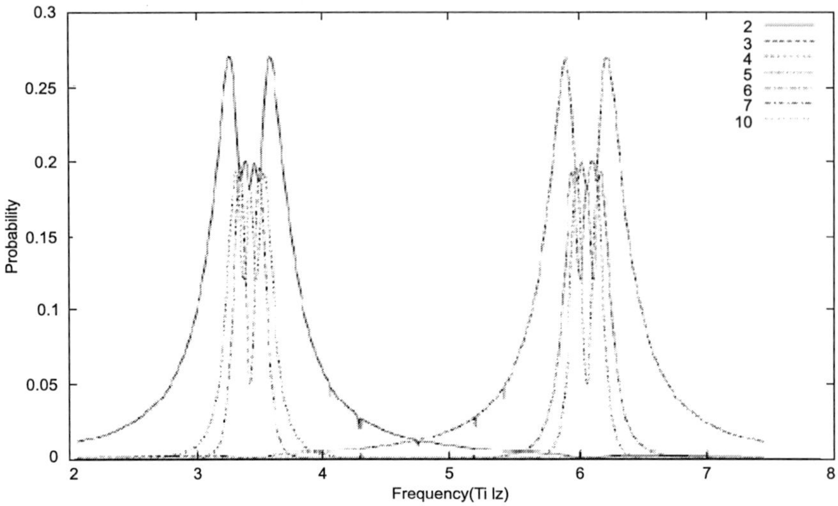

Figure 7.2 Variation of transition probability from the round state |1 > to the left ladder states namely |2 >,|4 >,|7 > etc and the right ladder states |3 >,|6 >,|10 > etc for magnetic field $B = 1.0$ T. Here 2 correspond to |1 > to |2 > transitions, etc.

For a given geometry, the onset of non-perturbative effects depends on laser intensity I. Increasing the laser intensity, the quasienergy level splitting also increases. Figure 7.3(a) and (b) shows the variation of probability from |1 > to |2 > transitions for $B = 0.0$ and 1.0 T, respectively. As the laser intensity increases, we clearly see the broadening and shifting of the curves showing power broadening and non-linear ac Stark effect. The visibly contrasting features between the figures are due to the modification in all the energy levels, due to the magnetic field contributing to the evaluation of the dynamics of the system. In the $B = 0.0$ T case, the subsequent degenerate states, being equi-spaced as depicted in Fig. 7.1, contributes to symmetric broadening of the |1 > to |2 > transition peak for two way shift from around $\omega = 4.55$ THz to $\omega = 2.42$ THz and $\omega = 6.68$ THz, respectively. The introduction of a magnetic field breaks the degeneracy and modifies the excited levels. This results in asymmetric broadening of the |1 > to |2 > transition line as shown in Fig. 7.3(b), making it equivalent to a red shift from around $\omega = 3.36$ THz to $\omega = 1.87$ THz. Power broadening will also be useful for bio-imaging as the system can be scanned for large frequency intervals to give photoluminescence spectra from QDs inserted in biological systems.

One can make a quantum optical device with a colloidal solution of QDs with high density ($\approx 10^{15}$ QD/cm^3). The transitions in QDs can be tuned using the external fields to yield efficient number of fluorescent photons. The biological specimen exposed to such a device is expected to give a high contrast image of its different components. The number of fluorescent photons F_p emitted by the specimen can be written as

$$F_p = \sigma^N_{(G)} N_F (I_p)^N, \tag{7.50}$$

where N_F is the number of QD per cm^3; I_p is the photon flux of the laser; $\sigma^N_{(G)}$ is the generalized N-photon cross-section (independent of photon flux) and is related to the total N-photon cross-section σ^N as $\sigma^N_{(G)} = \sigma^N/(I_p)^{N-1}$. As shown in Eq. (7.50), the fluorescent yield can be increased by optimizing the transition probability (cross-section) of multiphoton excitation using external fields. Thus, the tuning effects presented in this chapter will help in shaping up highly efficient future bio-imaging and photo-detecting devices.

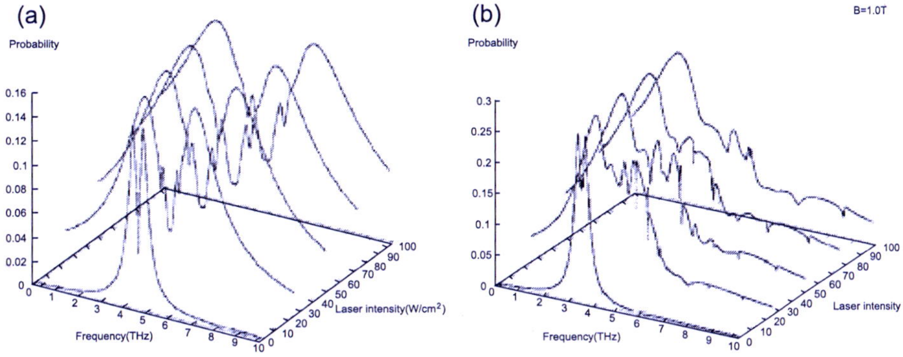

Figure 7.3 Variation of transition probability for transition |1 > to |2 > for different laser frequencies and intensities for (a) $B = 0.0$ T; (b) $B = 1.0$ T

Interband transition to the system with one level in the conduction band and the other in the valence band will now be discussed. The transition is determined by the overlap integral between the two levels. Figure 7.4(a) shows the transition probability from the ground state in the valence band to the excited state in the conduction band. The multiphoton block Floquet method is used to obtain the transition probability as well as the optical susceptibility.

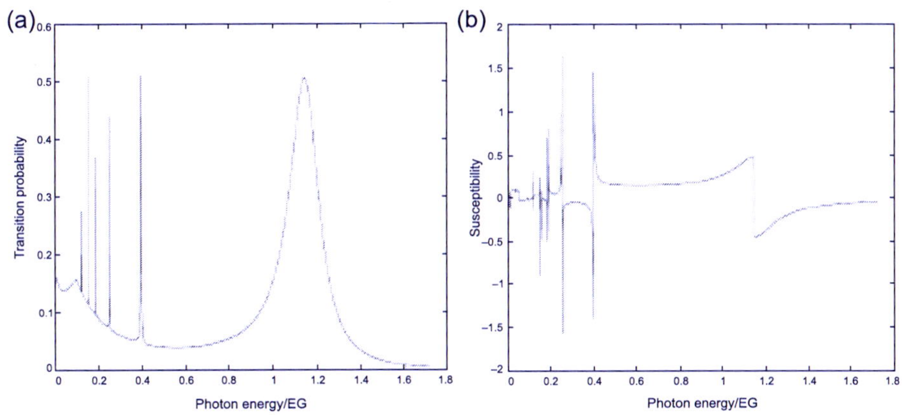

Figure 7.4 (a) The transition probability of the QD excited state for laser intensity 5.0 GW/cm^2. (b) Susceptibility (arbitrary unit) of the QD to the laser of the same intensity. Both the quantities are plotted against the laser frequency scaled down by the band gap energy of the QD

Figure 7.4(a) presents all the allowed transitions involving odd number of photons. For this very high intensity of the laser (5.0×10^9 W/cm^2), the non-perturbative Floquet method gives a very accurate result also showing the multiphoton transitions. In Fig. 7.4(b), the susceptibility of the QD is plotted against the laser frequency keeping the same intensity. Dispersions are obtained at the points of resonances. Very sharp dispersions occur at frequencies corresponding to higher photon transitions. The susceptibility is also observed to be decreasing with laser intensity corresponding to one and two photon transitions, whereas for transitions involving higher number of photons, the susceptibility of the QD to the laser increases with intensity.

Electron–hole pairs are generated by the interband transition, both due to single photon and multiphoton transition, corresponding to the transition peaks in Fig. 7.4(a) and also the regions of dispersion in Figs 7.4(b) and 7.5.

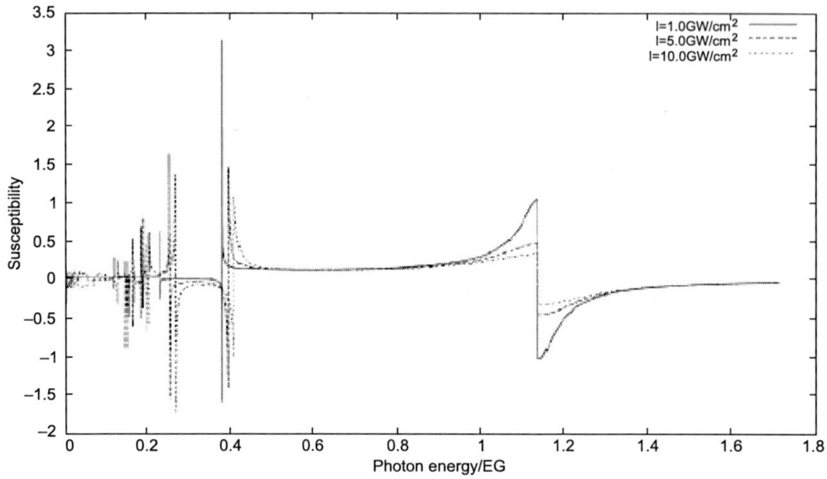

Figure 7.5 The susceptibility (in arbitrary unit) of the two level QD system interacting with lasers of different intensities

The maximum photocurrent is obtained when all the incident photons, N_{ph}, are absorbed to generate the same number of electron–hole pairs (N_{ph}/N for N-photon absorption). The number of electron–hole pairs collected to give photocurrent is

$$N_{eh} = \eta \frac{E_{abs}}{\hbar\omega}, \tag{7.51}$$

where η is the internal quantum effciency, E_{abs} is the energy absorbed by the QD. The electron–hole pairs thus generated from a number of QDs, embeded in a wetting layer, give rise to a photocurrent proportional to the number of the QDs at the regions corresponding to transition peaks. The photocurrent spectrum thus obtained involves a very straight forward experimental technique which gives the superposition of the absortion spectra or the transition spectra of the QDs on the linearly increasing wetting layer photocurrent (noise)[45, 46]. The absorption spectrum photocurrent (J)[47] from a QD is given as

$$J = e\frac{dN_{eh}}{dt} = \eta \frac{eP_{abs}}{\hbar\omega} \tag{7.52}$$

with P_{abs} being the total power absorbed by the QD involving multiphoton absorption and η is the internal quantum effciency. For $\eta = 3.0 \times 10^{-4}$ and absorbtion power of 5μ W/cm^2 and photon wavelength $\approx$ 1460 nm, the photocurrent is found to be $\approx$ 2 nA, which can be very easily experimentally measured[45, 46]. The power absorption occurs at the frequency ranges corresponding to the transition peaks in the transition probability curve plotted against photon energy. Hence, the results obtained from this theoretical study will enhance the sensitivity of a photo-detector based on the photocurrent spectra measurement. It will also explain the peaks of the photocurrent spectra corresponding to one photon and multiphoton absorptions.

7.4 Conclusion

In this chapter, the spectra of QD in the presence of magnetic and THz laser fields involving multiphoton processes was investigated. Floquet theory was used for solving time-dependent equations, which provides calculations with high accuracy compared to other theoretical methods in literature. Convergent results were obtained for long time average transition probabilities for various intersubband transitions involved. The breaking of degeneracy with magnetic field is revealed by the appearance of two sets of patterns (left and right) in the probability curves. Further, power broadening, ac Stark effect, and hole burning were also observed with changes in laser field intensities. A point worth mentioning is that the tuning of magnetic and laser fields in QD with the appearance of two seperate patterns and non-linear effects, especially power broadening, will be important for making devices for use in bio-imaging and photo-detection.

Bibliography

1. R. L. Liboff. 1980. *Introductory Quantum Mechanics*, Oakland: Holden-Day.
2. C. Cohen-Tannoudji, B. Diu and F. Laloe. 1977. *Quantum Mechanics*, New York: Wiley.
3. B. Sharma and M. Mohan. 1992. *J. Phys. B*. 25: 399.
4. N. Singhal, V. Prasad and M. Mohan. 2002. *Eur. Phys. J. D*. 21: 293.
5. H. Schneider, H. C. Liu, S. Winnerl, C. Y. Song, M. Walther and M. Helm. 2009. *Opt. Express* 17: 12279.
6. S. I Chu. 1985. *Adv. At. Mol. Phys*. 21: 197.
7. G. Floquet. 1883. *Ann. l Ecol. Norm. Sup*. 12: 47.
8. J. H. Poincare. 1892, 1893, 1899. *Les Methodes de la Mechanique Celeste*, I to IV, Paris: Gauthier-Villars.
9. S. H. Autler and C. H. Townes. 1955. *Phys. Rev*. 100: 703.
10. J. H. Shirley. 1965. *Phys. Rev., B*. 979: 138.
11. A. I. Ekimov, A. L. Efros and J. A. Onushchenko. 1985. *Solid State Commun*. 56: 921.
12. L. Banyai and S.W. Koch. 1993. *Semiconductor Quantum Dot*, Singapore: World Scientific.
13. S. M. Reimann and M. Mannien. 2002. *Rev. Mod. Phys*. 74: 1283.
14. L. Zhang and H. J. Xie. 2003. *Phys. Rev. B*. 68: 235315.
15. I. Karabulut, U. Atav and H. Safak. 2005. *Phys. Rev. B*. 72: 207301.
16. S. L. Chuang and D. Ahn. 1989. *J. Appl. Phys*. 65: 2822.
17. J. J. Shi and S. H. Pan. 1995. *Superlattice Nanostructures*. 17: 91.
18. M. M. Fejer. 1989. *Phys. Rev. Lett*. 62: 1041.

19. M. K. Gurnick and T. A. DeTemple. 1983. 'Quantum Electron.' *IEEE J.* 19: 791.
20. W. Xu and C Zhang. 1996. *Appl. Phys. Lett.* 68: 3305.
21. C. J. G. M. Langerak and B. N. Murdin. 1995. *Appl. Phys. Lett.* 67: 3453.
22. M. E. Donovan, A. Schulzgen, J. Lee, P. A. Blanche, N. Peyghambarian, G. Khitrova, H. M. Gibbs, I. Rumyantsev, N. H. Kwang, R. Takamaya, Z. Yang and R. Binder. 2001. *Phys. Rev. Lett.* 87: 237402.
23. T. Unold, K. Mueller, C. Lienau, T. Elsaesser and A. D. Wieck. 2005. *Phys. Rev. Lett.* 94: 137404.
24. J. W. M. Chon and M. Gu. 2004. *Appl. Opt.* 43: 1063.
25. N. Thantu. 2005. *J. Lumin.* 111: 17.
26. Y. Fu, Y. Luo and H. Agren. 2006. *Thin Solid Films* 515: 842.
27. L. Cao, X. Wang, M. J. Meziani, F. Lu, H. Wang, P. G. Luo, Y. Lin, B. A. Harruff, L. M. Veca, D. Murray, S. Y. Xie and Y. P. Sun. 2007. *J. Am. Chem. Soc.* 129: 11318.
28. S. Lahon, M. Gambhir, P. K. Jha and M. Mohan. 2010. *Physica Status Solidi B.* 247: 962.
29. S. Lahon, P. K. Jha and M. Mohan. 2011. *J. Appl. Phys.* 109: 054311.
30. H. Schneider and H. C. Liu. 2006. *Quantum Well Infrared Photodetectors: Physics and Applications*, Heidelberg: Springer.
31. M. Etienne, A. Biney, A. D. Walser, R. Dorsinville, D. L. V. Bauer and V. Balogh-Nair. 2005. *Appl. Phys. Lett.* 87: 181913.
32. S. M. Kim, F. Hatami, J. S. Harris, A. W. Kurian, J. Ford, D. King, G. Scalari, M. Giovannini, N. Hoyler, J. Faist and G. Harris. 2006. *Appl. Phys. Lett.* 88: 153903.
33. J. Darmo, V. Tamosiunas, G. Fasching, J. Kroll, K. Unterrainer, M. Beck, M. Giovannini, J. Faist, C. Kremser and P. Debbage. 2004. *Opt. Express* 12: 1879.
34. B. P. Holder and L. E. Reicjl. 2005. *Phys. Rev. A.* 72: 43408.
35. L. V. Vela-Arevalo and R. F. Fox. 2004. *Phys. Rev. A.* 69: 63409.
36. Y. B. Zeldovich. 1966. *Zh. Eksp. Teor. Fiz.* 51: 1492.
37. V. I. Ritus. 1966. *Zh. Eksp. Teor. Fiz.* 51: 1544.
38. H. Sambe. 1973. *Phys. Rev. A.* 7: 2203.
39. C. C. Tannoudji. *Frontiers of Laser Spectroscopy* 103. R. Ballian, S Haroche, S. Liberman, Eds. New York: Wiley.
40. Z-H. Zhang, K-X. Guo, B. Chen, R-Z. Wang and M-W. Kang. 2009. *Superlattices and Microstructures*. 46: 672.
41. V. Popsueva, R. Nepstad, T. Birkeland, M. Forre, J. P. Hansen, E. Lindroth and E. Waltersson. 2007. *Phys. Rev. B.* 76: 035303.
42. G. F. Quinteiro and P. I. Tamborenea. 2009. *Phys. Rev. B.* 79: 155450.
43. M. Wagner, U. Merkt and A. V. Chaplik. 1992. *Phys. Rev. B.* 45: 1951.
44. J. V. Moloney and F. H. M. Faisal. 1979. *J. Phys. B.* 12: 2829.

45. P. W. Fry, M. S. Skolnick, D. J. Mowbray, I. E. Itskevich, J. J. Finley, L. R. Wilson, K. L. Schumacher, J. A. Barker, E. P. O'Reilly, M. Al-Khafaji, A. G. Cullis, M. Hopkinson, J. C. Clark and G. Hill. 2000. *Physica E.* 9: 106.
46. P. W. Fry, I. E. Itskevich, S. R. Parnell, J. J. Finley, L. R. Wilson, K. L. Schumacher, D. J. Mowbray, M. S. Skolnick, M. Al-Khafaji, A. G. Cullis, M. Hopkinson, J. C. Clark and G. Hill. 2000. *Phys. Rev. B.* 62: 016784.
47. D. Duchesne, L. Razzari, L. Halloran, R. Morandotti, A. J. Spring Thorpe, D. N. Christodoulides and D. J. Moss. 2009. *Optics Express.* 17: 5298.

8 Optical Diagnostics of Singlet Oxygen for Chemical Oxygen-Iodine Laser

Mainuddin, Gaurav Singhal, A. K. Varshney, R. Dohare and R. K. Tyagi

8.1 Introduction

Chemical oxygen–iodine laser (COIL) was first demonstrated by McDermott et al. in 1978 and is the only chemical laser based on electronic transition. It operates on $^2P_{1/2} \rightarrow {}^2P_{3/2}$ transition of iodine atom at 1.315 μm and is pumped by the following reaction scheme,

$$O_2(^1\Delta) + I(^2P_{3/2}) \Leftrightarrow O_2(^3\Sigma) + I^*(^2P_{1/2}) \tag{8.1}$$

$$I^*(^2P_{1/2} + nh\upsilon \rightarrow I(^2P_{3/2}) + (n+1)h\upsilon \tag{8.2}$$

This laser due to its short wavelength, high efficiency and scalability has found application in various military and civilian scenarios in the last decade. The possibility of carrying high power beams via optical fibers makes it extremely useful for decommissioning and dismantling dangerous structures such as obsolete nuclear reactors through remotely operated mechanisms by Tei et al.

The pumping source, $O_2(^1\Delta)$ is an excited form of oxygen molecule with energy level very close to that of atomic iodine enabling near resonant energy transfer as shown in Fig. 8.1. Although, different techniques such as chemical, RF discharge, photo-sensitizer, have been reported for the production of $O_2(^1\Delta)$, the chemical method is the only one till date that has been successful for scaling up power. Therefore, amongst the various mechanisms studied for the production of $O_2(^1\Delta)$, the chemical method still remains at the forefront for the development of large-scale COIL systems.

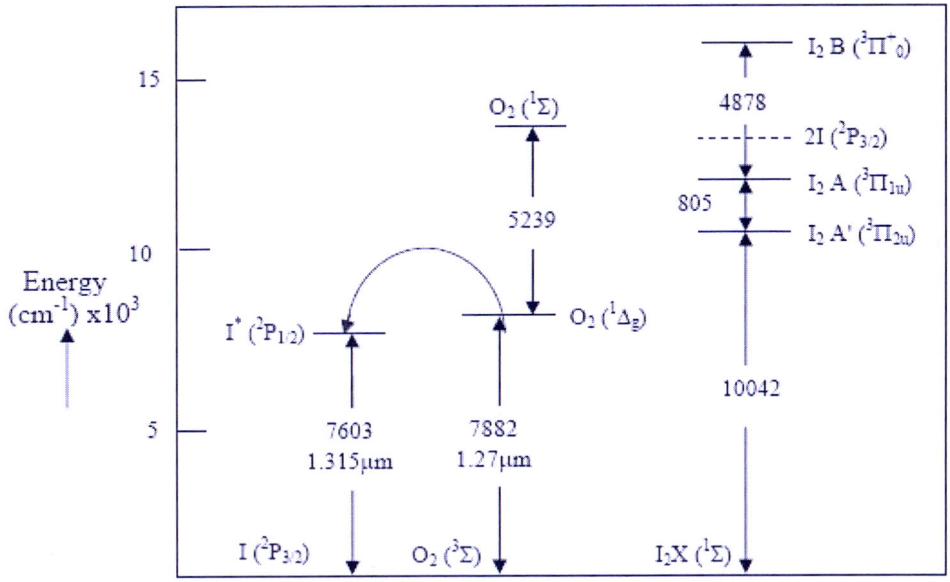

Figure 8.1 Energy level diagram of O_2, I_2 and I in COIL scheme

The chemical method is based on chemical reactions between chlorine gas and basic hydrogen per-oxide (BHP) solution resulting in production of singlet oxygen, which in turn dissociates iodine molecules into iodine atoms and also subsequently excites these atoms.

$$2KOH + H_2\,O_2 + Cl_2 \rightarrow 2KCl + 2H_2\,O + O_2(^1\Delta_g) \tag{8.3}$$

For an efficient production of singlet oxygen, jet type singlet oxygen generator (JSOG) by Rajesh et al. has proved its potential over the other techniques such as rotating disc or bubbler type. Other forthcoming generators such as twisted aerosol and centrifugal bubble/ flow type are yet to be proven for large-scale power levels.

Figure 8.2 shows the functional block diagram of chemical oxygen iodine laser[3, 9] which exhibits coupling of all subsystems. BHP solution is prepared in a separate preparation tank and supplied to the SOG reaction chamber in the form of jets. Chlorine is supplied to the SOG reaction chamber from the bottom in a counter-flow direction to the jets so that it reacts at the surface of the liquid jets to produce singlet oxygen molecules. The supplied BHP solution through the SOG is collected in the receiving tank attached to the bottom of the SOG reaction chamber. The BHP solution has to be maintained at a very low temperature, of the order of −20°C, from the point of view of safety and operational efficacy. Thus, interfacing of both the BHP preparation and receiving tanks with a cooling system is essential.

The singlet oxygen generator is made of plexiglas for the ease of viewing. The singlet oxygen produced by the reaction between chlorine and BHP at the surface of the jets is transported to the nozzle throat via the diagnostic duct. The diagnostics set up is placed at the exit of the singlet oxygen generator.

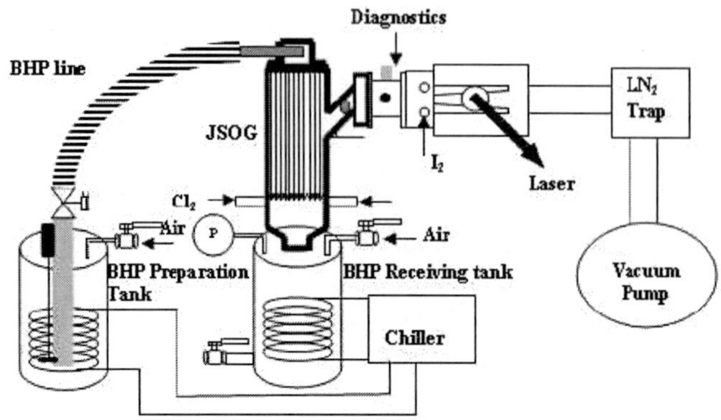

Figure 8.2 Functional block diagram of COIL

The basic lasing species, i.e., iodine is injected into the main flow at the nozzle throat, and the total medium is then supersonically expanded into the laser cavity. The total active medium passes through the LN_2 trap prior to being exhausted to the pump in order to arrest any unutilized chlorine or iodine present in the flow.

The iodine supply system has the prime function of generating the lasing species, i.e., iodine vapor in requisite quantities at the desired temperature. Iodine vapors are then injected into the lasing chamber through the supersonic nozzle bank system where the iodine molecules first get dissociated into iodine atoms, which are then pumped to the lasing level by the singlet oxygen molecules. Further, in order to carry heavier iodine molecules from the iodine supply system to the laser cavity and also to create desired pressure conditions in various subsystems, addition of buffer gas (nitrogen) is required which necessitates provision of its corresponding gas feed system. Figure 8.3 shows the photograph of a developed 500 W class COIL system.

It is apparent that the singlet oxygen generator forms the most important subsystem in the chemical oxygen–iodine laser. The design of SOG relevant for COIL applications requires thorough understanding of the physics, chemical kinetics and operational mechanics of the COIL system, which is imperative for achieving the optimum performance of the final system. The concept of jet type generators for the production of singlet oxygen with a view to use it for COIL applications was first demonstrated by Balan et al. and later implemented by Zagidullin et al. Basically, the generator consists of a large number of fine jets of basic hydrogen peroxide (BHP) flowing through an atmosphere of chlorine. The advantages of this technique can be summarized as follows:

1. Due to the large reaction surface, JSOG has much higher singlet oxygen yield even at higher operating pressures, which is one of the major requirements for supersonic operation of the COIL system.
2. Easy replenishment and recycling of BHP is possible, which not only makes the system compact but also allows it to run for longer duration.
3. High specific surface area resulting in relatively smaller generator volume because of the large reaction surface.
4. No heat accumulation in the reaction chamber because of the fast replenishment of the cold BHP and hence minimum water vapor production.

5. Self separation of large aerosol droplets.
6. Better option for power up-gradation.

Figure 8.3 Photograph of developed 500 W Class COIL system

In order to run the COIL at its optimal performance, some critical conditions in relation to the generator need to be satisfied. The most essential conditions are in terms of: (i) generator pressure, which is a function of the relative jet and gas momentum ratios and their mass fluxes, and specifically controls chlorine utilization and singlet oxygen yield, (ii) BHP temperature, which controls the water vapor fraction, large value of which can have detrimental effect on COIL performance, (iii) amount and manner of dilution of singlet oxygen with nitrogen buffer gases which controls the transport duct losses and to an extent both the generator and plenum pressures.

Another important feature which needs to be taken care of is jet instabilities, which may lead to liquid reagent carryover in terms of aerosol droplets or catastrophic carryover. The former to an extent are inevitable due to the multiphase nature of the SOG operation involving interaction of gas and liquid phases. However, the latter occur as a result of abnormal momentum ratio of liquid and gas flows or faulty jet orientations leading to merging of jets or early jet break up leading to massive aerosol spraying or catastrophic liquid reagent carryover into the flow medium. Jet stability is also a weak function of jet temperature which actually affects jet viscosity and hence its stability.

The chemical generation of singlet oxygen is a complex phenomenon involving a two phase interaction between chlorine gas and BHP liquid. Thus, the parameters in terms of generator pressure, relative momentum ratio of liquid reagent and gas, BHP temperature, geometry of flow extraction and subsequent transportation of the singlet oxygen pumping medium needs to be well controlled.

Hence, it is critical to accurately quantify the actual amount of singlet oxygen generated after the involved losses to be able to characterize the generator for achieving its optimal operation. Thus, there is a need to develop dedicated measurement systems for singlet oxygen using non-intrusive techniques.

In this chapter, the optical spectroscopic based singlet oxygen diagnostics system for COIL system optimization and performance analysis is presented.

8.2 System Description

A detachable diagnostics cell, made of plexiglas, having duct dimensions identical to the generator exit (75×10 mm^2) has been developed for measurement of singlet oxygen signal and pressure signal. The diagnostic duct is of 70 cm length and is situated at a distance of 80 mm from the generator exit. It has provisions for two optical ports with one-inch windows for $O_2(^1\Delta)$ emission measurements. Provisions for plenum pressure measurement and for probing the gas sample required for chlorine utilization measurement have also been included in the same diagnostic cell. The details regarding measurement of singlet oxygen are discussed below.

The diagnostic system for the measurement of singlet oxygen concentration is based on the principle of optical emission spectroscopy. The meta-stable singlet oxygen molecules $O_2(^1\Delta)$ emit at 1.27 μm[2]. The emitted photons which are collected using a one-inch lens fixed in the diagnostic cell are focused on to a liquid nitrogen cooled germanium photodiode of 5 mm diameter (M/s EG & G Judson, Model No. JD16-M204). An interference filter with $\lambda = 1270 \pm 5$ nm is introduced between the lens and the photodiode for reducing the unwanted photons reaching the photo-detector.

Figure 8.4 and 8.5 correspondingly show the schematic of singlet oxygen diagnostics setup and the developed hardware.

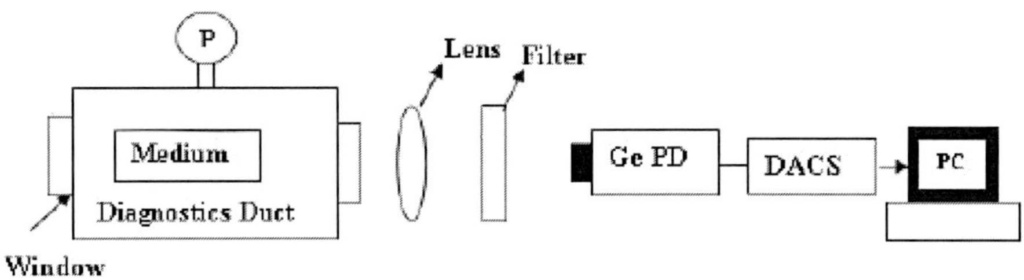

Figure 8.4 Schematic of the singlet oxygen diagnostics setup

Figure 8.5 Developed diagnostics system for singlet oxygen

8.2.1 Principle

The photodiode signal (S_Δ) is proportional to the singlet oxygen concentration $[O_2(^1\Delta)]$, sensitivity and collection efficiency,

$$S_\Delta \propto \left[O_2(^1\Delta)\right]\eta_\lambda T_\lambda d\Omega, \tag{8.4}$$

where, η_λ is the detector efficiency, T_λ is the resultant transmission and $d\Omega$ is the solid angle. The above relation can also be written as

$$S_\Delta = A\left[O_2(^1\Delta)\right], \tag{8.5}$$

$$S_\Delta = \frac{AP_\Delta}{kT}, \tag{8.6}$$

where, A is the calibration factor of the entire detection system, k is the Boltzmann constant, T is the gas temperature and $P\Delta$ is the singlet oxygen partial pressure. In our case, A is found to be 0.33 V/torr.

Therefore, Eq. (8.6) can be written as:

$$S_\Delta = \frac{0.33P_\Delta}{kT}. \tag{8.7}$$

From the measurement of the singlet oxygen concentration, the yield can be estimated as

$$Y_\Delta = \frac{\left[O_2(^1\Delta)\right]}{[O_2]_{\text{total}}} = \frac{P_\Delta}{P_0}, \tag{8.8}$$

where, P_0 is the oxygen partial pressure in the diagnostic duct and can be calculated for the known values of chlorine M_{C12} and buffer gas flow rates and cell pressure P in accordance with the relation:

$$P_0 = \frac{PM_{Cl_2}U_{Cl}}{M_{Cl_2} + M_{\text{buffer}}} \tag{8.9}$$

Equation (8.8) can be rewritten as,

$$Y_\Delta = kTS_\Delta\frac{(M_{Cl} + M_{\text{buffer}})}{APM_{C12}U_{Cl}} \tag{8.10}$$

where P' is the transport duct total pressure and U_{Cl} is measured using another diagnostics system based on optical absorption of chlorine at 330 nm.

8.2.2 Chlorine utilization estimation

Chlorine molecules have peak absorption at 330 nm[4, 7] with an absorption cross-section (σ_v) of 2.75×10^{-19}cm^2. Chlorine utilization measurement gives the percentage of chlorine molecules reacted with BHP for generation of laser pumping medium singlet oxygen. The hardware setup is shown in Fig. 8.6.

Figure 8.6 Developed system for chlorine utilization

A portion of the singlet oxygen generator (SOG) flow exit medium is passed through an optical cell of 25 cm length and 25.4 mm diameter. The flow medium from the optical cell is continuously pumped out during operation for maintaining identical conditions as in the main duct. Ultraviolet mercury lamp is used as a light source along with an interference filter with 330 ± 5 nm. Collimated light is passed through the cell and detected using silicon photodiode. The flow rate of exit chlorine $(M_{Cl2})_{exit}$ can be estimated using Eq. (8.11) and chlorine input flow rate $(M_{Cl2})_{input}$ is known using the principle of orifice under choke flow condition. The residual chlorine molar flow rate $(M_{Cl2})_{exit}$ is estimated by calculating the partial pressure of chlorine (using Beer Lambert's law and perfect gas law), and substituting the expression for the partial pressure in Dalton' law,

$$M_{Cl} = \left\{ \frac{M_c kT}{\sigma_\nu l} \right\} \frac{\ln(I_0/I_\nu)}{P_{tot} - (kT/\sigma_\nu l) l_m (I_0/I_\nu)}, \tag{8.11}$$

where I_ν and I_0 are the light intensity at frequency ν with and without chlorine respectively. σ_ν is the absorption cross-section (2.75×10^{-19}cm^2), l is length of cell, k is the Boltzmann constant, T is the temperature of the gas, P_{tot} is the measured total pressure, M_c is carrier/buffer gas molar flow rate.

Therefore, chlorine utilization has been estimated using Eq. (8.12).

$$U_{Cl} = 1 - \frac{(M_{Cl2})_{exit}}{(M_{Cl2})_{input}}. \tag{8.12}$$

Experiments have been conducted for large variation in chlorine flow rate conditions. It has been observed that the system is capable of measuring unutilized chlorine traces of the order of ~ 0.8 mmols^{-1} The detailed experimentation in the present case shows a typical chlorine utilization of nearly 94%[7].

8.2.3 Pressure measurement

The total pressure in the transport duct (P) is measured using a capacitance gauge (M/s Pfeiffer make Model No. CMR 262) due to its high accuracy (0.2%), However, these sensors require a warm up time of about one hour for achieving accuracies of this order and moreover, the water vapor and hazardous gases such as Cl_2 and I_2 adversely affect the performance. Therefore, diaphragm type pressures transmitters from M/s Metran of 'Metran-22 series' have been used. These employ metallic diaphragm, which are suitable for the corrosive environment encountered in COIL application.

In these diaphragm type sensors, there is a monocrystalline sapphire plate with silicon film tensoresistors (silicon-on-sapphire-structure), connected to the diaphragm of a tensotransducer. These tensoresistors are connected into the bridge circuit. The deformation of the measuring diaphragm (deformation of the tensotransducer's diaphragm) causes proportional resistance change of tensoresistors and misbalance of bridge circuit. Electrical signal from the bridge circuit output is fed to a microprocessor based electronic module, where it is transformed into a unified current output (4–20 mA). These operate at 24 V supply to produce 4–20 mA output current signal. The current signal is preferred because it is less likely to be affected by noise.

In order to estimate singlet oxygen yield, estimation of chlorine utilization is required. For estimation of chlorine utilization, again pressure measurement is done using the same diaphragm type pressures transmitters from M/s Metran of Metran-22 series.

8.2.4 Photo signal measurement

As already discussed, liquid nitrogen cooled germanium photodiode of 5 mm diameter (M/s EG & G Judson, Model No. JD16-M204) has been used for detecting emission-photo signal of singlet oxygen molecules $O_2(^1\Delta)$ at 1.27μm. For estimation of chlorine, a silicon photo detector (RS component stock no. 303–674) with amplifier has been used to detect the absorbed light signal of 330 nm wavelength. Figure 8.7 shows the photograph of a silicon detector on which amplifier circuit is directly mounted. The amplifier circuit has been developed using surface mounted miniature circuit components.

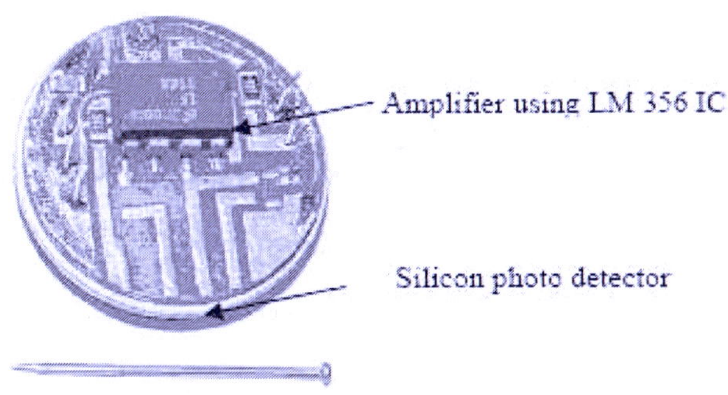

Figure 8.7 Electronics hardware for photodiode with amplifier

8.2.5 Temperature measurement

The temperature of the medium is measured using a resistance temperature detector (RTD Pt-100) as this sensor has good sensitivity, stability, and linearity. RTD resistance varies with the change in temperature. This circuit provides 4–20 mA output current signal, which can be converted, in the range of 0 to 10 V, as the data acquisition card can accept 0 to 10 V input only at its analog input channel. Thus, we need to have a constant current source so that one can get a variable output voltage in the range of 0 to 10 V with respect to changes in the resistance of RTD.

XTR-105 (Burr-Brown) is used to get 4–20 mA output current signal with respect to change in temperature of RTD sensor. It is a monolithic 4–20 mA, two-wire current transmitter with two precision current sources (800 μA each). It provides complete current excitation for platinum RTD temperature sensors. XTR-105 is a 14 pin IC and is specified for the −40°C to +85°C industrial temperature range with offset drift of 0.4 μV/°C. Figure 8.8 shows the circuit diagram of the temperature to voltage converter used for temperature measurement.

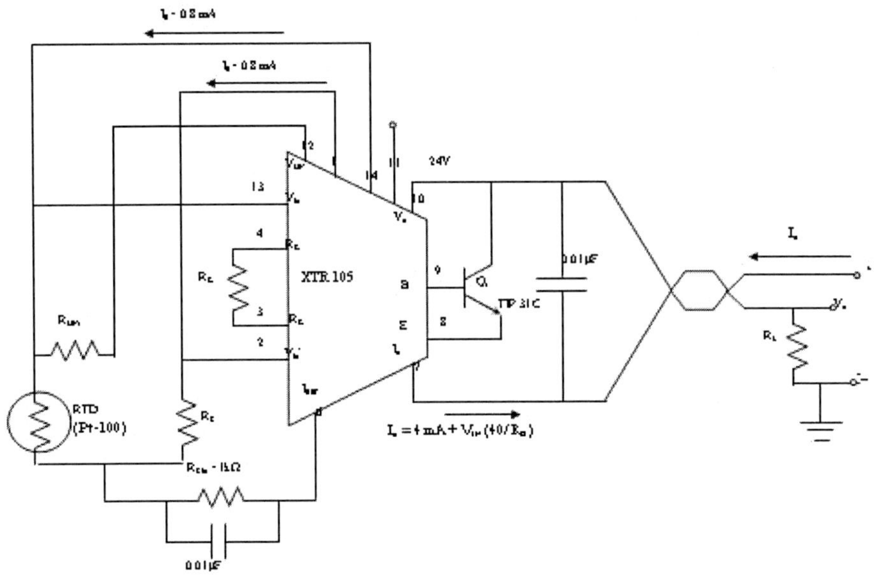

Figure 8.8 Electronics circuit diagram of temperature measurement

8.2.6 Data acquisition system

The continuous recording of the temporal variation of pressure and the photodiode signal is carried out using a dedicated PCI bus based data acquisition system (DAS). DAS[5] is not only required for diagnostics parameters estimation but also for real time operational control, acquisition, measurement, monitoring, display, storage and analysis of parameters. Since COIL[3] is one among many infrared gas lasers with flowing gas medium; the output performance of these lasers is strongly dependent on the composition of the lasing, pumping and other species present in the laser system along with the flow parameters like flow rates, pressures, temperatures, Mach number etc. There is a need for continuous monitoring of the concentration of involved gas species based on non-intrusive techniques so that the gas flow is not disturbed. The operation time of these lasers is limited to a few seconds for various reasons. Therefore, the sequential and switching control of fuel supplies (supply of BHP, chlorine and buffer gases) is also very critical for optimal performance of the system. Figure 8.9 shows the scheme of DAS which has been used for singlet oxygen estimation as well as for other operational requirements of COIL.

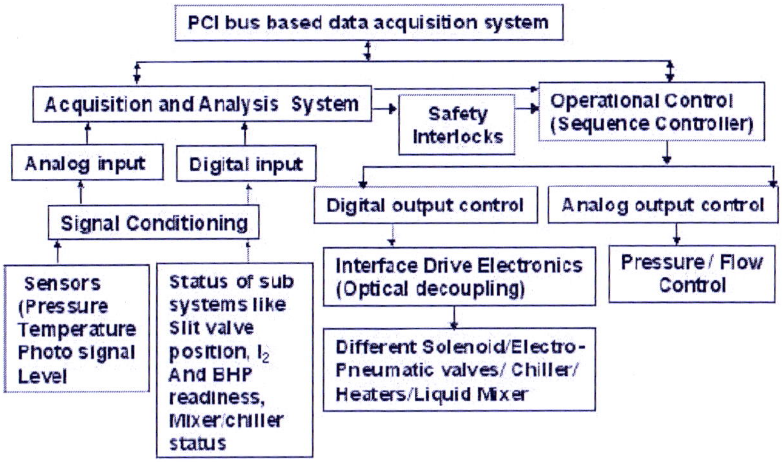

Figure 8.9 DAS scheme for estimation of $O_2(^1\Delta)$ as well as for other operational requirements of COIL

DAS has been realized using Advantech PCI-1716 multifunction data acquisition cards, custom built electronics circuits; various transducers, actuators and application software developed using programming language VC++. In order to fulfill all the requirements for COIL operation, the developed data acquisition system comprises 168 channels including 64 analog inputs (AI), 8 analog output (AO), 32 digital input (DI) and 64 digital output (DO) channels. Figure 8.9 shows the block diagram of the data acquisition system.

The design of DAS for COIL is based on the 24 V dc supply and accordingly, the selection of compatible transducers, actuators and other devices has been carried out. The reason for selection of the above-stated operating voltage is that low voltages are highly susceptible to interference whereas high voltages are prone to inducing interference. The medium DC voltages (specially 24 V) tend to be in the middle of the two extremes, thus being an ideal choice in the present application. The selection of transducers is such that either they produce 4–20 mA current output or are signal conditioned to do the same. The current output ensures low susceptibility to noise.

DAS has been divided into three modules: acquisition and analysis module, operational and sequential control module and safety interlocks module. The acquisition and analysis module is responsible for the acquisition of analog and digital parameters along with the storage and online estimation of diagnostic parameters in graphical form for analysis of the performance of different subsystems. The analog input channels are required for acquiring the signals from various sensors like pressure sensor, temperature sensor and photodiode. On the other hand, the digital input channels are used to indicate the status of various subsystems such as the system readiness status of subsystems for laser firing and on/off status of various electro-pneumatic/solenoid/slit valves.

The operational and sequential control is responsible for the sequential and switching control of electrical/electro-pneumatic valves/devices and online control and adjustment of physical parameters like flow rates of gases involved in lasing action. This system incorporates both analog as well as digital output channels. Analog output channels are required for the flow rate control of different gas feed lines, whereas digital output channels are required for performing switching and sequential operations (on/off control) of various subsystems/valves. The safety interlocks is an essential part of chemical lasers to avoid any accident or potential human hazards. This module utilizes input from different subsystems through acquisition and analysis module and interacts with the operational sequence control module for implementing necessary emergency actions accordingly.

8.2.7 Graphical user interfaces (GUI)

COIL is a large and complex system in which different subsystems are located at different positions. It is desired that a single person from the control room should operate this laser. The numbers of displayed/monitored parameters at a time are 150, which cannot be accommodated in a single window screen. Thus, management of the displayed parameters is essential. Moreover, it is required that the control windows should be so structured that the operator should develop a feel for the system sitting in the control room itself. Thus, to provide ease of operation, all the controls and crucial parameter display are incorporated into five user friendly bitmap control windows (graphical interface units) according to their operations. These windows are:

1. BHP (basic hydrogen peroxide) control bitmap
2. SOG (singlet oxygen generator) control bitmap
3. Gas feed control bitmap
4. Iodine evaporator control bitmap
5. Sequential setting control.

Among these GUIs, SOG window is responsible for display of singlet oxygen photo signal, pressure of SOG, chlorine absorption signal, temperature etc. This window is also responsible for starting the laser experiment using the start control button on this window. The bitmap of these graphical users's interfaces have been designed according to the subsystem's operational, monitoring, effective display and control requirements. All the control windows have toggling switch to switch over from one window to other windows. Figure 8.10 shows the SOG graphical user interface bitmap window.

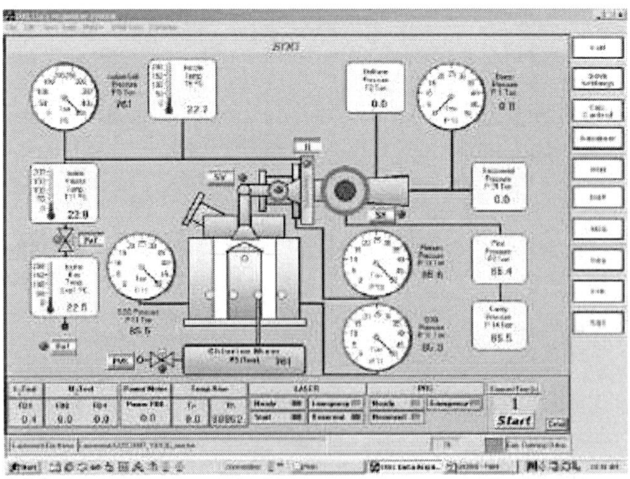

Figure 8.10 Graphical user interface for SOG

8.3 Results and Discussions

The developed system was calibrated prior to be used for measurement and optimization of singlet oxygen. The calibration of the singlet oxygen measurement system was carried out by the technique suggested by Zagidullin et al. A mixture of chlorine and nitrogen with a ratio of 1:9 was fed to the SOG through the chlorine supply line. The SOG pressure was maintained at about 30 torr by adjusting the slit valve position so that the gas velocity is about 15 ms^{-1} and hence the gas residence time of 6.7 ms in the reaction zone, which is within the required range for complete utilization of chlorine. Under these conditions, the partial pressure of chlorine inside the generator (3 torr) was low enough to neglect the pooling losses and a yield of almost 100% could be expected. Large number of experiments were carried out by varying the pressure in the diagnostic cell (by choking the pump line), while maintaining SOG conditions. Figure 8.11 shows the $O_2(^1\Delta)$ photodiode signal variation with the oxygen pressure in the diagnostic cell. The results show that the calibration factor (A) in the present setup is 0.33 V/torr of singlet oxygen. Several experiments have been conducted to confirm the repeatability.

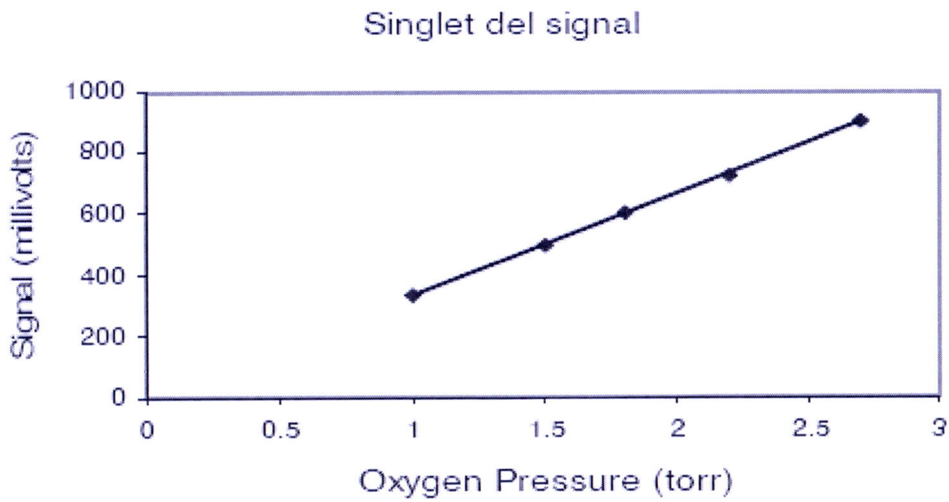

Figure 8.11 Singlet delta oxygen calibration curve

In order to be able to correctly estimate singlet oxygen, it is clear from Eq. (8.10) that we should have temporal variation of SOG pressure. The continuous monitoring of SOG pressure is essential so as to ascertain that the SOG pressure lies in the desired range of 20–35 torr for optimal functioning in terms of singlet oxygen generation. A pressure higher than 35 torr may also lead to a cascading pressure rise inside the generator which may pose a safety hazard. The measured plenum pressure (P) is utilized for determination of singlet oxygen yield Y_Δ using Eq. (8.10). Figure 8.12 shows the acquired SOG pressure (P) which is utilized by Eq. (8.10).

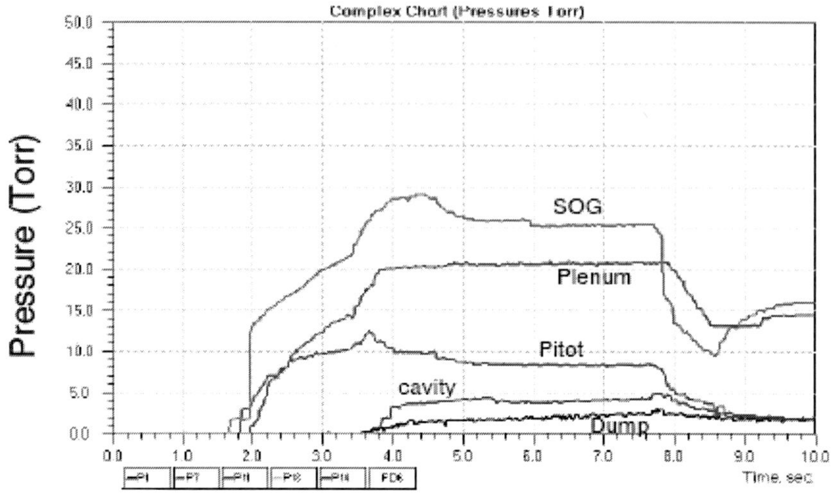

Figure 8.12 Temporal variation of various pressures along with SOG pressure

The estimation of partial pressure (P_Δ) of Eq. (8.6) is done using S_Δ signal whose temporal variation is shown in Fig. 8.13. The other values like calibration factor A, Boltzmann constant (k) and temperature (T) are already known.

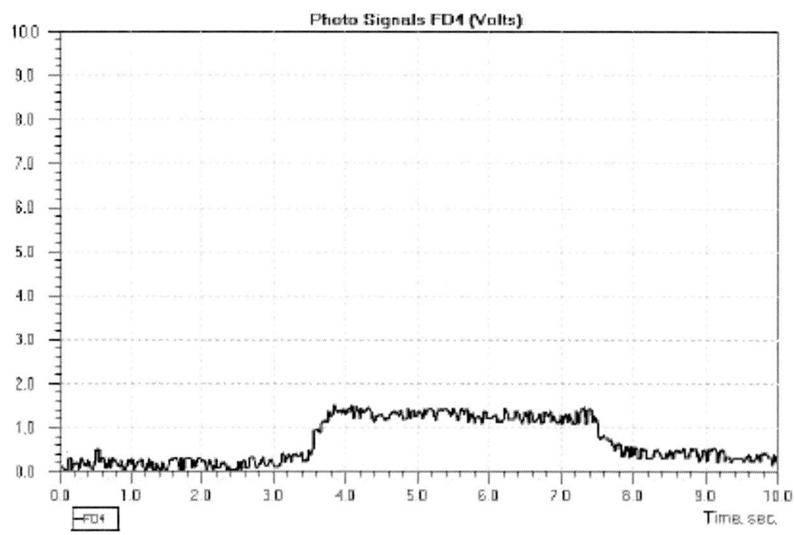

Figure 8.13 Temporal variation of singlet oxygen signal (S_Δ)

Figure 8.14 shows the typical power pulse in one of the experiments in which singlet oxygen yield of 64% for a typical chlorine utilization of 94% was obtained.

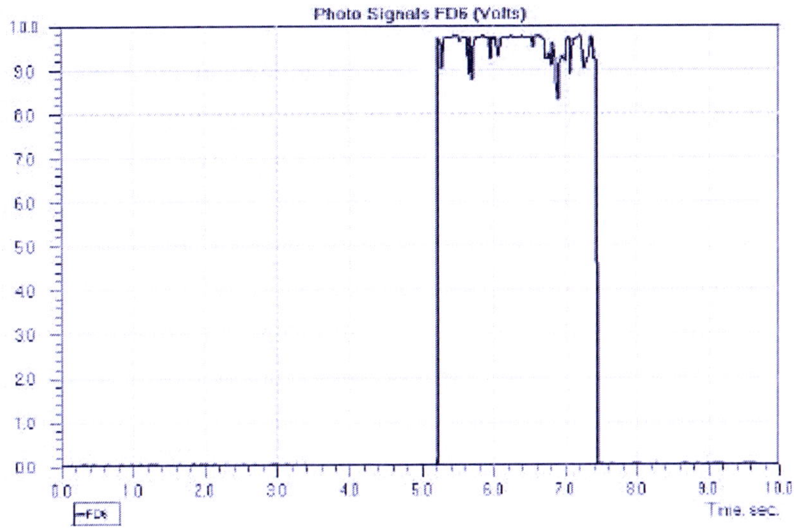

Figure 8.14 Temporal variation of laser power

8.4 Conclusions

An optical emission spectroscopic based singlet oxygen diagnostics system has been designed and developed for the optimization of COIL. This chapter discusses in detail the implementation of critical diagnostics in a COIL laser source starting from various kinds of transducer selection, data acquisition and processing, and calibration of results. In an optimal scenario, the estimated singlet oxygen yield at the exit of the generator inside the plenum is thus 64% which is as per the design requirement of greater than 60% yield. The developed system is being used routinely for various experiments of this laser.

Bibliography

1. N. F. Balan, R. M. Gizadulin, M. V. Zagidulin, A. Yu. Kurov, V. D. Nikolaev, V. M. Pichkasov and M. I. Svistin. 1989. 'Investigation of a jet generator of $O_2(^1\Delta_g)$.' *Sov. J. Quantum. Electron* 1412.
2. D. J. Benard and Pchelkin. 1978. 'Measurement of $O_2(^l\Delta_g)$ content in the gaseous effluents of a chemical generator.' *Review of Scientific Instruments* 49: 794.
3. D. L. Carroll. 2011. 'Overview of high energy lasers: Past, present and future.' *AIAA paper* 3102.
4. J. Kodymova and O. Spalek. 1998. 'Performance characteristics of jet type generator of singlet oxygen for supersonic chemical oxygen iodine laser.' *Japanese Journal of Applied Physics* 37: 117–21.
5. Mainuddin, R. K. Tyagi, R. Rajesh, G. Singhal and A. L. Dawar. 2003. 'Real-time data acquisition and control system for a chemical oxygen-iodine laser.' *Journal of Measurement Science and Technology* 14: 1364–1372.

6. W. E. McDermott, N. R. Pchelkin, D. J. Benard and R. R. Bonsek. 1978. 'An electronic transition chemical laser.' *Appl. Phys. Lett.* 32(8): 469–70.
7. R. Rajesh, G. Singhal, Mainuddin, R. K. Tyagi and A. L. Dawar. 2010. 'High throughput jet singlet oxygen generator for multi kilowatt SCOIL.' *Optics and Laser Technology* 42: 580–85.
8. K. Tei, D. Sugimoto, M. Endo, S. Takeda and T. Fujikova. 1999. 'Chemical oxygen iodine laser for decommissioning and dismantlement of nuclear facilities.' *SPIE* 3887: 162–69.
9. R. K. Tyagi, R. Rajesh, G. Singhal, Mainuddin, A. L. Dawar and M. Endo. 2003. 'Supersonic COIL with angular jet singlet oxygen generator.' *Optics and Laser Technology* 35: 395–99.
10. M. V. Zagidullin, V. D. Nikolaev, M. I. Svitin, V. S. Safonov and N. I. Ufimtsev. 1996. 'The study of buffer gas mixing with active gas on chemical oxygen iodine laser performance with jet SOG.' *SPIE* 2702: 310.

9 Raman Scattering in Crystalline, Amorphous and Nanocrystalline Silicon

Vivek Kumar and A. K. Shukla

9.1 Introduction

Silicon, the second most abundant element (after oxygen) in the earth's crust, making up 25.7% of the crust by mass, is one of most (probably the absolute) striking material for electronic and technological applications. It has thus become the leading and most prominent building-block material for electronics. Silicon has an indirect band gap of 1.12 eV that is ideal for room temperature operation. It allows the processing flexibility to place today more than 10^9 devices on a single chip. However, all the single transistors and electronic devices have transferred information to length scale which is relevant with respect to their nanometric scale. Pushing down the dimension of silicon-based devices toward the nanometric scale determines a high concentration of transistors on a single silicon crystal (around 200 mm for the moment) that have permitted high integration level high-speed device performances and unprecedented interconnection levels.

Amorphous semiconductors represent an important area in materials science, which is interesting both from the technological and the theoretical point of view. The simplest way to describe the structure of an amorphous material like a disordered solid is short-range coordination. Extended atomic network of amorphous silicon (a-Si) is random[1]. In other words, the local coordination is tetrahedral (sp^3 hybridization) as in crystalline Si (c-Si). The first nearest-neighbor distance, coordination number, or binding energy remains more or less same as in the amorphous and crystalline phases. Moreover, the covalent nature of the bonds means that short-range interactions play a prominent role. Density of electronic or vibrational states is similar in amorphous Si (a-Si) and c-Si[2, 3]. Semiconducting properties are also preserved in the a-Si. As a matter of fact, the presence of dangling bonds in the a-Si affects the performance of electronic devices. In practice, this can be circumvented by hydrogenation of the amorphous material, since hydrogen is a good terminator for insaturated bonds. From the theoretical point of view[2], the main features of the density of states can be accounted for by a molecular description (tight binding or valence force field models). Effect of the topological disorder mainly broadens linewidth of the Raman modes.

Semiconductor nanostructures have also been investigated in recent years due to their immense applications in electronic and optoelectronic devices[4, 5]. Electronic and optical properties of semiconductor nanostructures are strongly affected by the quantum confinement effect due to the reduced dimensions of these systems. In recent years, a great deal of research has gone into the optical, electrical and mechanical properties of silicon nanoparticles. The key motivation for such research is the discovery of luminescence in the visible spectral range of nanocrystalline silicon[6]. This property of nanocrystalline silicon can be exploited in the optoelectronics industry. High frequency optical phonons have mostly been studied on silicon nanoparticles[7].

Thermodynamically stable phases of silicon are found to be crystalline silicon and amorphous silicon at room temperature and atmospheric pressure. Various intermediate phases such as microcrystal and nanocrystal also exist in nature between the crystalline limit and the amorphous limit. Raman spectroscopy can reveal the crystalline phase, nanocrystalline (i.e., quantum confinement effect) and amorphous phase in silicon. This chapter is concerned with Raman spectra in the c-Si, nanocrystalline silicon and a-Si which are compared over a wide spectral range. Raman scattering of the c-Si takes place with either absorption or emission of optical phonons of zone center phonons ($\mathbf{q} = 0$). Raman spectrum from a-Si shows the same behaviour as the density of vibrational states (DVS) due to short range order, i.e., all phonons take part in Raman scattering. In silicon nanostructures (SiNS), the selection rule ($\mathbf{q} = 0$) of the first-order Raman scattering is relaxed. As a result, spatial confinement of acoustic phonons and optical phonons gives rise to low-frequency Raman active acoustic mode and optic mode respectively. A phenomenological quantum confinement model is discussed for the Raman active acoustic mode of silicon nanostructures at low laser power density. It is found here that large Raman shift of the acoustic mode is observed in comparison with the phonon softening of the optical mode for a given size of nanostructure.

9.2 Experimental Details

Amorphous silicon sample is prepared by the ion irradiation in the c-Si. Three samples (sample S1, sample S2 and sample S3) of the silicon nanostructures (SiNS) were fabricated by the laser induced etching (LIE) method[8]. The LIE was done by immersing n-Si wafer into HF acid (concentration 48%) and then focusing the argon ion laser beam (COHERENT, INNOVA 90, $\lambda = 514.5$ nm) on it. The n-Si wafers were etched to fabricate SiNS of different sizes using laser power density of 1.76 kW/cm^2. Etching time was 25 minutes, 45 minutes and 60 minutes for sample S1, sample S2 and sample S3, respectively. All Raman spectra of the c-Si, a-Si and SiNS were recorded by using micro-Raman setup of the triple monochromator (T64000 Jobin Horiba).

9.3 Results and Discussion

9.3.1 Raman scattering from semiconductors

When incident light of angular frequency ω_i interacts with a semiconductor crystal, scattered light of frequency ω_s may have values higher or lesser than the incident light frequency ω_i. The observation of higher or lower frequency in the scattered light is due to the creation or destruction of one or more lattice vibrational quanta, known as phonon. Energy and momentum conservation rules have to be followed during the first-order Raman scattering process, where one phonon participates:

$$\hbar\omega_i = \hbar\omega_s \pm \hbar\omega_0 \tag{9.1}$$

$$\hbar\vec{k}_i = \hbar\vec{k}_s \pm \hbar\vec{k} \tag{9.2}$$

$\vec{k}_i$, $\vec{k}_s$ and $\vec{k}$ are wave vectors of the incident light, scattered light and phonon respectively. ω_0 is the phonon frequency.

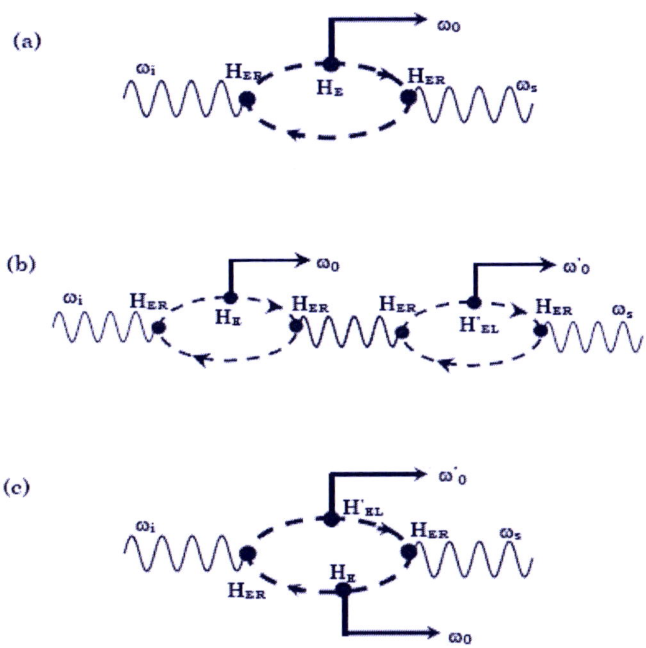

Figure 9.1 Schematic representation of Raman scattering processes in terms of elementary interactions H_{EL} and H_{ER}. (a) Shows first-order Raman scattering process whereas (b) and (c) show second-order Raman scattering

The process of Raman interaction between photon and phonon in a semiconductor has been known for very long time[9, 10]. The interaction is intermediated by a virtual electronic state. The complete interaction taking place in the whole Raman scattering process is summarized in Fig. 9.1. It shows electron radiation and electron lattice interaction as H_{ER} and H_{EL} respectively. First-order Raman scattering process is depicted in Fig. 9.1(a). For Si, the electron–lattice interaction is due to the excitation of phonons, which perturbs the periodic potential acting on the electrons by displacement of atoms. Since $\vec{k}_i$ and $\vec{k}_s$ (~ 10^5 cm^{-1}) are negligible in comparison to the width of the Brillouin zone (~ 3×10^8 cm^{-1}), momentum conservation allows the first-order Raman scattering for the phonon frequency at $\vec{k} \approx 0$.

The second-order Raman scattering results due to the involvement of two phonons in the Raman scattering process. Involvement of two phonons can be in two ways as shown in Figs 9.1(b) and 9.1(c). In one case, the second-order Raman spectra (Raman line spectrum) is due to two successive first-order Raman scatterings as depicted in Fig. 9.1(b). In this case, the frequency of the Raman scattered photon is the sum or the difference of shifts occurring in the first-order spectrum. The line nature of the second order scattering is the result of wave vector conservation in two successive first-order Raman scattering process[11]. In the other case, a couple of phonons participate as a single event as shown in Fig. 9.1(c). According to the conservation of wave vectors, the wave vectors of the two phonons should be equal and opposite because wave vectors of incident and scattered photons are negligible as compared to the dimension of the Brilloun zone dimension. As a result, the second-order continuous Raman spectrum reflects a combined density of phonon pairs with equal and opposite wave vectors. Frequency dependence of the interactions H_{EL} and H_{ER}[11] is also observed in this process. Therefore, zone edge phonons having high symmetry at the critical point (i.e., Δk is maximum) can participate in such second-order Raman scattering.

9.3.2 Raman spectrum of crystalline silicon

Raman spectrum from the c-Si shows only the zone-centered phonons due to the selection rules discussed in the above section. Phonons present in the diamond structured Si can be investigated by one-phonon and two-phonon Raman scattering. Figure 9.2 shows the Raman spectrum of the c-Si. The Raman scattering peak with Lorentzian line-shape has the maximum intensity at 520.5 cm^{-1}. This Lorentzian Raman line reflects the first-order Raman active phonon having Γ_{25} symmetry located at the zone center in Fig. 9.2. The zone center phonon (ZCP) has a tendency to decay to other sets of phonons[12]. The coupling of the ZCP with other sets of phonons is reflected as the width of this one-phonon Raman line-shape at 520.5 cm^{-1}. Natural line-width of a first-order Raman line is 3.5 cm^{-1}.

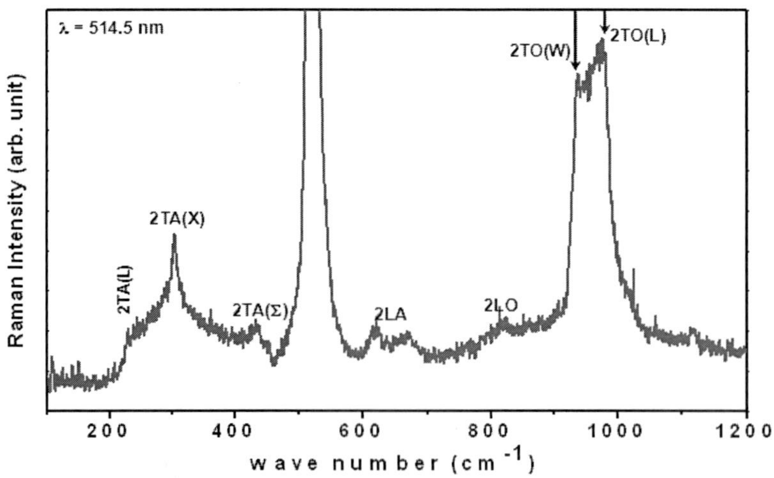

Figure 9.2 Raman spectrum from the c-Si, recorded using 514.5 nm excitation wavelengths

Second-order Raman scattering is shown in the Fig. 9.2 in the frequency range 200–500 cm^{-1} and 550–1000 cm^{-1}. Temple et al.[13] have observed peaks at 300 cm^{-1} and 435 cm^{-1} which were associated with the 2TA phonons from the critical points X and Σ respectively. The kink at 230 cm^{-1} is associated with the 2TA phonon from L point in the phonon dispersion curve. Weak scattering from the phonons along and near Σ direction is reflected as a weak peak at 460 cm^{-1}. The two-phonon Raman spectra associated with a combination of the acoustic and optic phonons in the Σ direction is seen near 610 cm^{-1} and 800 cm^{-1}. The scattering in the frequency range 900–1050 cm^{-1} has been identified with 2TO phonon overtones from the critical points at X. The shoulders at 940 cm^{-1} and 975 cm^{-1} are associated with 2TO phonon overtones at *W* and *L* points respectively[13].

9.3.3 Raman spectrum of amorphous silicon

As discussed above, the first-order Raman spectrum of the c-Si is a measure of the phonon frequency in the vicinity of $\vec{k} = 0$. Other phonons with non-zero $\vec{k}$ values are Raman forbidden as a result of the translational symmetry. However, these phonons can be observed as two phonon processes in the second-order Raman spectrum. In this case, only the sum of the $\vec{k}$ values of the two phonons must be vanishingly small and the optical phonons with $\vec{k} \neq 0$ may also contribute to the second-order Raman spectrum[13–15]. In the amorphous Si (a-Si), the translational invariance is broken due to lack of long-range order. As a result, the $\vec{k}$-conservation selection rule breaks down and all phonon modes become first-order Raman allowed[16, 17]. The tetrahedral short-range order is preserved in a-Si. Thus the density of the vibrational states of these materials is simply a broadened version of that of their amorphous counterparts. First-order Raman spectrum of the a-Si reflects the density of states (DOS) modulated by transition matrix elements and statistical factors[18]. Raman study by Wihl et al.[19] also establishes that the short-range order remaining in the a-Si material suffices to preserve the main features of the phonon density of states of the corresponding crystalline material. This type of finding is also mentioned by Smith et al.[20] in their study. Many authors[10, 21] have studied the gradual change in the Raman spectra of Si when a crystal is damaged and is transformed into the amorphous form.

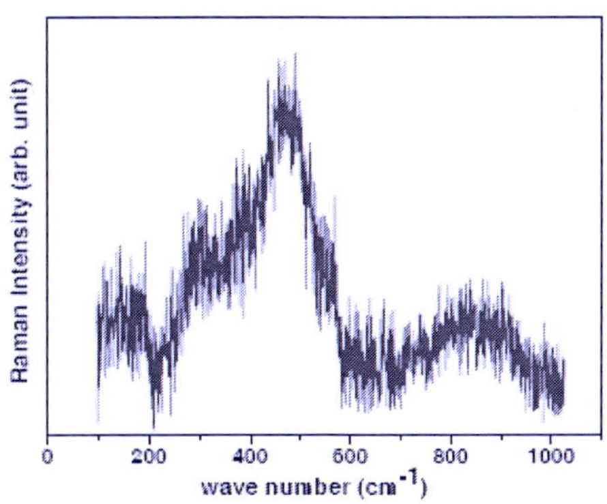

Figure 9.3 Raman spectrum from a-Si recorded using 514.5 nm excitation

Different theories are presented for the observed Raman spectrum of the a-Si[22–25]. As mentioned above, disorder in the a-Si relaxes the Raman selection rule for $\vec{k} = 0$, which allows first-order Raman scattering by phonons of larger $\vec{k}$ values. The maximum value of $\vec{k}$ of the participating phonon is of the order of $1/\Lambda_j$[16], where Λ_j is the coherence length of the normal mode 'j'. Room temperature Raman spectrum of the a-Si is shown in Fig. 9.3. Shuker and Gammon[25] attempted to quantify the Raman scattering from a-Si. Their analysis was based on following assumptions: (i) the vibrations are harmonic so that they can be analyzed into normal modes, and (ii) the vibrations couple to light through the displacement dependence of the electronic polarizability of the material. The following line-shape of the a-Si was proposed:

$$I(\omega) = \sum_b C_b \left(\frac{1}{\omega}\right) [1 + n(\omega)] g(\omega) \tag{9.3}$$

where, $n(\omega) = [\exp(\hbar\omega/k_B T) - 1]^{-1}$ (9.4)

with k_B as the Boltzman constant and $g(\omega)$ is the phonon DOS function. The C_b represents the coupling of different vibration bands to the radiation. It also contains the polarization dependence of Raman scattering from the a-Si. If all vibrational modes in a given band couple equally to the incident light, Eq. (9.3) can be used to obtain $g(\omega)$ from the Raman spectrum reduced by the term $[n(\omega) + 1]/\omega$. This technique shows that Raman scattering in amorphous solids can be used to obtain a semiquantitative picture of the phonon DOSs of the crystalline solids. Figure 9.4(b) depicts the one-phonon density of states calculated by Weber[26] using a bond-charge model. Figure 9.4(c) shows the Raman spectrum of the a-Si.

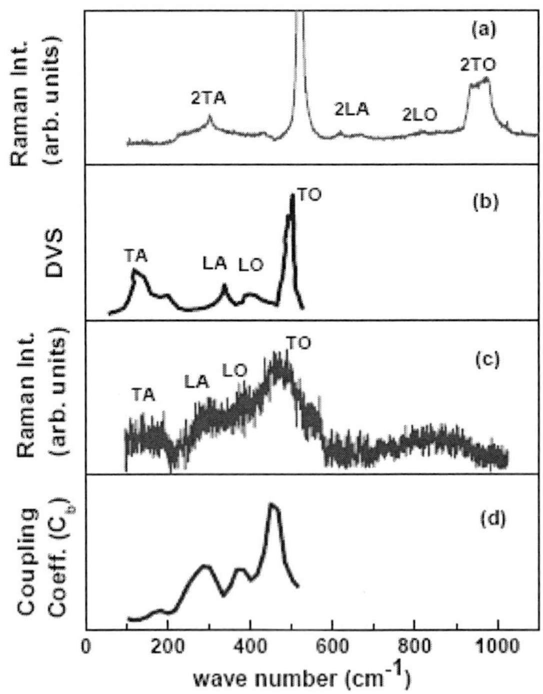

Figure 9.4 (a) Raman spectrum from c-Si. (b) Density of vibrational states of the c-Si from Weber[10]. (c) Raman spectrum of the a-Si. (d) Coupling coefficient of light interaction with the vibrational excitations

Disorder in the c-Si relaxes the selection rule and makes first-order Raman scattering by phonons of larger **q** possible. Theoretical Raman spectrum of the Stokes[25, 27] component for amorphous material is expressed as Eq. (9.3). The factors C_b, allow different vibrational bands to couple with varying strength to the radiation as shown in Fig. 9.4(d). In order to obtain a measure of the C_b, we have reduced the Raman spectrum of the a-Si in Fig. 9.1(c) by $[n(\omega)+1]^*\ g(\omega)/\omega$. It is shown in Fig. 9.4(d). Peaks at 150 cm^{-1}, 280 cm^{-1}, 380 cm^{-1} and 450 cm^{-1} observed in Fig. 9.4(d), where C_b is maximum can reveal overtones for 2TA, 2LA, 2LO and 2TO in the Fig. 9.4(a). The consistency of these curves shows the validity of interpretation of the results in terms of the C_b. The reduced Raman spectrum of the a-Si and $g(\omega)$ is very similar in peak location and in general shape for a-Si as shown in Figs 9.4(b) and 9.4(c). This confirms the view that the phonon spectrum of the crystal is largely determined by near -neighbor interactions, i.e., the density of states of solid predicted by Debye for calculating molar heat capacity is preserved here.

9.3.4 Raman scattering in nanocrystalline silicon

Figures 9.5(a)–9.5(c) show the Raman spectra of samples S1–S3 respectively. Raman spectra of Fig. 9.2 display the usual optical mode (i.e. 520 cm^{-1} for the c-Si) that is shifted to lower wavenumber with lineshape broadening and asymmetry towards lower wavenumber. Phonon softening, FWHM (full width at half maxima) and asymmetry of the optical mode are found to increase from Fig. 9.5(a) to Fig. 9.5(c) revealing smaller size nanostructures in the sample S3. The average size of nanostructures, found from optical phonon confinement model[28, 29] is 5.3 nm, 4.5 nm and 4.0 nm for sample S1, sample S2 and sample S3 respectively.

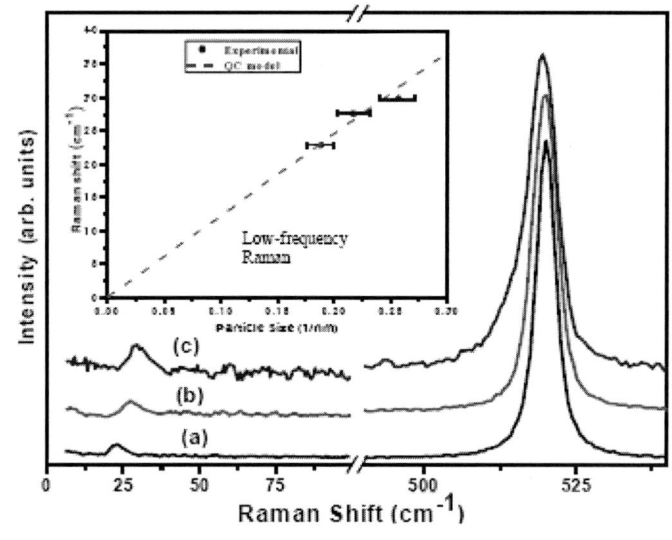

Figure 9.5 Raman spectra from samples S1, S2 and S3 are shown in (a), (b) and (c) respectively. Raman shift due to confined acoustic phonons has been plotted with the inverse of the diameter of the SiNS in the inset of the figure (Solid circles and dashed line show the experimentally observed Raman shifts and the Raman shifts calculated by the present QC model respectively

Furthermore, low-frequency Raman spectra of these samples (S1 – S3) are also shown in Fig. 9.5. An acoustic mode is observed at 22.9 cm^{-1}, 27.6 cm^{-1} and 29.8 cm^{-1} in Fig. 9.5 for sample S1, sample S2 and sample S3 respectively. Frequency of Raman active acoustic mode is increased with decreasing size of the SiNS. Frequency, FWHM and asymmetry towards higher wavenumber of the acoustic mode are increased from Fig. 9.5(a) to Fig. 9.5(c) with decreasing size of the SiNS. As average size of the SiNS is decreased from sample S1 to sample S3, oscillator strength of the Raman scattering due to acoustic mode is increased in Fig. 9.5.

In a perfect crystal, the first-order Raman scattering takes place due to participation of phonons at the **Γ** point (**q** = 0) of the Brillouin zone. Since there are no acoustic phonons at the **Γ** point in the phonon dispersion curve, first-order Raman scattering due to acoustic phonons is not observed for c-Si. Spatial confinement of phonons leads to the relaxation of the momentum conservation rule (i.e., $\mathbf{q} \geq 0$). It allows first-order Raman scattering due to acoustic phonons of non-zero wave vector ($q = 0$) in the SiNS. Therefore, frequency of the acoustic mode is increased with decreasing size of the SiNS. This can be explained by a dispersion curve where the acoustic phonons have increasing frequency with increasing wave vector **q** from the Γ point. Furthermore, localization of phonons (i.e., short correlation length) in the nanostructures is described by a Gaussian weighting function[28]. It is expressed as $W(r, L) = \exp(-cr^2/L_0^2)$. Fourier coefficient of this weighing function is given by $|F(q)|^2 = \exp(-cq^2L_0^2/a^2)$, where L_0 is the average size of the SiNS and 'c' is a constant. a is lattice parameter of the silicon.

As a result of the quantum confinement (QC) of acoustic phonons, Raman intensity for two-dimensionally confined SiNS can be described by[30]

$$I(\omega) = \int_0^1 \frac{\omega^3}{[\omega - \omega(q)]^2 + \left(\frac{\Gamma_0}{2}\right)^2} [n(\omega) + 1]|F(q)|^2 \, 2\pi q \, dq \tag{9.5}$$

where Γ_0 is inversely proportional to the life time. Limit of q in the integral is from 0 to 1 in the dimensionless unit (i.e., normalized by $2\pi/a$, where a is the lattice constant). Since short range order (i.e., diamond structure as well as identical nearest neighbour distances) is preserved in nanostructures as well as amorphous material[31], dispersion curve of the c-Si near the Γ point is good enough to map the acoustic phonons in the SiNS. Acoustic phonons can contribute appreciably in the integration of Eq. (9.5) for range $0 \leq q \leq 2\pi/L_0$. Therefore, anisotropic phonon dispersion may not be important in Eq. (9.2) due to the participation of phonons near the zone center (i.e., $q_{\max} \approx 2\pi/L_0$). Weighing function, $F(q)$ also localizes the acoustic phonons within the nanostructures. Phonon dispersion relation of the Si for TA branch[32], which is used here to explain our experimental Raman data with Eq. (9.5), is described by analytical fitting using the relation

$$\omega(q) = A\left[-\left(q\cos^{5/3}(\pi q/4) - 1\right)^2 + 1\right], \tag{9.6}$$

where $A = 182\ \text{cm}^{-1}$ is a constant. Raman shifts calculated by Eq. (9.5) are shown by dashed line for various sizes of the SiNS in the inset of Fig. 9.5. Equation (9.5) describes Raman lineshape (i.e., variation of Raman intensity with frequency shift) for acoustic mode. Frequency of Raman active acoustic mode (i.e. frequency at maximum Raman intensity) is plotted in the inset of Fig. 9.5 for varying sizes of nanostructures. Experimental Raman shifts observed in Fig. 9.5 are shown by dark circles in the inset of Fig. 9.5, where sizes L_0 and size distribution (i.e., σ) are calculated here by a phenomenological quantum confinement model[28, 29] for the Raman active optical mode observed in Fig. 9.5. Experimental Raman shifts fit well with Raman shifts calculated by Eq. (9.5). Frequency of the acoustic mode is inversely proportional to the size of the SiNS. Frequency of the acoustic mode approaches zero for the case of crystalline limit ($L_0 \rightarrow \infty$), i.e., first-order Raman scattering due to acoustic phonons is absent in the c-Si. It is emphasized here that TA like phonons participate in low-frequency Raman scattering. Asymmetry towards higher wavenumber of the acoustic mode in Fig. 9.5 reveals the participation of acoustic phonons having increasing frequencies with increasing '**q**' due to relaxation of the selection rule ($\mathbf{q} = 0$). Equation (9.6) ensures availability of such phonons.

9.4 Conclusions

In conclusion, the lack of translational invariance from the crystalline phase to the amorphous can be accounted for by simply relaxing the Raman selection rules. Raman spectrum of the a-Si is related to the density of vibrational states in the silicon. It is emphasized that the short range order is very nearly the same in the amorphous and crystalline phases of these materials. Quantum confinement model can also be utilized to determine the size of nanostructures using the Raman active acoustic mode. Short range order in the nanostructures and density of vibrational states are exploited in the quantum confinement model for acoustic mode. Agreement between our calculations and experimental data indicates that low-frequency Raman scattering is due to participation of the acoustic phonons in the SiNS. Frequency of the Raman active acoustic mode is found to be inversely proportional to size of the SiNS. Furthermore, oscillator strength of the Raman scattering due to acoustic mode is also increased with decreasing size of the SiNS. Furthermore, frequency of the Raman active optical mode is decreased (i.e., phonon softening) with decreasing size of nanostructures. More lineshape asymmetry of the first-order Raman scattering is found towards higher wavenumber and lower wavenumber for acoustic mode and optical mode respectively.

Bibliography

1. D. E. Polk. 1971. *J. Non-Cryst. Solids* 5: 365.
2. S. R. Elliot. 1983. *Physics of Amorphous Materials*. New York: Longman Inc.
3. E. Smith, Jr., M. H. Brodsky, B. L. Crowder, M. I. Nathan and A. Pinczuk. 1973. *Phys. Rev. Lett.* 30: 1141.
4. J. Xiang, W. Lu, Y. J. Hu, Y. Wu, H. Yan and C. M. Lieber. 2006. *Nature* (London) 441: 489.
5. S. Lu, Z. Lingley, T. Asano, D. Harris, T. Barwicz, S. Guha and A. Madhukar. 2009. *Nano Lett.* 9: 4548.
6. L. Pavesi, L. Dal Negro, C. Mazzoleni, G. Franco and F. Priolo. 2000. *Nature* (London) 408: 440.
7. R. Kumar, H. S. Mavi and A. K. Shukla. 2008. *Micron* 39: 287.
8. H. S. Mavi, A. K. Shukla, R. Kumar, S. Rath, B. Joshi and S. S. Islam. 2006. *Semicond. Sci. Technol.* 21: 1627.
9. R. Loudon. 1963. *Proc. Royal Soc.* 275: 218. London.

10. A. K. Shukla, Ph. D. Thesis. 1986. 'Raman study of ion implanted and laser annealed silicon'. Delhi: Indian Institute of Technology.
11. R. Loudon. 1964. *Adv. in Phys*. 13: 423.
12. T. R. Hart, R. L. Aggarwal and B. Lax. 1970. *Phys. Rev. B*. 01: 638.
13. P. A. Temple and C. E. Hathway. 1971. *Phys. Rev. B*. 07: 3685.
14. M. A. Renucci, J. B. Renucci, R. Zeyher and M. Cardona. 1974. *Phys. Rev. B*. 10: 4309.
15. J. B. Renucci, R. N. Tyte and M. Cardona. 1975. *Phys. Rev. B*. 11: 3885.
16. D. Bermajo and M. cardona. 1979. *J. Non-Cryst. Solids* 32: 405.
17. M. Brodsky. 1975. *Light Scattering in Solids*, 205. Edited by M. Cardona. Verlag: Springer.
18. J. S. Lannin. 1973. *Solid State Commun*. 12: 947.
19. M. Wihl, M. Cardona and J. Tauc. 1972. *J. Non-Cryst. Solids* 172: 810.
20. J. E. Smith Jr., M. H. Brodsky, B. L. Crowder, M. I. Nathan and A. Pinczuk. 1971. *Phys. Rev. Lett*. 26: 642.
21. C. Thomsen and E. Bustarret. 1992. *J. Non-Cryst. Solids* 141: 265.
22. R. L. C. Vink, G. T. Barkema and W. F. van der Weg. 2001. *Phys. Rev. B*. 63: 115210.
23. M. Marinov and N. Zotov. 1997. *Phys. Rev. B*. 55: 2938.
24. D. Beeman, R. Tsu and M. F. Thorpe. 1985. *Phys. Rev. B*. 32: 874.
25. R. Shuker and R. W. Gammon. 1970. *Phys. Rev. Lett*. 25: 222.
26. W. Weber. 1977. *Phys. Rev. B*. 15: 4789.
27. Y. M. Yang, L. W. Yang and P. K. Chu. 2007. *Appl. Phys. Lett*. 90: 081909.
28. H. S. Mavi, A. K. Shukla, R. Kumar, S. Rath, B. Joshi and S. S. Islam. 2006. *Semi cond. Sci. Technol*. 21: 1627.
29. H. Richter, Z. P. Wang and L. Ley. 1981. *Solid State Commun*. 39: 625.
30. A. K. Shukla and Vivek Kumar. 2011. *J. Appl. Phys*. 110: 064317.
31. K. P. Jain, A. K. Shukla, R. Ashokan, S. C. Abbi and M. Balkanski. 1985. *Phys. Rev. B*. 32: 6688.
32. R. Tubino, L. Piseri and G. Zerbi. 1972. *J. Chem. Phys*. 56: 1022.

III
Laser Cooling and Trapping

10 Ultra-Stable Coherent Matter-Wave Gyroscope Exploiting Quantized Vortices in Bose–Einstein Condensates Generated from Orbital Angular Momentum Beams of Light

Sulakshana Thanvanthri, Jonathan P. Dowling and Kishore T. Kapale

10.1 Introduction

The interaction of atomic or molecular matter with light carrying quantized orbital angular momentum (OAM)[1], is of considerable research interest[2, 3, 4]. New tools are becoming increasingly available for creation[5, 6, 7], manipulation[8], detection[9, 10], and application[11] of the orbital angular momentum (OAM) states of light. However, there is a caveat—due to the size mismatch between the atoms and the spatial features of the phase structures in the OAM beams of light—it is difficult to couple the OAM degrees of freedom of light to the internal states of single atoms. As a result, one needs to look for macroscopic coherent objects that could interact with the OAM carrying beams of light. In this context, Bose–Einstein condensates (BEC)[12, 13, 14] come to mind for the obvious reason that they are macroscopic coherent objects that are obtained by trapping atoms in particular, internal states, and cooling them further so that all the atoms have the same motional and internal states. The excitation of vortices in BECs, that have the phase structure similar to light carrying OAM, have been traditionally achieved via stirring of the BEC cloud with a laser beam[15]. These vortex states are fairly stable and could be candidates for qubits in quantum information[16], if appropriate means to manipulate them are developed.

To recall, the OAM states of light have unique amplitude and phase structures. Monochromatic OAM beams have an azimuthal phase dependence of the type $\exp(il\phi)$. Laguerre–Gaussian (LG) laser modes are an example of such OAM states[17]. The normalized LG mode at the beam waist ($z = 0$) and beam size w_0 at the waist is given in cylindrical coordinates (ρ, ϕ, z) by

$$\mathrm{LG}_p^l(\rho,\phi) = \sqrt{\frac{2p!}{\pi(|\ell|+p)!}}\frac{1}{w_0}\left(\frac{\sqrt{2}\rho}{w_0}\right)^{|\ell|} L_p^{|\ell|}\left(\frac{2\rho^2}{w_0^2}\right)\exp(-\rho^2/w_0^2)\exp(i\ell\phi), \tag{10.1}$$

where $L_p^l(\rho)$ are the associated Laguerre polynomials,

$$L_p^{|\ell|}(\rho) = \sum_{m=0}^{p}(-1)^m \frac{(|\ell|+p)!}{(p-m)!(|\ell|+m)!m!}\rho^m; \tag{10.2}$$

w_0 is the beam width, p is the number of non-axial radial nodes of the mode, and the index ℓ, referred to as the winding number, describes the helical structure of the wave front around a phase dislocation. For further discussion, we consider only pure LG modes with charge ℓ and $p = 0$. We denote such a state of the light field by $|\ell\rangle$ such that $\langle \mathbf{r}|l\rangle = \mathrm{LG}_0^{\ell}(\rho, \phi)$. Thus it can be easily seen that the states $|+\ell\rangle$ and $|-\ell\rangle$, with ℓ being a whole number, differ only in the sense of the winding of the phase either clockwise or counter-clockwise.

There exist several proposals to transfer pure OAM states of light to the BECs[18, 19]; there is also a proposal to use BECs for storage of the OAM state of light[20]. This transfer of OAM of light to BEC is possible due to the coherent nature of the BEC and because the size of the BEC cloud is about the same as the distance over which the intensity variations occur in the beam of light carrying OAM. Schemes have been proposed for creating superposition of two counter-rotating vortex states (with same absolute value of the charge) in BECs via coupling of the OAM degrees of freedom of incident beam to the center-of-mass motion of the atoms[21, 22]. To illustrate, transfer of a pure OAM state of light to a BEC cloud is an experimental reality[23, 24, 25]; nevertheless the experimental system involves different linear momentum for the initial and final states of the BEC as opposed to the different internal states in the scheme discussed in our works[21, 22]. It is clear that imparting linear momentum to the BEC cloud along with the orbital angular momentum would impose several practical constraints for further applications. Furthermore, the experiment involves transfer of only pure OAM states to the BEC as opposed to a OAM superposition as we proposed.

The counter-rotating vortex superposition generated in BEC is analogous to the counter-rotating persistent currents in superconducting circuits[26, 27, 28]. These counter-rotating persistent currents in the superconducting systems have already been recognized as candidates for qubits for quantum information processing. Thus the study presented here may have useful applications to the area of quantum information. Further applications of the light–matter interaction between BEC and OAM states of light could make a quantum memory device for such superposition states possible. This memory device could be useful in the quantum communication networks using OAM states of light[29]. OAM states of light offer a higher dimensional Hilbert space to obtain extra security and dense coding of quantum information. The memory application would deem it necessary to have a time-reversal symmetric procedure to write the superposition in OAM states of light to BEC so that the superposition can be extracted at a later stage. In this context, it is instructive to note that along with our frequency chirp based earlier proposal[21], we have also worked out a population transfer between internal states of the trapping potential through the stimulated Raman adiabatic passage (STIRAP)[30] based time-reversal symmetric mechanism for transfer of special OAM states of light to BEC[22]. Here we review only important details of a specific trapping potential, which can support toroid-shaped condensate cloud, in order to increase the stability of the generated vortex superposition.

This chapter is organized as follows. In Section 10.2, we discuss detailed dynamics of optically coupled three component BEC and STIRAP based methods of transferring the OAM superposition from the light beam to the vortex states of BEC in complete detail. As an example, we consider the so-called Mexican hat trap that can sustain large charge vortices. In fact, we provide a very general set of equations to study the transfer of OAM superposition from light to BEC for any kind of trapping potential. In Section 10.4 we discuss the effect of rotation on the optically generated vortex superposition. Later, in Section 10.5 we discuss how one can detect the Sagnac phase shift and through it the rotational speed of the trapping potential. We provide a summary in Section 10.6.

10.2 Coupling of Optical Vortex Beams to BEC

In this section, we discuss the STIRAP-like time-reversal symmetric method to transfer superposition of optical orbital angular momentum states to the states of atoms. We would like to refer the interested reader to our earlier works[21, 22] for further technical details on how to generate arbitrary superpositions of two counter-rotating, and carrying the same absolute value of charge, OAM states of light such as: $(a_+|\ell\rangle + a_-|-\ell\rangle)$, with $|a_+|^2 + |a_-|^2 = 1$.

10.2.1 Basic equations

The methods for transfer of pure OAM states of light to BEC has been studied by Marzlin et al.[18] and Nandi et al.[19]. Nevertheless, for the transfer of OAM superpositions, one needs a special transfer scheme. The internal level scheme of the atoms and the transitions of interest are shown in Fig. 10.1. The internal hyperfine quantum number of the initial non-rotating state $|0\rangle$ is $m_F = 0$ and that of the final states $|+\rangle$ and $|-\rangle$ is $m_F = 2$. Thus, the internal quantum number of both the final states is chosen to be the same to obtain a pure superposition of the vortex states while the rest of the quantum numbers needed to describe the two states are exactly identical. The intermediate states, $|i\rangle$ and $|i'\rangle$, through which Raman coupling between the initial and final states occurs, have the hyperfine quantum number $m_F = 1$. The two components of the optical OAM states correspond to the Rabi frequencies Ω_+ and Ω_- and both are σ_+ polarized. The coupling field is designated by the Rabi frequency Ω_c and it is σ_- polarized. The OAM of the state $|0\rangle$ is zero and that of $|i\rangle$ and $|+\rangle$ is $+l$ and that of $|i'\rangle$ and $|-\rangle$ is $-l$, where l is a positive integer. Even though we are targeting the generation of superposition of two counter-rotating components, it can be quickly seen that this scheme can also be used as it is for transfer of arbitrary superposition of two arbitrary OAM states of light to the BEC cloud.

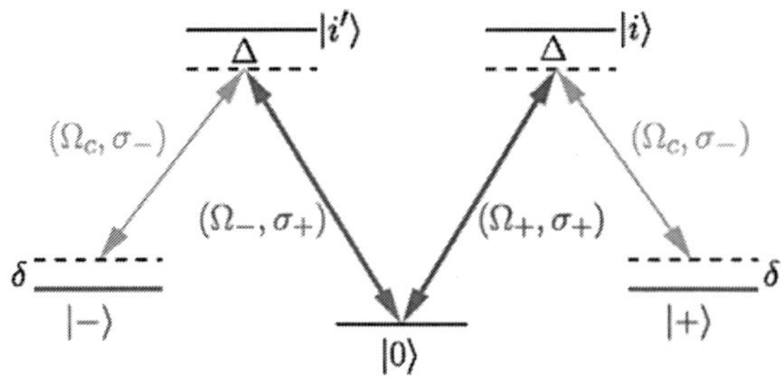

Figure 10.1 Internal level scheme of the trapped BEC atoms and the applied fields. The fields are specified through the corresponding Rabi frequencies and the polarization state. The fields are highly detuned from the excited states to avoid populating the excited states and ensure the stability of the condensate. The two-photon detunings δ are needed for one method of transfer of population from the non-rotating state to the vortex states but needs to be zero for the STIRAP-like transfer mechanism. The internal quantum numbers of the states $|0\rangle$, $|\pm\rangle$ are exactly the same. This makes specific polarization modes necessary for the coupling fields as shown. The intermediate states may have different internal quantum numbers, but they are not populated

Due to the various possibilities available for the internal state quantum numbers and the OAM quantum numbers, there are five distinct states that the atoms in the BEC can have as a result of their interaction with the optical fields. Thus, the BEC cloud can be described through a five-component spinor $\{\Psi_0, \Psi_+, \Psi_-, \Psi_i, \Psi_{i'}\}$. All the internal states are assumed to be trapped by the trapping potential and thus there are mutual interactions among all the components. The optical fields couple various components of the spinor BEC, as shown in Fig. 10.1, which lead to the evolution equations for the spinor components:

$$
\begin{aligned}
i\dot{\Psi}_0 &= \Omega_+^* \Psi_i + \Omega_-^* \Psi_{i'} + (\mathcal{H}_I/\hbar)\Psi_0, \\
i\dot{\Psi}_+ &= \Omega_c^* \Psi_i + (\mathcal{H}_I/\hbar)\Psi_+, \\
i\dot{\Psi}_- &= \Omega_c^* \Psi_{i'} + (\mathcal{H}_I/\hbar)\Psi_-, \\
i\dot{\Psi}_i &= -\Delta\Psi_i + \Omega_c\Psi_+ + \Omega_+\Psi_0 + (\mathcal{H}_I/\hbar)\Psi_i, \\
i\dot{\Psi}_{i'} &= -\Delta\Psi_{i'} + \Omega_c\Psi_- + \Omega_-\Psi_0 + (\mathcal{H}_I/\hbar)\Psi_{i'}.
\end{aligned} \tag{10.3}
$$

It is important to note the time derivatives appearing on the left-hand side of the equations. In general, the spinor components have both spatial and time dependence; nevertheless, for optical evolution, the time dependence is of prime importance. The optical coupling is governed by the Rabi frequencies Ω_c (for the coupling field) and $\Omega_\pm$ (for the OAM superposition). Here, Δ is the single-photon detuning. In setting up the above equations, we have assumed that the two fields are in two photon resonance with the transitions $|0\rangle - |+\rangle$ and $|0\rangle - |-\rangle$, meaning that the parameter δ appearing in Fig. 10.1 is zero. This detuning can be added later on, if needed, as will become clear later. Furthermore,

$$
\mathcal{H}_I = (\mathcal{I} + \mathcal{V} - \mu) + \eta(|\Psi_0|^2 + |\Psi_+|^2 + |\Psi_-|^2 + |\Psi_i|^2 + |\Psi_{i'}|^2), \tag{10.4}
$$

where $\mathcal{I}$ is the kinetic energy operator, $\mathcal{V}$ is the trapping potential of the BEC and μ is the chemical potential. The interatomic interaction strength is given by $\eta = 4\pi\hbar^2 \mathrm{aN/m} = NU$, where a is the s-wave scattering length, N is the mean number of atoms in the cloud, and m is the atomic mass of the atom. The absolute square terms in Eq. (10.4) thus correspond to the self-energy arising due to the inter-particle interaction.

Noting the spatial profile of the LG beams Eq. (10.1), we obtain

$$\Omega_{\pm}(\mathbf{r}) = a_{\pm}\Omega_0 e^{-r^2/w^2}\left(\frac{\sqrt{2}r}{w}\right)^{|\ell|} e^{\pm i|\ell|\phi}e^{ikz}, \tag{10.5}$$

with Ω_0 is the Rabi frequency of the atom–field interaction that could be time dependent if necessary. $a_{\pm}$ are the probability amplitudes of the $\pm|\ell|$ vortices in the superposition state we want to create. We assume that the Gaussian fall-off of the light intensity is over a length larger than the transverse size of the BEC cloud. As a result, we can ignore the gaussian term in the LG beam profile to arrive at

$$\Omega_{\pm} = a_{\pm}(\sqrt{2})^{|\ell|}\Omega_0\left(\tfrac{x\pm iy}{w}\right)^{|\ell|} e^{ikz}. \tag{10.6}$$

Note that we have incorporated the phase term inside the complex quantity $x \pm iy = r\cos\phi \pm i\sin\phi = re^{\pm i\phi}$. Similarly, the Gaussian fall-off of the intensity of the coupling beam (or Ω_c) along the transverse direction is ignored as the BEC cloud size is much smaller than the transverse beam profile.

The next step is to perform complete adiabatic elimination of the excited levels $|i\rangle$ and $|i'\rangle$. It is completely justified to use the adiabatic elimination procedure as: (i) the optical evolution happens on a time scale much faster than other processes in the BEC; (ii) the optical fields are so detuned from the excited states that the excited states are never occupied. Thus, by substituting $\dot{\Psi}_i = \dot{\Psi}_{i'} = 0$ and eliminating Ψ_i and $\Psi_{i'}$ from Eq. (10.3), we arrive at

$$\begin{aligned} i\dot{\Psi}_0 &= \frac{1}{\hbar}(\mathcal{T} + \mathcal{V} - \mu)\Psi_0 + \frac{\eta}{\hbar}(|\Psi_0|^2 + |\Psi_+|^2 + |\Psi_-|^2)\Psi_0 \\ &+ \frac{1}{\Delta}(|\Omega_+|^2 + |\Omega_-|^2)\Psi_0 + \frac{1}{\Delta}\Omega_+^*\Omega_c\Psi_+ + \frac{1}{\Delta}\Omega_-^*\Omega_c\Psi_-, \end{aligned} \tag{10.7a}$$

$$\begin{aligned} i\dot{\Psi}_+ &= \frac{1}{\hbar}(\mathcal{T} + \mathcal{V} - \mu)\Psi_+ + \frac{\eta}{\hbar}(\Psi_+|^2 + |\Psi_-|^2 + |\Psi_0|^2)\Psi_+ \\ &+ \frac{1}{\Delta}|\Omega_c|^2\Psi_+ + \frac{1}{\Delta}\Omega_+\Omega_c^*\Psi_0, \end{aligned} \tag{10.7b}$$

$$\begin{aligned} i\dot{\Psi}_- &= \frac{1}{\hbar}(\mathcal{T} + \mathcal{V} - \mu)\Psi_- + \frac{\eta}{\hbar}(|\Psi_+|^2 + |\Psi_-|^2 + |\Psi_0|^2)\Psi_- \\ &+ \frac{1}{\Delta}|\Omega_c|^2\Psi_- + \frac{1}{\Delta}\Omega_-\Omega_c^*\Psi_0. \end{aligned} \tag{10.7c}$$

Our ultimate interest lies in the temporal dynamics of the populations of various components of the BEC. Thus, the spatial part of the condensate is to be integrated out of the above equations. To deal with the spatial profiles of the spinor components of the BEC explicitly, we note that the general state of the BEC cloud can be written as

$$\langle \mathbf{r}|\Psi\rangle = \alpha(t)\langle \mathbf{r}|0 + \beta(t)\langle \mathbf{r}|+\rangle + \gamma(t)\langle \mathbf{r}|-\rangle, \tag{10.8}$$

where the spatio-temporal projections are given by

$$\begin{aligned}
\Psi_0(\mathbf{r},t) &= \alpha(t)\langle \mathbf{r}|0\rangle = \alpha(t)\exp[i(\mu/\hbar - \kappa)t]\psi_{\text{g}}(\mathbf{r}),\\
\Psi_+(\mathbf{r},t) &= \beta(t)\langle \mathbf{r}|+\rangle = \beta(t)\exp[i(\delta+\mu/\hbar - \kappa)t]\psi_{\text{v}+}(\ell,\mathbf{r}),\\
\Psi_-(\mathbf{r},t) &= \gamma(t)\langle \mathbf{r}|-\rangle = \gamma(t)\exp[i(\delta+\mu/\hbar - \kappa)t]\psi_{\text{v}-}(-\ell,\mathbf{r}).
\end{aligned} \tag{10.9}$$

The two-photon detuning δ introduced in the phase factors is defined as $\delta = \nu_\pm - \nu_c - \omega_{0\pm}$, where $\nu_\pm$ and ν_c are the angular frequencies of the optical field carrying OAM and the coupling field respectively, and $\omega_{0\pm}$ is the energy level difference between the states $|0\rangle$ and $|\pm\rangle$ expressed as an angular frequency. Furthermore, the parameter κ, in general, depends on the interaction strength η between the BEC atoms and the dimensional parameters of the BEC cloud, which in turn depend on the trapping potential. It is instructive to note that when one is dealing with condensates consisting of single species, as we are, κ need not be explicitly incorporated in the phase factors as it just causes a shift in the energy equally for all the component states of the BEC cloud.

The ansatz for the spatial profiles of the non-rotating ground state and the vortex states depend on the chosen trapping potential. After substituting the ansatz, Eq. (10.9) into the Eq. (10.7), we obtain the temporal dynamics of the populations of the BEC components of interest. We can then formally perform the coordinate integrals to remove the spatial dependence from the rate equations. In terms of the resulting spatial integrals, the rate equations are:

$$\begin{aligned}
i\dot{\alpha}(t) &= (T_g + V_g)\alpha(t) + (I_{gg}|\alpha(t)|^2 + I_{g+}(\ell)|\beta(t)|^2 + I_{g-}(\ell)|\gamma(t)|^2)\alpha(t)\\
&\quad + \frac{|\Omega_0|^2}{\Delta} I_{gg}^{(2l)}(\ell)\alpha(t) + \frac{\Omega_0^*\Omega_c}{\Delta}(I_{g+}^{(\ell)}(\ell)a_+^*\beta(t) + I_{g-}^{(\ell)}(\ell)a_-^*\gamma(t)),\\
i\dot{\beta}(t) &= [T_+ + V_+(\ell) + \delta + I_{g+}(\ell)|\alpha(t)|^2 + I_{++}(\ell)\,|\beta(t)|^2 + I_{+-}(\ell)|\gamma(t)|^2]\beta(t)\\
&\quad + \frac{|\Omega_c|^2}{\Delta}\beta(t) + \frac{\Omega_0\Omega_c^*}{\Delta} I_{+g}^{(\ell)}(\ell)a_+\alpha(t),\\
i\dot{\gamma}(t) &= [T_- + V_-(\ell) + \delta + I_{g-}(\ell)|\alpha(t)|^2 + I_{+-}(\ell)|\beta(t)|^2 + I_{--}|\gamma(t)|^2]\gamma(t)\\
&\quad + \frac{|\Omega_c|^2}{\Delta}\gamma(t) + \frac{\Omega_0\Omega_c^*}{\Delta} I_{-g}^{(\ell)}(\ell)a_-\alpha(t).
\end{aligned} \tag{10.10}$$

Note that the spatial integrals are denoted by the letter I with different subscripts, superscripts and in most cases, with the argument of ℓ to properly distinguish them. These spatial integrals are explicitly defined below:

$$T_{\text{g}} = \frac{1}{\hbar}\int \psi_{\text{g}}^*(\mathbf{r})\mathcal{T}\,\psi_{\text{g}}(\mathbf{r})d^3\mathbf{r},$$

$$V_{\rm g} = \frac{1}{\hbar}\int \psi_{\rm g}^{*}(\mathbf{r})\mathcal{V}\psi_{\rm g}(\mathbf{r})d^{3}\mathbf{r},$$

$$T_{\pm}(\ell) = \frac{1}{\hbar}\int \psi_{\rm v\pm}^{*}(\pm\ell,\mathbf{r})\mathcal{T}\psi_{\rm v\pm}(\pm\ell,\mathbf{r})d^{3}\mathbf{r},$$

$$V_{\pm}(\ell) = \frac{1}{\hbar}\int \psi_{\rm v\pm}^{*}(\pm\ell,\mathbf{r})\mathcal{V}\psi_{\rm v\pm}(\pm\ell,\mathbf{r})d^{3}\mathbf{r},$$

$$I_{\rm gg} = \frac{\eta}{\hbar}\int |\psi_{\rm g}(\mathbf{r})|^{2}|\psi_{\rm g}(\mathbf{r})|^{2}d^{3}\mathbf{r},$$

$$I_{\rm g\pm}(\ell) = \frac{\eta}{\hbar}\int |\psi_{\rm v\pm}(\ell,\mathbf{r})|^{2}|\psi_{\rm g}(\mathbf{r})|^{2}d^{3}\mathbf{r},$$

$$I_{++}(\ell) = \frac{\eta}{\hbar}\int |\psi_{\rm v+}(\ell,\mathbf{r})|^{2}|\psi_{\rm v+}(\ell,\mathbf{r})|^{2}d^{3}\mathbf{r},$$

$$I_{--}(\ell) = \frac{\eta}{\hbar}\int |\psi_{\rm v-}(\ell,\mathbf{r})|^{2}|\psi_{\rm v-}(\ell,\mathbf{r})|^{2}d^{3}\mathbf{r},$$

$$I_{\rm gg}^{(2\ell)}(\ell) = \int \psi_{\rm g}^{*}(\mathbf{r})\left(\frac{\sqrt{2}r}{w}\right)^{2|\ell|}\psi_{\rm g}(\mathbf{r})d^{3}\mathbf{r},$$

$$I_{\rm g\pm}^{(\ell)}(\ell) = \int \psi_{\rm g}^{*}(\mathbf{r})\,e^{\mp I\ell\phi}\left(\frac{\sqrt{2}r}{w}\right)^{|\ell|}\psi_{\rm v\pm}(\ell,\mathbf{r})d^{3}\mathbf{r},$$

$$I_{\pm\rm g}^{(\ell)}(\ell) = \int \psi_{\rm v\pm}^{*}(\ell,\mathbf{r})\,e^{\pm I\ell\phi}\left(\frac{\sqrt{2}r}{w}\right)^{|\ell|}\psi_{\rm g}(\mathbf{r})d^{3}\mathbf{r}. \qquad (10.11)$$

The definitions given above are independent of the trapping potential or the ansatz used for the spatial profile of the wavefunction. As a result, the above set of rate equations [Eq. (10.10)], specified in terms of the spatial integrals, is very general and applicable to a wide variety of trapping potentials, provided the integrals are evaluated by appropriately taking the spatial forms of the wavefunctions, $\psi_g(\mathbf{r}), \psi_{v\pm}(\mathbf{r})$, suitable for the trapping potential. The compact analytical expressions of these integrals in the case of a three-dimensional harmonic trapping potential are given in[22]. To note, in case the spatial ansatz for the BEC wavefunction is to be obtained in the Thomas–Fermi limit, then the kinetic energy terms can be ignored. Furthermore, the equations can also be applied to situations where the applied OAM and coupling fields have a certain time profile.

10.2.2 BEC vortex superposition via STIRAP-like pulse sequence

As it turns out, to obtain a complete transfer of population from a non-rotating BEC state to the vortex superposition, it is necessary to employ either a frequency chirp on the two-photon transition via a time-dependent detuning parameter δ as depicted in Fig. 10.1 or via a STIRAP-like scheme where $\delta = 0$ and simple time dependance is imprinted on the Rabi frequencies[22].

The STIRAP-like counter-intuitive pulse sequence is attained by adding appropriate time dependence to the Rabi frequencies as given below:

$$\Omega_0 = |\Omega_0| f(t) \quad \text{and} \quad \Omega_c = |\Omega_0| g(t), \tag{10.12}$$

where the temporal profiles of the beams are taken to be of the form:

$$f(t) = f_0 \, e^{-\left(\frac{t-t_1}{\sigma_1}\right)^2}, \quad g(t) = g_0 \, e^{-\left(\frac{t-t_2}{\sigma_2}\right)^2}. \tag{10.13}$$

Any arbitrary superpositions of the vortex states can be generated via the counter-intuitive pulse sequence—the coupling field pulse (time profile $g(t)$ and pulse center t_2) needs to come before the OAM superposition pulse (time profile $f(t)$ and pulse center t_1). We have shown that this scheme works immaterial of the chosen trapping potential[22].

10.2.3 Mexican hat trapping potential

Here we review the transfer of OAM to a BEC cloud when the trapping potential is shaped like a Mexican hat (Sombrero) in the x-y plane (See Fig. 10.2). The advantage of this potential is the toroidal symmetry it offers. The toroidal trap configurations naturally support a toroid shaped BEC cloud and as a result can sustain large vortices without disintegrating into several single charge vortices. This offers stability to the vortex superpositions we are aiming to generate. The trapping potential, in this case, is of the form [31]

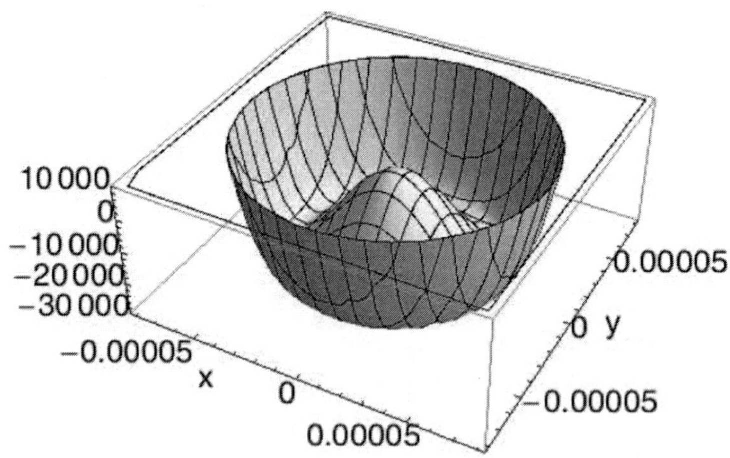

Figure 10.2 The spatial profile of the Mexican hat potential plotted in arbitrary units in two dimensions with $\sigma = 2.0$ and $\lambda = 0.005$

$$V(\rho, z) = -\frac{1}{2}\sigma m\omega_\perp^2 \rho^2 + \frac{1}{4}\lambda\left(\frac{m^2\omega_\perp^3}{\hbar}\right)\rho^4 + \frac{1}{2}m\omega_z^2 z^2, \tag{10.14}$$

where σ and λ are dimensionless parameters. The potential is harmonic in the z-direction. The dynamics of the OAM transfer to BEC can be studied in the Thomas–Fermi (TF) approximation where the kinetic energy of the BEC cloud is neglected. In this approximation, the spatial part of the BEC wavefunction is given by

$$\psi(\ell,\rho,\phi,z) = \left(\frac{1}{L_\perp \sqrt{L_z}}\right)\left(\frac{1}{\sqrt{|\ell|!}}\right)\left(\frac{\rho}{L_\perp}\right)^{|\ell|} e^{-\frac{z^2}{2L_z^2}} \mathrm{Max}\left[\mathbf{Re}\left(\sqrt{\frac{\mu - V(\rho,0)}{\eta}}\right),0\right] e^{i\ell\phi}. \tag{10.15}$$

Here we have assumed that the wave function in the z-direction has a Gaussian form and has little effect on the vortex dynamics of interest in the transverse direction[32]. The spatial profile of the above wavefunction is shaped like a toroid (or a donut) with a hole in the center. Thus, the two radii (inner and outer) are necessary to describe the shape of the BEC cloud. The radii can be found by setting $|\psi(\ell,\rho,\phi,z)|^2 = 0$ to determine where the particle density goes to zero. The two real solutions for the radii are given by

$$R_\pm^2 = \frac{\sigma M \pm \sqrt{\sigma^2 M^2 + 4\lambda g \mu}}{\lambda g}, \tag{10.16}$$

where $M = m\omega_\perp^2$ and $g = m^2\omega_\perp^3/\hbar^2$. A typical spatial form of the ground state BEC cloud in the mexican hat potential is shown in Fig. 10.3 along one of the transverse direction ($\hat{x}$ or $\hat{y}$). The toroid shaped particle distribution should be clear from the figure as it can be easily obtained by rotating the distribution around the central vertical line.

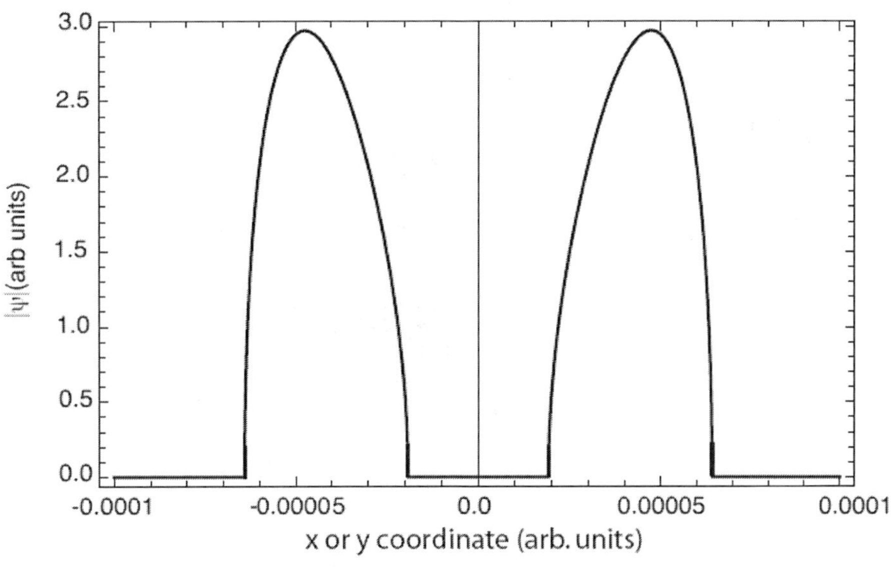

Figure 10.3 The Thomas–Fermi spatial wavefunction of the BEC in a Mexican hat potential plotted in arbitrary units versus one spatial dimension. The donut like particle density distribution is quite clear

The spatial part of the BEC wavefunction for different components is taken to be

$$\psi_g(\mathbf{r}) = \psi(0,\rho,\phi,z),$$

$$\psi_{v\pm}(\mathbf{r}) = \psi(\pm\ell, \rho, \phi, z). \tag{10.17}$$

Using the above, one can evaluate the spatial integrals appearing in Eq. (10.10) governing the population dynamics. We, furthermore, ignore kinetic energy of the system as we are working in the TF approximation, in this case and choose the optical field frequencies so that there is two-photon resonance, i.e., the two-photon detuning $\delta = 0$. The light fields Ω_c and $\Omega_\pm$ have the same saptio-temporal form as discussed in section 10.2.2. Numerically solving the evolution equations, we get the transfer curves for the Mexican hat potential depicted in Fig. 10.4. We observe that the time-scale of transfer is about the same as in the case of harmonic trap. The figure depicts generation of a 60:40 superposition of $|+2\rangle$ and $|-2\rangle$ vortex states. Furthermore, it is important to note that any arbitrary superposition could be obtained in the Mexican hat trap; the generated vortex state would be naturally stable and would not disintegrate into single charge vortices.

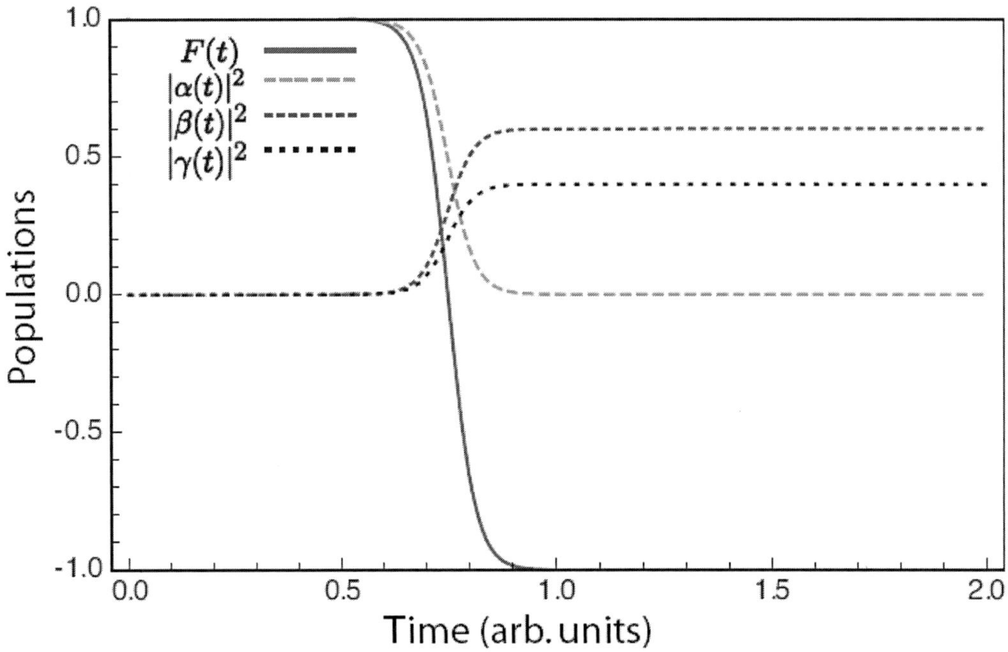

Figure 10.4 Generation of the vortex state superposition in a Mexican hat trap: Results of the numerical solutions of the equations for STIRAP scheme shows the superposition 60:40 of the $|+\rangle$ and $|-\rangle$ vortex states. The Rabi frequency $\Omega_0 = 1$ kHz and $\Delta = 100\Omega_0$

From the magnitude of the spatial integrals, it can be easily seen that the change in population of the levels over time does not affect the transfer significantly as the atom–light interaction has a much smaller timescale (μs) than the atom–atom interaction time scale (ms).

10.3 Introduction to Matter–Wave Interferometry

Matter–wave interferometers[33], which use atomic beams in place of coherent light beams, have been investigated extensively in the past decade in the context of metrology. The main advantage of matter–wave interferometers (employing beam of atoms with atomic mass m) stems from the fact that they offer sensitivities higher by a factor of about $mc^2/\hbar\omega \approx 10^{10}$ in comparison with optical interferometers (employing laser light of angular frequency ω)[34], where c is the speed of light and $\hbar$ is Planck's constant divided by 2π. The matter–wave interferometers naturally require mirrors and beam splitters for atomic beams[35], which are much more complex devices compared to their optical counterparts, resulting in added complications to the interferometric setup. Thus better schemes for interferometric metrology are actively being pursued.

In particular for gyroscopy, which deals with measurement of angular velocity of the laboratory frame of reference, the interferometer is known as Sagnac interferometer[36, 37, 38]; it employs two counter-propagating waves (light or matter) that experience a phase shift, termed as Sagnac phase shift proportional to the area of the interferometer and the angular velocity of the lab frame. Recently, a hybrid scheme has been proposed that uses alkali atom vapor cells, acting as a slow light medium, in the path of an optical Sagnac interferometer to obtain higher sensitivity for gyroscopy[39]. There is also another proposal employing entangled, parametrically down-converted photons to attain the Heisenberg limit of sensitivity for Sagnac interferometry[40]. The vortex superpositions in Bose–Einstein condensates (BEC) can be used as a Sagnac interferometer. This scheme provides a compact and robust scheme for gyroscopy without the need of traditional interferometers employing mirrors, beam splitters and single photon or atomic state-selective detectors.

Two important benchmarks for any gyroscope are its sensitivity and stability. The atomic beam gyroscopes, while having high sensitivity ($10^{-9}\text{s}^{-1}\ \text{Hz}^{-1/2}$), also suffer from instability due to a drift caused by fluctuations in the beam direction and thermal expansion of the beam diameter. Since the sensitivity of the Sagnac-effect based gyroscope is directly proportional to the area of the interferometer, there is a trade-off between sensitivity and stability. We show that employing BEC vortex superposition leads to a highly stable gyroscope.

10.4 Effect of Rotation on the Vortex Superpositions in a BEC

As discussed in Section 10.2 in detail, in the long time limit $t \to \infty$, it is possible to generate an arbitrary two-component vortex superposition state, in a Mexican hat trapping potential, given by:

$$\Psi(\mathbf{r},t) = b_+\Psi_+(\mathbf{r}) + b_-\Psi_-(\mathbf{r}), \tag{10.18}$$

as $\alpha(t) \to 0$, $\beta(t)/\gamma(t) \to \Omega_+/\Omega_- = b_+/b_-$, where $b_\pm$ are the amplitudes corresponding to the two components of the OAM superposition of light coupled to the BEC.

This superposition amounts to counter-rotating matter–wave currents causing the density profile of the BEC to be the interference pattern given by:

$$\Sigma(\mathbf{r},t) = [1+\cos(2|\ell|\phi)]|\psi(\ell,\rho,z)|^2, \tag{10.19}$$

where $\psi(\ell,\rho,z)$ is the azimuthally symmetric spatial profile which is common to the two counter-rotating vortex components when they share a common vortex-charge magnitude. This density distribution at $z=0$ is plotted in Fig. 10.5. When the trap (or laboratory) itself rotates in time about the z-axis, the two counter-rotating matter waves pick up different phases and Eq. (10.18) is modified to $\Psi_\Omega(\mathbf{r},t) = b_+\Psi_+(\mathbf{r})e^{i\phi(t)} + b_-\Psi_-(\mathbf{r})e^{-i\phi(t)}$.

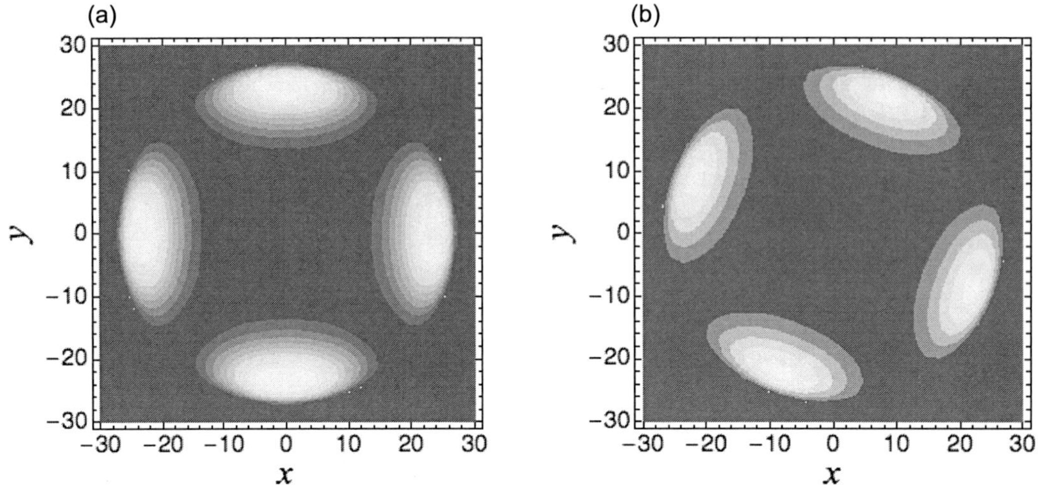

Figure 10.5 Interference pattern in the atomic density produced by the OAM superposition state $(|+2\rangle + |-2\rangle)/\sqrt{2}$. The pattern is in the transverse plane for $z = 0$: (a) with no rotation of the trap and (b) when angular velocity of the trap is $\Omega = 0.05\ \mathrm{s}^{-1}$. The lengths x and y are normalized with respect to the trap dimension $L_x = L_y = 2.4\ \mu\mathrm{m}$

The phase $\phi(t)$ accumulated over a given time duration can be calculated using the prescription of the optical Sagnac effect, in which two optical beams counter-propagate in a closed loop. In an optical interferometer with the light of wavelength λ, the phase in one round trip of the interferometer by each component wave, is given by $\phi_{\rm opt} = 4\mathbf{A}\cdot\Omega/(\lambda c)$, where c is the velocity of light, $\mathbf{A}$ is the area and Ω, which is treated as a vector here, is the angular velocity of the interferometer[38]. For a BEC composed of atoms with atomic mass m, $\lambda c \to \hbar/m$, the area of the interferometer is given by the cross-sectional area of the trapped BEC cloud. Assuming that the interferometer rotates about a z-axis, the Sagnac phase accumulated is given by[34]

$$\phi_\Omega(t) = N(t)\frac{4Am}{\hbar}\Omega, \tag{10.20}$$

where $N(t)$ is the number of times an atom goes around the trap center in time t, and $\hbar$ is the Dirac's constant. Noting that each atom in the BEC cloud has the same magnitude of the angular momentum, we can determine the linear velocity, period (T) and as a result $N(t) = t/T$, as a function of time and ℓ to arrive at $N(t) = t\hbar\ell/(2m\pi r^2) = t\hbar\ell/(mA)$ and thus

$$\phi_\Omega(t) = 2\ell\Omega t. \tag{10.21}$$

Therefore, the accumulated phase in time t is proportional to the time, angular velocity of the laboratory frame and the magnitude of the vortex charge in the superposition. Even though the area dependence is not explicit in the above formula, the dependence on ℓ implicitly includes the number of round trips performed by each atom in unit time and the area of the interferometer. In this case, the transverse density distribution (at $z = 0$) is modified to

$$\Sigma_\Omega(\rho,\phi,t) = \{1 + \cos[2|\ell|\phi + \phi_\Omega(t)]\}\,|\psi(\ell,\rho,z)|^2, \tag{10.22}$$

which when compared with Eq. (10.19) signifies rotation of the interference pattern by an angle $\phi_\Omega(t)/(2|\ell|)$ as the Sagnac phase accumulates over time. Following the rotation of an imaginary line drawn in the dark region of the interference pattern, the density profile allows determination of the angular velocity Ω.

10.5 Detection of the Sagnac Phase

The precision in measuring the angular velocity depends on the sensitivity of the BEC density measurement. The phase contrast optical imaging of BEC clouds as a non-destructive technique has been extensively used for *in situ* imaging of trapped BECs[41, 42] for imaging the rotating BEC cloud. In this technique, images are formed by photons scattered coherently from a condensate. The phase modulation of the probe through the BEC cloud is converted to amplitude modulation by retarding the transmitted light. In the following paragraph, we discuss the signal-to-noise ratio (SNR) and the sensitivity achievable in such a detection scheme.

The SNR is given by the ratio of the rotational phase shift to the shot-noise that depends on the number of scattered photons N_{SC} arriving at the detector per second[39],

$$\text{SNR} = \frac{\phi_\Omega}{\sqrt{1/N_{sc}}}, \tag{10.23}$$

where N_{sc} is the number of scattered probe photons arriving at the detector per second from the condensate. From Eq. (10.23), it appears that the SNR can be indefinitely increased by using a higher probe-field Rabi frequency of the probe field in the phase contrast imaging scheme. However, increasing the probe intensity enhances the risk of losing atoms from the BEC and thereby reducing the number of scattered photons from the BEC cloud and effectively reducing the SNR. An optimum balance can be reached for an atomic cloud of a given density as depicted in Fig. 10.6. For our current analysis, we assume that the condensate holds 10^6 atoms in a pancake shaped trap with dimensions $L_x = L_y = 2.4\ \mu\text{m}$ and $L_z = 0.8\ \mu\text{m}$. On average, the detector receives in the order of 10^6 scattered photons per 0.1 ms, i.e., 10^{10} photon per second from the BEC. In Fig. 10.6, the SNR is plotted as a function of the number of photons detected. The loss of atoms from the BEC with increasing photon scattering with an exponential dependence of the SNR on the photon number is modeled. The final SNR has a maximum at a certain rate of photon detection. The sensitivity Ω_{min} is calculated by setting SNR = 1.

The sensitivity of this device strongly depends on the value of the vortex charge and the general formula can be obtained by fixing SNR = 1 to obtain $\Omega_{\text{min}} = 1/(2\ell t\sqrt{N_{\text{sc}}})$. A conservative value is chosen by allowing the magnitude of the vortex charge to be $\ell = 100$; such a high charge vortex can be accommodated in the Sombrero trap with appropriate choice of the trap parameters. Thus, the numerical value for the smallest rotational speed measurable is calculated to be $\Omega_{\text{min}} = 5 \times 10^{-8}$ rad s^{-1} $\text{Hz}^{-1/2}$ (for $t = 1$ s and $N_{\text{sc}} = 10^{10}\ \text{s}^{-1}$) which is slightly worse than the sensitivity of about $10^{-9}\ \text{s}^{-1}\ \text{Hz}^{-1/2}$ reported for the state-of-the-art atomic beam gyroscopes[43, 44, 39]. Despite the slightly lower sensitivity, the present scheme offers tremendous simplicity and robustness in comparison with the other matter–wave interferometers.

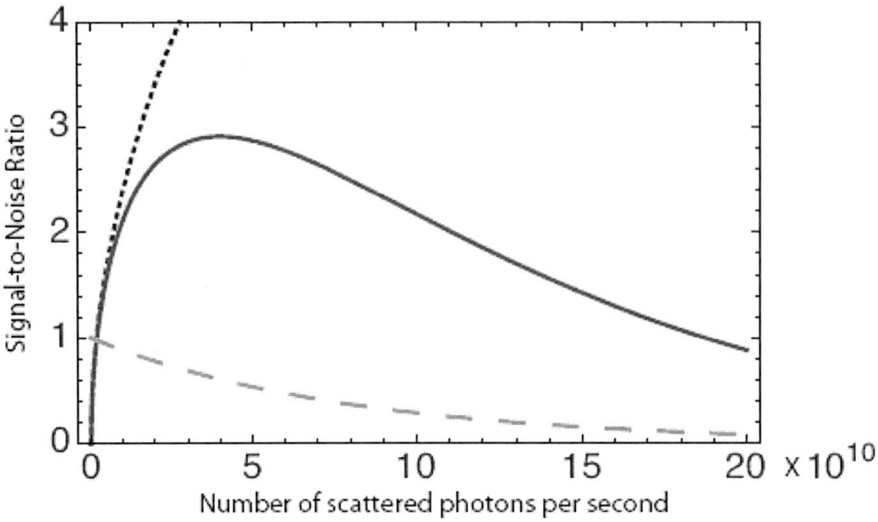

Figure 10.6 Signal-to-noise ratio (SNR) as a function of the number of scattered photons arriving at the detector per second. The dotted curve gives the $1/\sqrt{N_{sc}}$ behavior, the dashed curve models the exponential decay of the BEC with increase in number of probe photons. The solid curve gives the effective SNR. Here the rotational velocity of the trap is taken to be $\Omega = 6 \times 10^{-6}$ rad s^{-1}

To address the stability of the proposed gyroscope, it helps to note that the beam drift is an important source of instability in the long integration time gyroscopy employing traditional atomic or optical beams. In BEC vortices, angular momentum is quantized and all the atoms move with a common angular velocity and besides having to maintain the trapping potential, no complicated maneuvering of the atomic beams is required. Moreover, thermal expansion over long beam trajectories is not an issue for a trapped BEC. Thus, the proposal eliminates the drift problem commonly faced by current atomic gyroscopes with drifts in the order of 68 μdeg/h after about four hours of operation[45].

The area of the interferometer is another important parameter governing the sensitivity; in the present scheme, the size of the BEC trap, which is a few mm^2, is the area of the interferometer. On the other hand, the typical area of an atom beam gyroscope is about 4 m^2[45]. Thus, despite such a large difference in the area, current numbers for sensitivity of the vortex-based gyroscope are only slightly worse than that of the state-of-the-art atomic gyroscopes. Several methods are already available for improving the sensitivity of the vortex-based gyroscope such as using higher density BEC clouds, using optical lattice of BECs[46] and improving the sensitivity and speed of the BEC imaging techniques. An important point to note is that for the BEC vortex gyroscope, an increase in sensitivity (by increasing the area) does not correspond to a decrease in stability. The stability can be maintained and enhanced by manipulating the geometry of the BEC trap and optical beams used to create the vortex superposition. Further, if it is necessary to replenish the trap with a fresh BEC cloud, it can be noted that creating and reloading BECs has become a routine procedure with improvements in the experimental techniques coming regularly.

10.6 Conclusion

To summarize, we have discussed the mechanism for coupling of optical angular momentum of light to vortices in BEC. We gave a complete formalism that goes beyond the Mexican hat type potential considered in detail here. We discussed how a STIRAP like population transfer mechanism can be employed to transfer the superposition of optical angular momentum of light to vortices in BEC. We also discussed how the superposition of counter-rotating currents in the BEC could also be used for inertial sensing, especially as a gyroscope. Several advantages offered by a gyroscope based on superpositions of counter-rotating vortex structures would be the tunability of the effective de-Broglie wavelength by choosing appropriate atomic masses and the angular velocity and phase sensitivity via the choice of the quantized angular momentum of the atoms.

Bibliography

1. M. Padgett, J. Courtial and L. Allen. 2004. *Physics Today* 57, 5: 35.
2. A. Muthikrishnan and C. R. Stroud. 2002. *J. Opt. B*. 4: S73.
3. M. Babiker, C. R. Bennett, D. L. Andrews and L. C. D. Romero. 2002. *Phys. Rev. Lett.* 89, 14: 143601.
4. A. Alexandrescu, D. Cojoc and E. D. Fabrizio. 2006. *Phys. Rev. Lett.* 96: 243001.
5. N. R. Heckenberg, R. McDuff, C. P. Smith and A. G. White. 1992. *Opt. Lett.* 17, 3: 221.
6. J. Arlt, K. Dholakia, L. Allen, M. J. Padgett and J. Mod. 1998. *Opt.* 45, 6: 1231.
7. K. Sueda, G. Miyaji, N. Miyanaga and N. Nakatsuka. 2004. *Opt. Exp.* 12, 15: 3548.
8. D. Akamatsu and M. Kozuma. 2003. *Phys. Rev. A*. 67: 023803.
9. G. Molina-Terriza, J. P. Torres and L. Torner. 2002. *Phys. Rev. Lett.* 88: 013601.
10. J. Leach, M. J. Padgett, S. M. Barnett, S. Franke-Arnold and J. Courtial. 2002. *Phys. Rev. Lett.* 88, 25: 257901.
11. D. G. Grier. 2003. *Nature* (London) 424: 810.
12. M. H. Anderson, J. R. Ensher, M. R. Matthews, C. E. Wieman and E. A. Cornell. 1995. *Science* 269: 198.
13. C. C. Bradley, C. A. Sackett, J. J. Tollett and R. G. Hulet. 1995. *Phys. Rev. Lett.* 75, 9: 1687.
14. M. O. Mewes, M. R. Andrews, N. J. van Druten, D. M. Kurn, D. S. Durfee and W. Ketterle. 1996. *Phys. Rev. Lett.* 77, 3: 416.
15. J. R. Abo-Shaeer, C. Raman, J. M. Vogels and W. Ketterle. 2001. *Science* 292: 476.
16. M. A. Nielsen and I. L. Chuang. 2000. *Quantum Computation and Quantum Information*. Cambridge: Cambridge University Press.
17. L. Allen, M. W. Beijersbergen, R. J. C. Spreeuw and J. P. Woerdman. 1992. *Phys. Rev. A*. 45, 11: 8185.
18. K. P. Marzlin, W. Zhang and E. M. Wright. 1997. *Phys. Rev. Lett.* 79, 24: 4728.
19. G. Nandi, R. Walser and W. P. Schleich. 2004. *Phys. Rev. A*. 69: 063606.
20. Z. Dutton and J. Ruostekoski. 2004. *Phys. Rev. Lett.* 93, 19: 193602.

21. K. T. Kapale and J. P. Dowling. 2005. *Phys. Rev. Lett.* 95: 173601.
22. S. Thanvanthri, K. T. Kapale and J. P. Dowling. 2008. *Phys. Lett. A.* 77: 053825.
23. M. F. Andersen, C. Ryu, P. Clade, V. Natrajan, A. Vaziri. K. Helmerson and W. D. Phillips. 2006. *Phys. Rev. Lett.* 97: 170406.
24. K. Helmerson, M. F. Andersen, P. C. C. Ryu, V. Natarajan, A. Vaziri, A. Hansen and W. Phillips. 2006. *Proc. of SPIE.* 6326: 632603.
25. K. C. Wright, L. S. Leslie and N. P. Bigelow. 2008. *Phys. Rev. A.* 77: 041601.
26. Y. Nakamura, Y. A. Pashkin and J. S. Tsai. 1999. *Nature* (London) 398: 786.
27. J. R. Friedman, V. Patel, W. Chen, S. K. Tolpygo and J. E. Lukens. 2000. *Nature* (London) 406: 43.
28. C. H. van der Wal, A. C. J. ter Haar, F. K. Wilhelm, R. N. Schouten, C. J. P. M. Harmans, T. P. Orlando, S. Lloyd and J. E. Mooji. 2000. *Science* 290: 773.
29. F. M. Spedalieri. 2006. *Opt. Commun.* 260(1): 340–346.
30. B. W. Shore, K. Bergmann, A. Kuhn, S. Schiemann, J. Oreg and J. H. Eberly. 1992. *Phys. Rev. A.* 45: 5297.
31. M. Cozzini, B. Jackson and S. Stringari. 2006. *Phys. Rev. A.* 73: 013603.
32. A. Görlitz, J. M. Vogels, A. E. Leanhardt, C. Raman, T. L. Gustavson, J. R. Abo-Shaeer, A. P. Chikkatur, S. Gupta, S. Inouye, T. Rosenband and W. Ketterle. 2001. *Phys. Rev. Lett.* 87: 130402.
33. P. R. Berman Eds. 1996. *Atom Interferometry*. New York: Academic Press.
34. J. P. Dowling. 1998. *Phys. Rev. A.* 57, 6: 4736.
35. T. L. Gustavson, A. Landragin and M. A. Kasevich. 2000. *Class. Quantum Grav.* 17: 2385.
36. G. Sagnac. 1913. *Comptes Rendus* 157: 708.
37. G. Sagnac. 1913. *Comptes Rendus* 157: 1410.
38. M. O. Scully and M. S. Zubairy. 1997. *Quantum Optics*. Cambridge: Cambridge University Press.
39. F. E. Zimmer and M. Fleischhauer. 2006. *Phys. Rev. A.* 74: 063609.
40. A. Kolkiran and G. S. Agarwal. 2007. *Opt. Exp.* 15, 11: 6798.
41. M. R. Andrews, C. G. Townsend, H. J. Miesner, D. S. Durfee, D. M. Kurn and W. Ketterle. 1997. *Science* 275: 637.
42. M. R. Andrews, D. M. Kurn, H. J. Miesner, D. S. Durfee, C. G. Townsend, S. Inouye and W. Ketterle. 1997. *Phys. Rev. A.* 79: 553.
43. A. S. Arnold, C. S. Garvie and E. Riis. 2006. *Phys. Rev. A.* 73: 041606R.
44. S. Gupta, K. W. Murch, K. L. Moore, T. P. Purdy and D. M. Stamper-Kurn. 2005. *Phys. Rev. Lett.* 95: 143201.
45. D. S. Durfee, Y. K. Shaham and M. A. Kasevich. 2006. *Phys. Rev. Lett.* 97: 240801.
46. W. Li, A. K. Tuchman, H. Chien and M. A. Kasevich. 2007. *Phys. Rev. Lett.* 98: 040402.

11 Laser Induced Optical Lattices in BEC with Superfluid Properties

Priyanka Verma, Aranya B. Bhattacherjee and Man Mohan

11.1 Introduction

Bose–Einstein condensates, which are ensembles of atoms in a single quantum state with long coherence time, offer the possibility of studying quantum mechanics in a completely new regime. Ultracold atoms in optical lattices exhibit formation of Bloch energy bands, tunneling effects, Bloch oscillations, Bragg scattering and Josephson effect. Many of these properties are subject to extensive experimental investigations[1]. The fusion of ultracold atoms and the optical lattices has made significant experimental progress in the past years[2–6] by loading the Bose-condensed atoms in the optical lattices and the optical lattices inside the cavities. Therefore, many solid state phenomenon can be explored in these systems as it is possible for the experimentalists to handle both external and internal degrees of freedom of this system. The Hubbard model[7–8] for bosons and fermions have been studied within the framework of ultracold atoms trapped in optical cavities. Ground-breaking experiments[9–10] have also been performed to realize the Bose–Hubbard model in these systems.

Optical potential with and without the cavity are not the same. It is the presence of ultracold atoms inside the cavity which modifies the resonance frequency of the optical cavity. This in turn modifies the amplitude of the field prevailing inside the cavity responsible for the atomic localization. This kind of periodic potential always develops a band structure in the energy spectrum of the condensate. Thus, if the atom–field coupling inside the cavity is strong, this band structure will also lead to band structure of the field inside the cavity. Then this significantly modifies the Bloch energies, effective mass, Bogoliubov excitations and the superfluid fraction of the BEC[11] inside a cavity. Motivated by such interesting developments in the field of cavity opto-mechanics, the work in this chapter emphasizes on the effects of a movable cavity mirror on superfluid properties, tunneling parameters, the Bloch energy spectrum of a BEC confined in an optical cavity.

11.2 Optical Lattice

An optical lattice is basically a set of standing wave pattern buildup when two or more counter-propagated laser beams overlap. These OLs can be created in 1D, 2D or 3D depending upon the number of laser beams used. For example, when two identical laser beams of wavelength λ_L with parallel polarization are counter-propagated such that they entirely overlap, a spatially periodic interference pattern is created which is an OL in 1D. The distance between any two maximas (minimas) in the interference pattern is $d = \lambda_L/2$ which is the spacing of the OL. Instead of taking two laser beams, a single linearly polarized laser beam is sometimes retro-reflected with a high-Q mirror to create the simplest 1D OL. In order to create the 2D OL, one more set of laser beams in a direction perpendicular to the first pair is added resulting in a 2D periodic interference pattern. Similarly, three pairs of counter-propagating laser beams along the three mutually perpendicular axis results in a 3D OL resembling the simple lattice. In fact, by changing the orientation, relative phase, or the polarization direction between the laser beams, more complex kind of OLs can be created.

Optical lattices have given a big motivation to the researchers in trapping and coherently controlling a large number of ultracold atoms. They or the other type of optical dipole traps uses the principle of ac Strak shift to trap the ultracold atoms in them. We consider an EM field in the form of a standing wave which can be created by superimposing two counter-propagating laser beams traveling aligned on the same axis. The spatially modulated electric field of these lasers are given by, $E_{-}(x,t) = E_0 e^{i(-kx-\omega t)}\hat{e}_x$ and $E_{-}(x,t) = E_0 e^{i(-kx-\omega t)}\hat{e}$, where E_0 is the maximum strength of the electric field, $\hat{e}_x$ is the unit polarization vector, $k = \omega/c$ is the magnitude of the wave vector of the laser, ω is the frequency of laser beams used. The standing wave field can be calculated as

$$\begin{aligned} E(x,t) &= E_0 e^{i(-kx-\omega t)}\hat{e}_x + E_0 e^{i(-kx-\omega t)}\hat{e}_x \\ &\cong 2E_0 \cos(kx)\cos(\omega t)\hat{e}_x. \end{aligned} \tag{11.1}$$

Considering a two-level atom interacting with the above field, the ground state energy of the atom is shifted due to ac Stark shift. It is assumed that there is coupling only between the atomic ground state and the single excited state due to lasers. Under this assumption, the ground state shift is given by

$$\delta E_g(x) = -1/2\alpha \left\langle E^2(x,t) \right\rangle_t, \tag{11.2}$$

where α is the atomic polarizability being given by $\alpha \approx -\dfrac{|\langle 2|(\vec{\mathbf{D}} \cdot \hat{e}_x)|1\rangle|^2}{\hbar(\omega-\omega_0)}$. Here $\hbar\omega_0$ represents the excitation energy of the atom from ground state $|1\rangle$ to the excited state $|2\rangle$ with $\vec{\mathbf{D}}$ being the atomic dipole moment operator. Substituting Eq. (11.1) in Eq. (11.2) the following optical lattice potential periodic in space seen by the atom can be obtained

$$V_{OL}(x) = \delta E_g(x) = -\alpha E_0^2 \sin^2\left(\frac{\pi x}{d}\right) = V_0 \sin^2\left(\frac{\pi x}{d}\right), \tag{11.3}$$

where, V_0 is the OL depth. Depending upon the laser frequency for blue detuning $\Delta > 0$, the $V_0 > 0$ and for red detuning $\Delta < 0$, the $V_0 < 0$. For a red detuned laser beam, the standing wave of light traps the atom at the anti-node, and for a blue detuning case at the nodes. In this way, a number of ultracold atoms congregate in the potential minima or maxima of OL. In case of a 3D lattice, full OL potential can be written as

$$V(x,y,z) = V_1 \cos^2\left(\frac{\pi x}{d_1}\right) + V_2 \cos^2\left(\frac{\pi y}{d_2}\right) + V_3 \cos^2\left(\frac{\pi z}{d_3}\right), \tag{11.4}$$

where lattice depth and wave vectors can be changed independently if required. Thus, in short, the above can be described action as a process in which an oscillating dipole moment is induced in the atom which at the same time interacts with light creating the trapping potential. Apart from the above described conservative way of light–atom interaction due to induced dipole moment, the neutral atoms can also interact via a dissipative way with light. This kind of interaction is the outcome of absorption of photons followed by spontaneous emission. In this process, there is transfer of momentum from the absorbed and emitted photons which results in a dissipative force on atoms which is used in laser cooling and MOT. Generally, in the experiments, detuning is large enough to neglect spontaneous emission of photons from the atoms so that only a conservative way of interaction is used to create the trapping potential for the atoms. Hence, optical lattice is basically a landscape of energy where local potential energy is comparative to the local intensity of light.

11.3 Theoretical Description of the System

11.3.1 Model

The model considered in this chapter consists of an elongated cigar shaped BEC of N two-level ^{87}Rb atoms interacting with a single standing wave cavity mode of frequency ω_c. The one-dimensional optical lattice potential formed inside the optical cavity is the result of a standing wave in the cavity. The ^{87}Rb atoms are assumed to be in the $|F = 1\rangle$ state, each with mass m. The transition frequency of an ^{87}Rb atom from $|F = 1\rangle$ to $|F' = 2\rangle$ transition of D_2 line is symbolized as ω_a. To study the opto-mechanical effects on the superfluid properties of BEC, one of the optical cavity mirror is allowed to move with a frequency Ω_m. The frequency of the coherent laser beam is represented by ω_p and its amplitude by η. The symbol ε denotes the coupling strength between the movable mirror and the optical mode. Here ε is a dimensionless quantity represented by $\varepsilon = (x_o/\Omega_m)\frac{d\omega_p}{dx}\Big|_{x=0}$, where $x|_o$ is the zero point motion of the mechanical mode. $x_o = \sqrt{\hbar/(2m_{\text{eff}}\Omega_m)}$ and m_{eff} is the effective mass of the movable mirror.

11.3.2 Single particle Hamiltonian

The system described in the above section can also be explained by the following opto-mechanical type Hamiltonian in the rotating frame of reference for a single particle using rotating wave and dipole approximation as

$$H_{OM} = H_A + H_{\text{Cav.Pht.}} + H_I + H_P + H_M + H_{\text{M-Cav.}} \tag{11.5}$$

where,

$$H_A = \frac{P^2}{2m} - \hbar\Delta_a\sigma^+\sigma^-.$$

This term is the sum of the kinetic energy of the atom and electronic transition energy of atom. $\frac{P^2}{2m}$ is the kinetic energy of the atom with P and m as the momentum and mass of the atom. The $-\hbar\Delta_a\sigma^+\sigma^-$ is the electronic atomic energy part of the atom in terms of atomic transition operators, and is the pseudo-spin operator which excites and de-excites the atom from ground state to upper and from upper to lower level of the atom. The atom–pump detuning is written as $\Delta_a = \omega_p - \omega_a$ which is kept large so that the spontaneous emission process can be neglected.

$$H_{\text{Cav.Pht.}} = -\hbar\Delta_c a^\dagger a.$$

This term represents the energy of intra-cavity photons where $\Delta_c = \omega_p - \omega_c$ is the cavity pump detuning. a, $a^\dagger$ are the annihilation and creation operators for the photo mode.

$$H_I = -i\hbar g(x)(\sigma^+ a - \sigma^- a^\dagger).$$

This term represents the interaction energy between atom and photons. It is space dependent because we assume the coupling between atom–field to be $g(x) = g_o\cos(kx)$. Here, g_o is the maximum value of atom–photon coupling.

$$H_P = -i\eta\hbar(a - a^\dagger).$$

This is the energy of the laser pump incident on the stationary mirror where η is the intensity of the pump.

$$H_M = \hbar\Omega_m a_m^\dagger a_m.$$

H_M describes the mirror energy where a_m, $a_m^\dagger$ are the annihilation operators for the mirror mode. This is the energy of a single phonon being $\hbar\Omega_m$ multiplied by the total number of phonons $a_m^\dagger a_m$.

$$H_{\text{M-Cav.}} = \hbar\varepsilon\Omega_m a^\dagger a(a_m + a_m^\dagger).$$

The term $H_{\text{M-Cav.}}$ describes the dispersive interaction between cavity field and the mirror. This term is the outcome of modification in the cavity resonance frequency which changes with the motion of the mirror.

Large atom–pump detuning allows us to adiabatically eliminate the excited states. To do this, the Heisenberg equation of motion will be used. The Heisenberg equation of motion for any operator $\hat{A}$ is

$$i\hbar\dot{\hat{A}} = [\hat{A}, H]$$

Substituting the H_{OM} in place of H in the above equation, the equation of motion for the atomic degrees of freedom σ_+ and σ_- can be evaluated and in the steady state configuration of the system, time dependence of σ_- and σ_+ should be removed. Therefore, substituting $\dot{\sigma}_- = 0$ and $\dot{\sigma}_+ = 0$ in the equations of motion of these operators, the following expressions for pseudo-spin operators can be obtained

$$\sigma_- = \frac{g(x)a}{i\Delta_a}, \sigma_+ = -\frac{g(x)a^\dagger}{i\Delta_a}. \tag{11.6}$$

Using $U_0 = g_0^2/\Delta_a$ as the height of the optical lattice potential per photon which also implies the atomic backaction on the field. Depending upon the sign of U_0, the strength of coupling between the condensate atom and cavity can be increased or decreased. If the value of U_0 is less than zero, the system is highly non-linear[12]. In this case, the atomic density and optical lattice are highly dependent on each other, which increases the coupling between condensate and cavity field. However, in the case of $U_0 > 0$, the lowest bound-state is attracted towards the nodes of light field. In this condition also, atomic density and optical lattice depend on each other but very weakly as compared to the previous case. Therefore, the non-linearity present in the system is almost negligible. In order to minimize the non-linearity in it, the assumption is $U_0 > 0$. Substituting Eq. (11.6) in Eq. (11.5), and introducing V_{cl} as an external added classical potential, an expression for the single particle Hamiltonian can be derived as

$$H_0 = \frac{p^2}{2m} + \hbar U_0 \cos^2(kx)\left(V_{cl} + a^\dagger a\right) - i\eta\hbar\left(a - a^\dagger\right) - \hbar\Delta_c a^\dagger a + \hbar\Omega_m a_m^\dagger a_m$$
$$+\hbar\varepsilon\Omega_m a^\dagger a\left(a_m + a_m^\dagger\right). \tag{11.7}$$

In the above Hamiltonian, $\hbar U_0(V_{cl} + a^\dagger a)$ represents the optical lattice potential whose period is $\lambda/2$.

11.3.3 Opto-mechanical Bose–Hubbard–Hamiltonian

The Hamiltonian in Eq. (11.7) is the single particle Hamiltonian which will be converted into a many-body Hamiltonian in this section. Considering the second quantization formulation, the many-body Hamiltonian can be written as a combination of a two-body interaction term and free evolution of the atomic variables as

$$H = \int d^3x \Psi^\dagger(\vec{r}) H_0 \Psi(\vec{r}) + \frac{1}{2}\frac{4\pi a_{sc}\hbar^2}{m}\int d^3x \Psi^\dagger(\vec{r})\Psi^\dagger(\vec{r})\Psi(\vec{r})\Psi(\vec{r}). \tag{11.8}$$

In the above Hamiltonian, the second term represents the two-body interaction energy of the atoms where a_{sc} is the s-wave scattering length. The field annihilation and creation operators are represented by $\Psi(\vec{r})$ and $\Psi^\dagger(\vec{r})$. The atomic field operators can be expanded in the Wannier function basis as, $\Psi(\vec{r}) = \sum_j b_j w(\vec{r} - \vec{r}_j)$. Here b_j represents the annihilation operator for the bosonic atom at the jth site of optical lattice. Substituting atomic operators in the wannier function basis in Eq. (11.8) and separating the onsite and offsite, the following OMBH is obtained keeping only the onsite and nearest neighbor terms as

$$H = \sum_j b_j^\dagger b_j\left(E_0 + \hbar U_0(V_{cl} + a^\dagger a)J_0\right) + \sum_j \left(b_{j+1}^\dagger b_j + b_{j+1} b_j^\dagger\right)\left(E + \hbar U_0(V_{cl} + a^\dagger a)J\right)$$
$$+\left\{-i\eta\hbar\left(a - a^\dagger\right) - \hbar\Delta_c a^\dagger a + \hbar\Omega_m a_m^\dagger a_m + \hbar\varepsilon\Omega_m a^\dagger a\left(a_m + a_m^\dagger\right)\right\} + \frac{U}{2}\sum_j b_j^\dagger b_j^\dagger b_j b_j. \tag{11.9}$$

The summation $\langle i, j\rangle$ in the above Hamiltonian represents the offsite terms only. where it has been assumed,

$$U = \frac{4\pi\hbar^2 a_s}{m}\int d^3x\left|w(\vec{r})\right|^4 \Rightarrow \text{Effective onsite interaction energy}$$

$$E_0 = \int d^3xw(\vec{r}-\vec{r}_j)\left(\frac{-\hbar^2\nabla^2}{2m}\right)w(\vec{r}-\vec{r}_j) \Rightarrow \text{Onsite kinetic energy (K.E.).}$$

$$J_0 = \int d^3xw(\vec{r}-\vec{r}_j)\left(\cos^2(kx)\right)w(\vec{r}-\vec{r}_j) \Rightarrow \text{Onsite potential energy (P.E.).}$$

$$E = \int d^3xw(\vec{r}-\vec{r}_j)\left(\frac{-\hbar^2\nabla^2}{2m}\right)w(\vec{r}-\vec{r}_{j\pm1}) \Rightarrow \text{Nearest neighbor K.E. contribution to tunneling}$$

$$J = \int d^3xw(\vec{r}-\vec{r}_j)\left(\cos^2(kx)\right)w(\vec{r}-\vec{r}_{j\pm1}) \Rightarrow \text{Nearest neighbor P.E. contribution to tunneling}$$

To study the opto-mechanical effects on the superfluid properties of the system, the onsite energies E_0, J_0 are set equal to zero.

11.4 Heisenberg–Langevin Equation of Motion

In this section, the Heisenberg–Langevin equations of motion for a, b_j and a_m operators are derived so that the dynamics of the system can be investigated. Using the commutation relations, $[a,\ a] = \left[a^\dagger,\ a^\dagger\right] = 0$ and $\left[a,\ a^\dagger\right] = 1$, $[b_i,\ b_i] = \left[b_i^\dagger,\ b_i^\dagger\right] = 0$ and $\left[b_i,\ b_j^\dagger\right] = \delta_{i,\ j}$, $[a_m,\ a_m] = \left[a_m^\dagger,\ a_m^\dagger\right] = 0$ and $\left[a_m,\ a_m^\dagger\right] = 1$, and adding the damping term with γ_1 as the damping coefficient of the cavity photon operator, γ_3 as the damping coefficient of the mirror operator, the following equation of motion can be obtained

$$\dot{a}(t) = -iU_0\left(J_0\sum_j b_j^\dagger b_j + J\sum_j (b_{j+1}^\dagger b_j + b_{j+1}b_j^\dagger)\right)a + i\Delta_c a + \eta - i\Omega_m\varepsilon(a_m + a_m^\dagger)a - \gamma_1 a. \tag{11.10}$$

$$\dot{b}_j = -\frac{i}{\hbar}Ub_j^\dagger b_j b_j - \frac{i}{\hbar}\left(E_0 + \hbar J_0 U_0(V_{cl} + a^\dagger a)\right)b_j - \frac{i}{\hbar}\left(E + \hbar J U_0(V_{cl} + a^\dagger a)\right)\left(b_{j-1} + b_{j+1}\right). \tag{11.11}$$

$$\dot{a}_m = -i\Omega_m(a_m + \varepsilon a^\dagger a) - \gamma_3 a_m. \tag{11.12}$$

11.5 Opto-Mechanical Effects on Superfluid Properties within the Mean Field Approximation

In this section, the various superfluid properties (superfluid fraction, effective mass, Bloch spectrum, elementary excitation spectrum etc.) of BEC and the opto-mechanical effects on them are explained. First the Heisenberg–Langevin equations of motion discussed in the previous section in their steady states are solved. It is assumed that $\gamma_3 < \gamma_1$ and γ_1 are the fastest time scale. This implies that the cavity decay rate is very much larger than the oscillation frequency of bound atoms in the optical lattice of the cavity. The condensate is assumed to remain coherent and robust in the duration of the measurement process. In this limit, the steady state values of the a, a_m operators can be evaluated and symbolized by a_s, it being the steady state photon annihilation operator, and $a_{m,s}$ as being the steady state phonon annihilation operator. Therefore, putting $\dot{a}(t) = 0$, $\dot{a}_m(t) = 0$ in Eq. (11.11) and Eq. (11.12), the following steady state values can be obtained

$$a_s = \frac{-\eta}{-iU_0\left(J_0\sum_j b_j^\dagger b_j + J\sum_j (b_{j+1}^\dagger b_j + b_{j+1} b_j^\dagger)\right) + i\Delta_c - i\varepsilon\Omega_m(a_{m,s} + a_{m,s}^\dagger) - \gamma_1}. \tag{11.13}$$

$$a_{m,s} = \frac{-i\varepsilon\Omega_m}{\gamma_3 + i\Omega_m} a_s^\dagger a_s. \tag{11.14}$$

It is assumed that the number of atoms in the condensate is very large so that the steady state photon operator can be replaced by a complex number α and the tight binding model can be applied. In the tight binding model, the b_j operator can be replaced with ϕ_j and the solutions evaluated in the form of Bloch waves.

$$\phi_j = u_k \exp\left(i\left(kjd - \frac{\mu t}{\hbar}\right)\right), \tag{11.15}$$

where μ is the chemical potential of the system, d being the periodicity of the lattice, n_0 being atomic number density, I being total number of lattice sites, N being total number of atoms and F, is the steady state photon number, $F = |\alpha|^2$.

Substituting Eq. (11.15) in Eq. (11.13) the following cubic equation in F can be obtained

$$A_1' F^3 + A_2' F^2 + A_3' F + A_4' = 0, \tag{11.16}$$

where the coefficients are defined as,

$$A_1' = \left(\frac{2\varepsilon\Omega_m^3}{\gamma_3^2 + \Omega_m^2}\right)^2,$$

$$A_2' = 2\left(\frac{2\varepsilon\Omega_m^3}{\gamma_3^2 + \Omega_m^2}\right)[\Delta_c - U_0(J_0 N + 2JN\cos(kd))],$$

$$A_3' = \gamma_1^2 + [\Delta_c - U_0(J_0 N + 2JN\cos(kd))],$$

$$A_4' = -\eta^2. \tag{11.17}$$

Equation (11.16) is numerically evaluated and the value of F used in further calculations.

11.5.1 Effective optical lattice potential

Opto-mechanical effects on the effective optical lattice potential is given as

$$V_{\text{op}} = \hbar U_0 \left(V_{cl} + |\alpha|^2 \right) \tag{11.18}$$

Equation (11.18) is rewritten in terms of γ_1 variable with $\hbar = 1$ above potential as

$$\frac{V_{\text{op}}}{\gamma_1} = \frac{U_0}{\gamma_1}(V_{cl} + F). \tag{11.19}$$

The cubic equation, Eq. (11.16) is numerically solved and the result substituted in Eq. (11.19) to evaluate the optical lattice potential. Effective optical lattice potential as a function of quasi momentum kd is plotted in Fig. 11.1(a). The effective optical lattice potential is maximum in case of no mirror motion. As the mirror photon coupling is increased from 0 to 0.004, the effective optical lattice barrier height decreases which results in easy tunneling of atoms across the optical lattice sites. The effective optical lattice potential is plotted as a function of cavity pump detuning in Fig. 11.1(a). The effective optical lattice potential is modified due to the atomic backaction and the motion of the mirror. As the value of the mirror photon coupling is increased, the optical lattice potential starts moving towards the negative side of cavity pump detuning. As the mirror photon coupling is increased from $\varepsilon = 0$ to $\varepsilon = 0.004$ in Fig. 11.1(b) the potential shows a tendency to move towards the optical bistability regime on negative Δ_c, but optical lattice is always single valued when the cavity pump detuning is positive. For positive Δ_c, for a fixed value of cavity pump detuning, the barrier height decreases with the increasing value of mirror photon coupling. The optical lattice potential is continuously modified as the condensate moves across the Brillouin zone due to change in atom–field interaction. The maximas in V_{op} are determined by the condition

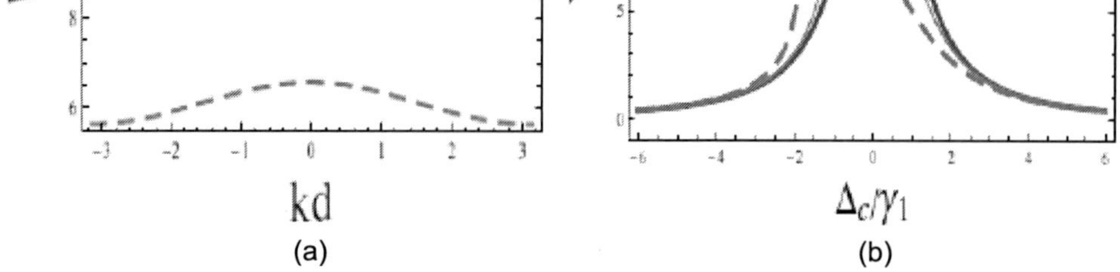

Figure 11.1 The effective optical lattice potential **(a)** as a function of the quasi-momentum kd **(b)** as a function of the detuning Δ_c/γ_1 for three different values of the $\varepsilon = 0.0$ (thick line), $\varepsilon = 0.002$ (thin line), $\varepsilon = 0.004$ (dashed line). The pump amplitude is $\eta/\gamma_1 = 40$, mechanical mirror frequency $\Omega_m/\gamma_1 = 40$, mirror damping $\gamma_3/\gamma_1 = 0.001$, number of particles $N = 100, U_0J/\gamma_1 = 0.0007, E/\gamma_1 = 2.0$ and $\Delta_c/\gamma_1 = 0.5$

$$\Delta_c - U(J_0 N + 2JN\cos(kd)) + \frac{2\varepsilon^2\Omega_m^3}{\gamma_3^2 + \Omega_m^2} = 0. \tag{11.20}$$

With increasing ε, the maxima also shifts towards negative Δ_c.

11.5.2 Bloch energy spectrum

In this section, the Bloch spectrum of the system will be evaluated. For this, the mean field solution in the form of Bloch waves is substituted in Eq. (11.11),

$$b_j\left(-i\frac{\mu}{\hbar}\right) = -\frac{i}{\hbar}Ub_j^\dagger b_j b_j - \frac{i}{\hbar}\left(E_0 + \hbar J_0 U_0(V_{cl} + |\alpha|^2)\right)b_j$$

$$-\frac{i}{\hbar}\left(E + \hbar J U_0(V_{cl} + |\alpha|^2)\right)\left(b_j e^{-ikd} + b_j e^{ikd}\right)$$

$$\Rightarrow \mu = Un_0 + (E_0 + J_0\hbar U_0[V_{cl} + F]) - 2\cos(kd)J_{\text{eff}}. \tag{11.21}$$

In the above, the tunneling parameter has been defined as $J_{\text{eff}} = -E - \hbar U_0 J(V_{cl} + F)$. The tunneling parameter depends on k and F. The effective tunneling in Fig. 11.2(a) has been plotted as a function of quasi momentum kd. The effective tunneling of atoms between the neighboring wells increases as the value of mirror photon coupling increases due to corresponding decrease in optical lattice height. This in turn increases the superfluid character of the BEC. The general expression for the energy per particle is given as

$$\varepsilon(k) = \frac{1}{n_0}\int \mu \, dn_0. \tag{11.22}$$

The Bloch energy of the system can be numerically evaluated by substituting Eq. (11.21) in Eq. (11.22). The Bloch spectrum is plotted as a function of quasi momentum kd in Fig. 11.2(b) after subtracting the ground state energy of the system. In numerically solving the ground state energy, the value of chemical potential for the ground state of translationally invariant lattice at $k = 0$ have been used

$$\mu_0 = Un_0 - K_{\text{eff}}^0 - 2J_{\text{eff}}^0 \tag{11.23}$$

where, $K_{\text{eff}}^0 = -E_0 - \hbar U_0 J_0 [1 + F_0]$ (11.24)

$$J_{\text{eff}}^0 = -E - \hbar U_0 J [1 + F_0]. \tag{11.25}$$

Here F_0 is the solution of the cubic equation, Eq. (11.16) at $k = 0$. It is evident from Fig. 11.2(b) that the Bloch energy of the system increases as the mirror photon coupling is increased because the corresponding tunneling of the atoms increases across the optical lattice sites. Mirror photon coupling plays a significant role in the superfluid properties of the system.

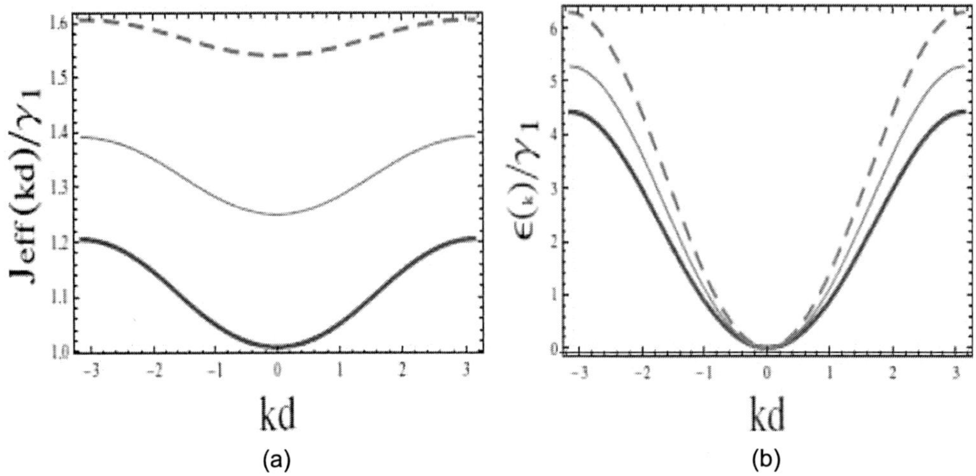

Figure 11.2 (a) Effective tunneling (b) Bloch spectrum as a function of the quasi-momentum kd for three different values: $\varepsilon = 0.0$ (thick line), $\varepsilon = 0.002$ (thin line), $\varepsilon = 0.004$ (dashed line). The pump amplitude is $\eta/\gamma_1 = 40$, mechanical mirror frequency $\Omega_m/\gamma_1 = 40$, mirror damping $\gamma_3/\gamma_1 = 0.001$, number of particles $N = 100$, $U_0 J/\gamma_1 = 0.0007$, $E/\gamma_1 = 2.0$ and $\Delta_c/\gamma_1 = 0.5$

11.5.3 Effective mass

In this section, the opto-mechanical effects on the effective mass of a condensate atom in the lowest energy band is determined. The general expression for the effective mass is given as

$$m^* = \frac{\hbar^2}{\left(\frac{\partial^2 \varepsilon(k)}{\partial k^2}\Big|_{k=0}\right)}. \tag{11.26}$$

The expression of Bloch energy as obtained in the previous section, is substituted in above the equation. Then, the above equation is numerically solved to evaluate the effective mass of the atom. The plot of the ratio of m^*/m of the effective mass to the bare mass $(m^*|_{U_0=0})$ as a function of $U_0 J/\gamma_1$ is shown in Fig. 11.3(a). As the magnitude of mirror photon coupling increases, the effective mass of the atom starts decreasing which is the outcome of the corresponding increase in tunneling of atoms. Apart from the decrease in effective mass, it is also observed that as the optical lattice vanishes in the limit $U_0 J/\gamma_1 \to 0$, the effective mass becomes equal to bare mass and the ratio $\frac{m^*}{m} \to 1$ which is expected to occur. The ratio of m^*/m is plotted as a function of cavity pump detuning Δ_c for three different values of the mirror photon coupling shown in Fig. 11.3(b). For positive value of cavity pump detuning, the effective mass increases with the increase in mirror photon coupling, but a peculiar behavior of effective mass becoming less than the bare mass is observed in the negative detuning side. The maximas in Fig. 11.3(b) corresponds to maximas in V_{op} as shown in Fig. 11.1(b). As the optical lattice height increases, the ground state wavefunction becomes more localized. Therefore in deeper lattices, the effective mass increases with a corresponding decrease in sound velocity. Thus it is clear than an opto-mechanical system can greatly affect the properties of a superfluid BEC.

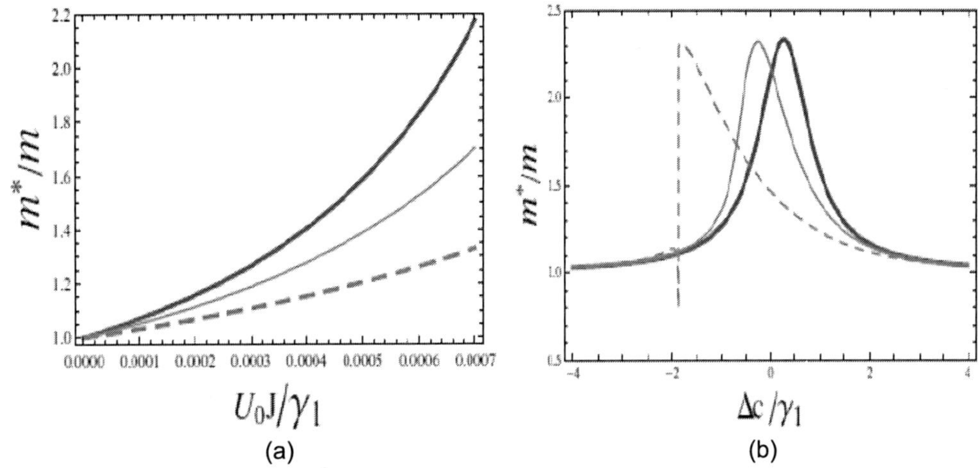

Figure 11.3 The ratio of effective mass to the bare mass as a function of (a) U_0J/γ_1. (b) Δ_c/γ_1 for three different values of $\varepsilon = 0.0$ (thick line), $\varepsilon = 0.002$ (thin line), $\varepsilon = 0.004$ (dashed line). The pump amplitude is $\eta/\gamma_1 = 40$, mechanical mirror frequency $\Omega_m/\gamma_1 = 40$, mirror damping $\gamma_3/\gamma_1 = 0.001$, number of particles $N = 100$, $U_0J/\gamma_1 = 0.0007$, $E/\gamma_1 = 2.0$, $U/\gamma_1 = 15$, $\Delta_c/\gamma_1 = 0.5$ (for right plot) and $kd = 0$

11.5.4 Spectrum of elementary excitations

In this section, the opto-mechanical effects on the spectrum of elementary excitations in the system using the Bogoliubov approach is explained. The ground state of the system corresponds to $k = 0$ and the excitation is considered respect to the ground state only. The Bogoliubov spectrum of elementary excitations can explain the energy of small perturbations with q as quasi-momentum. Assuming δ_j to be the fluctuation operator and considering the effect of quantum fluctuations, the annihilation operator is replaced as

$$b_j \to (\phi_j + \delta_j)e^{-i\mu t/\hbar}. \tag{11.27}$$

Substituting the above in Eq. (11.11), the following equation for δ_j is arrived at, where only terms linear in the fluctuations at k = 0. We get the Bogoliubov equation for the lattice

$$i\hbar\partial_t\delta_j = (2Un_0 - \mu)\delta_j + Un_0\delta_j^\dagger - J_{\text{eff}}^0(\delta_{j-1} + \delta_{j+1}) - K_{\text{eff}}^0\delta_j. \tag{11.28}$$

The Bogoliubov equation for the lattice using the quasi-particles which diagonalize the Hamiltonian is now solved. The following quasi-particles

$$\delta_j = \frac{1}{\sqrt{I}}\sum_q \left[u^q\alpha_q e^{i(qjd-\omega_q t)} - v^{q*}\alpha_q^\dagger e^{-i(qjd-\omega_q t)}\right], \tag{11.29}$$

$$\delta_j^\dagger = \frac{1}{\sqrt{I}}\sum_q \left[u^{q*}\alpha_q^\dagger e^{-i(qjd-\omega_q t)} - v^q\alpha_q e^{-i(qjd-\omega_q t)}\right], \tag{11.30}$$

where u^q and v^q are the Bogoliubov amplitudes which obey the commutation relations $[\alpha_q, \alpha_q^\dagger] = \delta_{qq'}$ Substituting these operators in Eq. (11.28) and comparing the coefficients of $e^{i(qjd-\omega_q t)}$ on both the sides of the equation,

$$\hbar\omega_q u^q = \left(2Un_0 - \mu_0 - K^0_{\text{eff}}\right)u^q + Un_0(-v^q) + Un_0(-v^q) + \left(-J^0_{\text{eff}}\right)u^q 2\cos(qd)$$
$$= u^q\left(2Un_0 - \mu_0 - K^0_{\text{eff}} - 2J^0_{\text{eff}} + 4J^0_{\text{eff}}\sin^2\left(\frac{qd}{2}\right)\right) + -v^q(-Un_0). \tag{11.31}$$

Similarly, comparing the coefficients of $e^{-i(qjd-\omega_q t)}$ on both the sides of the equation,

$$-\hbar\omega_q v^{q*} = v^{q*}\left(2Un_0 - \mu_0 - K^0_{\text{eff}} - 2J^0_{\text{eff}} + 2J^0_{\text{eff}}\sin^2\left(\frac{qd}{2}\right)\right) + u^{q*}(-Un_0). \tag{11.32}$$

Taking complex conjugate of Eq. (11.32),

$$-\hbar\omega_q v^q = v^q\left(2Un_0 - \mu_0 - K^0_{\text{eff}} - 2J^0_{\text{eff}} + 2J^0_{\text{eff}}\sin^2\left(\frac{qd}{2}\right)\right) + u^q(-Un_0). \tag{11.33}$$

The solution of Eqs (11.31) and (11.33) after imposing the condition that $(u^q)^2 - (v^q)^2 = 1$, is

$$(u^q)^2 = \frac{\hbar\omega_q + A}{2\hbar\omega_q}, (v^q)^2 = \frac{-\hbar\omega_q + A}{2\hbar\omega_q}. \tag{11.34}$$

The Bogoliubov spectrum as is

$$\hbar\omega_q = \sqrt{\chi(q)(\chi(q) + 2Un_0)}, \tag{11.35}$$

where

$$\chi(q) = 4J^0_{\text{eff}}\sin^2\left(\frac{qd}{2}\right), \tag{11.36}$$

$$A = Un_0 + 4J^0_{\text{eff}}\sin^2\left(\frac{qd}{2}\right). \tag{11.37}$$

Equation (11.35) is numerically solved and the Bogoliubov spectrum plotted as a function of quasi-momentum qd shown in Fig. 11.7. The positive cavity pump detuning is taken. As the mirror photon coupling is increased from 0 to 0.004, the Bogoliubov spectrum is enhanced which is due to a corresponding decrease in the effective mass. Moreover, the phononic linear regime is seen to increase for higher values of ε. Optical lattice potential in the cavity is dependent on the BEC wavefunction, hence if there is any variation in the wavefunction, it will also affect the optical lattice potential. However, the variations considered in this section are very small, therefore, it is assumed that there is no variation in the optical lattice potential due to it. This is true only for $\Delta_c > 0$. Therefore, the opto-mechanical effects can be used to control the Bogoliubov spectrum also.

11.5.5 Superfluid fraction

This section is devoted to the opto-mechanical effects on the superfluid fraction[14, 15] present in the system. The interacting many-body state of the BEC is the gesture of superfluidity associated with it. The expression for the superfluid fraction is given as

$$fs = \frac{1}{2N} J_{\text{eff}}^{0} \sum_{j=1}^{I} \langle \Psi_0 | b_{j+1}^{\dagger} b_j + b_j^{\dagger} b_{j+1} | \Psi_0 \rangle. \tag{11.38}$$

Using, $\langle \delta\phi \rangle = \langle \Psi_0 | \delta\phi | \Psi_0 \rangle = 0$, $\phi_{j+1}^{\dagger}\phi_j = \phi_j \phi_{j+1}^{\dagger} = \phi_j \phi_j = |\phi_j|^2$ and simplifying the right-hand side of Eq. (11.38) as

$$\sum_{j=1}^{I} \left\langle \Psi_0 \left| b_{j+1}^{\dagger} b_j + b_j^{\dagger} b_{j+1} \right| \Psi_0 \right\rangle = \sum_{j=1}^{I} \left\langle \Psi_0 \left| \left(2|\phi_j|^2 + \delta_{j+1}^{\dagger}\delta_j + \delta_j^{\dagger}\delta_{j+1} \right) \right| \Psi_0 \right\rangle$$

$$= \sum_{j=1}^{I} x1 + x2 + x3 + x4 + x5, \tag{11.39}$$

where,

$$x1 = \langle \Psi_0 | (2|\phi_j|^2) | \Psi_0 \rangle.$$

$$x2 = \langle \Psi_0 | \frac{1}{I} \sum_q 2\cos(qd) |u^q|^2 \alpha_q^{\dagger} \alpha_q | \Psi_0 \rangle.$$

$$x3 = \langle \Psi_0 | \frac{1}{I} \sum_q 2\cos(qd) |v^q|^2 \alpha_q \alpha_q^{\dagger} \Psi_0 \rangle.$$

$$x4 = \langle \Psi_0 | \frac{1}{I} \sum_q \left(-2u^q v^q \alpha_q \alpha_q \right) e^{2i(qjd - \omega_q t)} e^{iqd} | \Psi_0 \rangle = 0.$$

$$x5 = \langle \Psi_0 | \frac{1}{I} \sum_q \left(-2u^{q*} v^{q*} \alpha_q^{\dagger} \alpha_q^{\dagger} \right) e^{-2i(qjd - \omega_q t)} e^{-iqd} \Psi_0 \rangle = 0. \tag{11.40}$$

On solving Eq. (11.39) and substituting $\left\langle \alpha_q^{\dagger} \alpha_q \right\rangle \rightarrow 0$ in the limit of $T \rightarrow 0$,

$$\langle \Psi_0 | \left(2|\phi_j|^2 + \delta_{j+1}^{\dagger}\delta_j + \delta_j^{\dagger}\delta_{j+1} \right) | \Psi_0 \rangle = 2\phi_j^2 + \frac{1}{I} \sum_q 2\cos(qd) |v^q|^2. \tag{11.41}$$

Substituting Eq. (11.41) in Eq. (11.39), the superfluid fraction at zero temperature is evaluated as

$$fs = \frac{1}{2N} \left(\sum_{j=1}^{I} 2\phi_j^2 + \frac{1}{I} \sum_q 2\cos(qd) |v^q|^2 \right) = \frac{I}{N} \left(\phi^2 + \frac{1}{I} \sum_q \cos(qd) |v^q|^2 \right), \tag{11.42}$$

where, $q = \frac{2\pi}{Id} j$ with $j = 1, \ldots\ldots\ldots, (I-1)$ and $\phi_j = \phi$ for any value of j. Now, the superfluid fraction can be numerically evaluated in the system using Eq. (11.43) in the limit $d \to 0$ with q being finite. Under this limit, the normalization condition can be used as

$$I\phi^2 + \sum_q |v^q|^2 = N, \tag{11.43}$$

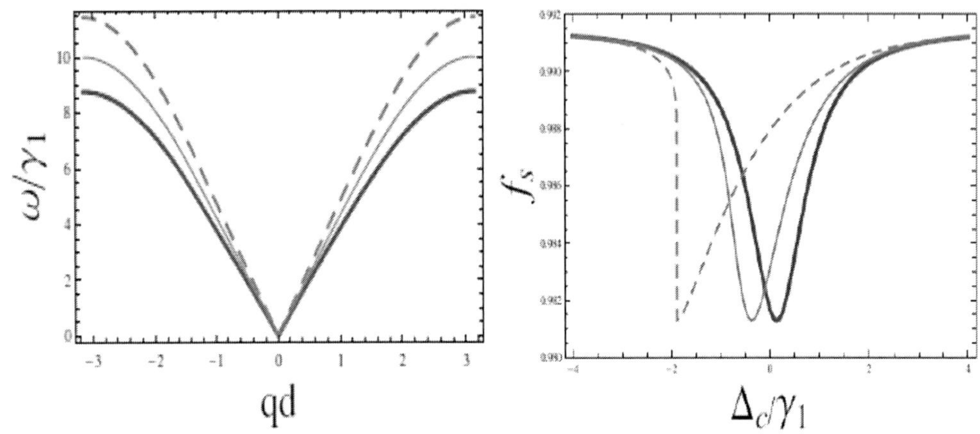

Figure 11.4 (a) The Bogoliubov spectrum as a function of the quasi-momentum qd (b) The superfluid fraction as a function of the Δ_c/γ_1 or three different values of $\varepsilon = 0.0$ (thick line), $\varepsilon = 0.002$ (thin line), $\varepsilon = 0.004$ (dashed line).The pump amplitude is $\eta/\gamma_1 = 40$, mechanical mirror frequency $\Omega_m/\gamma_1 = 40$, mirror damping $\gamma_3/\gamma_1 = 0.001$, number of particles $N = 100$, $U_0J/\gamma_1 = 0.0007$, $E/\gamma_1 = 2.0$, $U/\gamma_1 = 15$, $\Delta_c/\gamma_1 = 0.5$

which is substituted in Eq. (11.43) to numerically evaluate the f_S. Thus, the superfluid fraction is numerically evaluated as a function of Δ_c; it is shown it in Fig. 11.4(b). When the cavity pump detuning is positive, the superfluid fraction increases with the increase in mirror photon coupling. This is the outcome of the corresponding decrease in the optical lattice depth. As the value of ε is increased, the mirror shakes more rigorously which in turn shakes the optical lattice and transfers the atoms between the wells. As a result of this, there arise large atom number fluctuations in the system depending upon the value of ε accompanied by an increase in the superfluid fraction. From the experimental point of view also, when a BEC is released from an optical trap, its interference pattern can indicate about the superfluid properties.

11.6 Conclusion

In this chapter, a new tool has been developed for controlling the superfluid properties of a condensate loaded in an optical lattice inside a cavity. It has been shown that the motion of a cavity mirror gives a powerful insight into the way superfluid features of the BEC can be enhanced or diminished. Opto-mechanical effects (effects of moving one of the cavity mirror) on the Bloch energy, the effective mass, the Bogoliubov excitation spectrum, superfluid fraction of Bose–Einstein condensate confined in an optical cavity have been studied. The cavity light field develops a photonic band structure due to the strong coupling with the condensate. Since the moving mirror is coupled to the cavity photons, the photonic band structure is modified due to the mirror motion. In case of positive detuning, the mirror motion reduces the optical lattice depth and it keeps on decreasing with the increasing mirror photon coupling. A precise control over the mirror motion achieves the desired rate for the tunneling of atoms across the lattice wells by adjusting the desired optical lattice depth inside the cavity. Such a precise control might not be possible by changing the intensity of the laser beam used. The significance of reducing the lattice depth is to increase the Bloch and the Bogoliubov energies. There is also a decrease in the effective mass of the atoms accompanied by increase in the superfluid fraction. Thus the moving mirror could serve as a wonderful device in the future for investigating many properties of the BEC.

Bibliography

1. O. Morsch and M. Oberthaler. 2006. *Rev. Mod. Phys.* 78: 179.
2. B. Nagorny, Tn. Elssser and A. Hemmerich. 2003. *Phys. Rev. Lett.* 91: 153003.
3. J. A. Sauer, K. M. Fortier, M. S. Chang, C. D. Hamley and M. S. Chapman. 2004. *Phys. Rev. A.* 69: 051804(R).
4. A. ttl, S. Ritter, M. Khl and T. Esslinger. 2005. *Phys. Rev. Lett.* 95: 090404.
5. F. Brennecke, T. Donner, S. Ritter, T. Bourdel, M. Khl and T. Esslinger. 2007. *Nature* 450: 268.
6. Y. Colombe, T. Steinmetz, G. Dubois, F. Linke, D. Hunger and J. Reichel. 2007. *Nature* 450: 272.
7. M. P. A. Fisher, P. B. Weichman, G. Grinstein and D. S. Fisher. 1989. *Phys. Rev. B.* 40: 546.
8. D. Jaksch, C. Bruder, J. I. Circa, C. W. Gardiner and P. Zoller. 1998. *Phys. Rev. Letts.* 81: 31083111.
9. G. G. Batrouni and V. Rousseau. 2002. *Phy. Rev. Let.* 89: 117203.
10. G. G. Batrouni, F. F. Assaad, R. T. Scalettar, and P. J. H. Denteneer. *Phy. Rev. A.* 72: 031601 (R).
11. A. Bhattacherjee. 2008. *Optics Communications* 281: 3004.
12. D. Kleckner and D. Bouwmeester. 2006. *Nature* 444: 75.
13. J. Larson, S. Fernández-Vidal, G. Morigi and M. Lewenstein. 2008. *New J. Phys.* 10: 045002.
14. A. M. Rey, K. Burnett, R. Roth, M. Edwards, C. J. Williams and C. W. Clark. 2003. *J. Phys. B: At. Mol. Opt. Phys.* 36: 825.
15. R. Roth and K. Burnett. 2003. *Phys. Rev. A.* 67: 031602.

IV
Quantum Optics

12 Degenerate Four-Wave Mixing in DASPB Dye-Doped Polymer Film

S. Aithal, P. S. Aithal and N. G. Bhat

12.1 Introduction

Non-linear optical phase conjugation by degenerate four-wave mixing (DFWM) is an important technique with applications in many fields of science and technology including image transmission, optical image processing, optical filtering, and laser resonators[14, 29, 8]. When two counter-propagating and intense light beams interact with a non-linear medium, together with a less intense third one, a fourth beam is generated from the medium, which will be the phase conjugation of the third beam. This technique is called four-wave mixing. The unique feature of a pair of phase conjugate beams is that the aberration influence imposed on the forward (signal) beam passed through an inhomogeneous or disturbing medium can be automatically removed from the backward (phase conjugated) beam passed through the same disturbing medium[11]. The main applications of degenerate DFWM techniques are non-linear spectroscopy, real time holography, and phase conjugation. Phase conjugation by DFWM has been demonstrated in many organic and inorganic materials using pulsed or continuous-wave (CW) lasers[27, 12].

The phase conjugate light has a variety of characteristics and properties not associated with normal light and they can be enumerated as follows:

(1) *Phase compenzation effect:* Waves that pass through an aberrating medium are injected into a phase conjugate mirror. If the outgoing wave is re-propagated through the medium, the waves phase distortion is compensated.

(2) *Space domain multiplicative interaction:* Since phase conjugate waves are produced by interaction in a non-linear medium, the spatial multiplicative effect among the light waves appears in the phase conjugate waves.

(3) *Intensity dependant phase shift:* Due to the optical Kerr effect in a non-linear medium, the phase shift of the electric field depends on the wave intensity.

(4) *Detuning dependence:* The reflectance of a phase conjugate mirror strongly depends on the detuning between the pump and probe wave frequencies.

(5) *Time inversion:* A phase conjugate wave can be regarded as a time inverted wave since its direction of propagation is exactly opposite that of the probe wave and the wave front is identical to that of the probe wave.

(6) *Time domain multiplicative interaction:* When pulsed waves are used in the production of phase conjugate light, the resulting polarization, that is, the generalized phase conjugate wave intensity depends on the product of the interacting waves in the waves' temporal domains.

(7) *Quantum correlation:* When waves are treated quantum theoretically, the probe and phase conjugate waves correspond the photon annihilation and creation operators respectively.

Organic molecules exhibit large polarizabilities because exited π-bond electrons are delocalized and hence easily polarizable[19]. Saturable absorption plays a very important role when dyes are used for the production of phase conjugation light, because χ^3 is inversely proportional to the saturation intensity[16]. These systems exhibit large third-order susceptibilities. OPC has been reported in glasses and other solid matrices doped by organic dyes emerging as promising materials for OPC because of their large third-order non-linear susceptibilities $\chi^{(3)}$. In these materials, the phase conjugate wave can be generated at low light intensities provided by the continuous wave lasers. Moreover, these materials can be easily prepared in the laboratories.

Figure 12.1 shows the energy level diagram of dyes. These materials can be explained by a three-level system[24]. When light is injected, electrons excite from the S_0 level to the S_1 level. The dye molecules will switch over to the triplet state. The number of electrons at the S_0 level is reduced only by the number of electrons occupying the T_1 level. Therefore, the number of excitable electrons at the S_0 level is dramatically reduced if the intensity of the injected waves is high at a frequency corresponding to the S_0 and S_1 levels energy; the rate of absorption is thus reduced[8]. Thus, absorption becomes a function of intensity. The gratings produced by interference between the probe and pump wave are such that the period can be determined using the following relation :

$$\Lambda = \frac{2n}{K_G} = \frac{\lambda}{2n\sin\theta/2}, \tag{12.1}$$

where λ is the laser wavelength. At a small angle θ, the spacing is wider for a transmission grating than for a reflection grating[6, 5]. The theory of DFWM was formulated for a two-level saturable absorber by Abraham and Lind[1] and was modified to describe a three-level absorber (Fig. 12.1) by Silberberg and Bar-Joseph[25]. The phase conjugated reflectivity is then given by

$$R = \frac{|k|^2}{[a + w\cot(wL)]^2}, \tag{12.2}$$

where $w = (|k|^2 - a^2)^{\frac{1}{2}}$ and a is the saturated absorption co-efficient given by

$$a = a_0 \frac{1 + I_1 + I_2}{[1 + 2(I_1 + I_2) + (I_1 - I_2)^2]^{3/2}} \tag{12.3}$$

and the coupling coefficient k is given by

$$k^* = 2a_0 i \frac{(I_1.I_2)}{[1 + 2(I_1 + I_2) + (I_1 - I_2)^2]^{3/2}}. \tag{12.4}$$

I_1 and I_2 are two pump beam intensities normalized by the saturation intensity I_s[25].

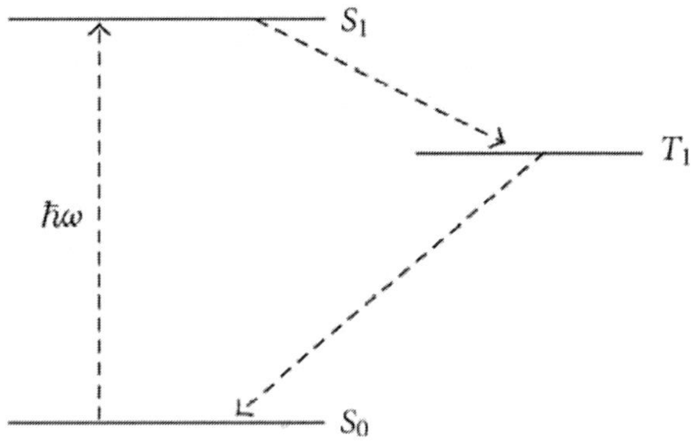

Figure 12.1 Energy level diagram of a three-level model for organic dyes

12.2 Non-linear Processes

Non-linear processes like non-linear refraction, thermal grating, saturation and reverse saturation absorption, two photon induced fluorescence, photorefraction, and stimulated Brillouin scattering lead to the formation of a laser-induced grating in the medium are associated with the generation of phase conjugated wave.

12.2.1 Non-linear refraction

In non-linear refraction, the refractive index of the material varies with light intensity. Materials with positive non-linear refractive index shows self-focusing and materials with negative non-linear refractive index gives rise to self-defocusing. For a purely third order response with input irradiance I_0,

$$\Delta n = n_2 I_0, \tag{12.5}$$

where n_2 is non-linear refractive index and is positive for self-focusing materials and is negative for self-defocusing materials. The non-linear optical properties of materials range from the index change due to the ultrafast interaction of light with bound electrons to the index change caused by the relatively slow thermal expansion of a liquid due to linear absorption. The effects caused by these non-linear interactions with matter range from the reduction of transmittance from increasing absorption with increasing irradiance (e.g., two-photon absorption) to beam spreading from self-defocusing to the ultimate non-linear interaction of laser-induced damage.

12.2.2 Thermal gratings

Thermal effect of illuminated light beam may create large non-linear optical effects and hence the temperature of the illuminated portion of the material increases; the refractive index invariably decreases with increasing temperature[4]. This intensity-dependent change of the refractive index in liquids or solids may result in very significant self-diffraction effects for phase conjugate beam generation. From early works on these mechanisms, it is well established from heat equations that the steady-state index modulation varies as the square of the grating period and is inversely proportional to the thermal conductivity of the material. Therefore, for small grating periods, thermal diffusion tends to reduce both the photoinduced index modulation amplitude and the diffraction efficiency of thermal holograms. For that reason, wave-mixing experiments based on thermal non-linearities will benefit from using a small pump-probe beam angle or a longer wavelength up to 10 μm.

Under steady-state conditions, the heat transport equation reduces to

$$\frac{\Delta(nL)}{nL} = \frac{1}{n}\frac{\partial n}{\partial T}\Delta T + \frac{1}{L}\frac{\partial L}{\partial L}\Delta T, \tag{12.6}$$

where n is refractive index and L is sample thickness[18, 5]. To perform efficient wave mixing, phase conjugation materials that display high thermal coefficient must be used, and this condition can be achieved in semiconductors like HgCdTe where $\partial n/\partial T$ is 10^{-3} at $\lambda = 10.6$ μm, or in liquid crystals where $\partial n/\partial T = 10^{-3}$ at room temperature and $\partial n/\partial T = 10^{-2}$ near the transition temperature[15].

12.2.3 Saturation and reverse saturation absorption

Saturable absorption was recognized early as an efficient mechanism for efficient phase conjugation. It is based on the dependence of the absorption and refractive index of two-level or three-level systems on the incident average intensity due to pump and probe beams. It is expected that transitions with a large cross-section and a long relaxation time are more easily saturable. The basic absorption processes in dyes can be divided into linear and non-linear absorption. Non-linear absorption is a phenomenon defined as a non-linear change (increase or decrease) in absorption with increasing intensity. This can be of either two types: saturable absorption (SA) and reverse saturable absorption (RSA). Depending on the pump intensity and on the absorption cross-section at the excitation wavelength, most molecules show non-linear absorption. With increasing intensity, if the excited states show saturation owing to their long life times, the transmission will show SA characteristics. If, however, the excited state has strong absorption compared with that of the ground state, the transmission will show RSA characteristics. These absorption characteristics are highly dependent on the wavelength, intensity and excited state lifetime. With the availability of intense laser sources, it is essential to have perfect knowledge of the absorption characteristics of the medium. SA is vital for use of dyes in mode locking. The most important application of RSA is in an optical limiting device[25] that protects sensitive optical components, including the human eye, from laser-induced damage. RSA is observed as a result of excited state absorption (ESA), TPA, or both.

Non-linear absorption parameters were therefore calculated for an effective three-level system with S_0, S_1, and S_n states. For the saturation of a homogeneously broadened line, the dependence of measured absorption coefficient α on intensity I of the incident laser radiation is given by the expression[23]

$$\alpha = \frac{\alpha_0}{1} + \frac{I}{I_s}, \qquad (12.7)$$

where α_0 is the low-intensity absorption coefficient and I_S is the saturation intensity.

A strong, narrow band laser field saturates one segment of the inhomogeneous absorption profile (e.g., Doppler broadening), leading to decreased absorption. Malcuit et al.[17] proposed the following equation for this kind of inhomogeneous saturation:

$$\alpha = \frac{\alpha_0}{1 + (I/I_s)^{1/2}}. \qquad (12.8)$$

Samoc et al.[22] observed SA in poly(indenofluorene), and they found that the experimental curves were not well reproduced by Eqs (12.7) and (12.8), proposed the following *ad hoc* formula, and obtained a satisfactory fit:

$$\alpha = \frac{\alpha_0}{1 + (I/I_s)^{1/2}}. \qquad (12.9)$$

12.2.4 Two photon induced florescence

According to the basic theoretical consideration, the TPA induced decrease of transmissivity can be expressed as

$$I(L) = \frac{I_0}{(1 + I_0 L\beta)}, \qquad (12.10)$$

where $I(L)$ is the transmitted beam intensity, I_0, is the incident beam intensity, L is the thickness of the sample, and β is the TPA coefficient of the sample medium. In the derivation of Eq. (12.10), it is assumed that the linear attenuation of the medium can be neglected and the beam has a nearly uniform transverse intensity distribution within the medium. Using Eq. (12.10), the value of β can be determined by measuring the transmitted intensity versus the incident intensity for a sample medium with a given L value. Furthermore, the TPA coefficient (in units of cm/GW) of a given sample is determined by

$$\beta = \sigma_2 N_o = \sigma_2 N_A d_o \times 10^{-3} \qquad (12.11)$$

Here, N_o is the molecular density of the dopant (in units of 1 cm^3), σ_2 is the molecular TPA coefficient (or cross-section) of the same dopant (in units of cm^4/GW), d_o is the concentration of the dopant compound in the matrix (in units of M), and finally N_A is Avogadro's number. For known values of β and d_o, the value of σ_2 can be calculated from Eq. (12.11).

Non-linear absorption coefficient is also calculated for an effective three-level system with S_0, S_1, and S_n states by using the β value determined from the theoretical fits to calculate the TPA cross-section (σ_{TPA}) from the following relation[13]:

$$\beta = \frac{N_0}{h_\vartheta}\sigma_{TPA}, \qquad (12.12)$$

where h is Planck's constant, ϑ is the frequency of light, and N_0 is the number of molecules per unit volume.

12.2.5 Photorefraction

Photorefraction is a particular type of non-linearity which arises in materials that exhibit linear (or quadratic) electro-optic (EO) effects. The illumination of the material by a two-beam interference pattern generates a photoinduced charge distribution. The photogenerated carriers (electrons or holes) are trapped, and thus it results in a space charge field in the volume of the material which modulates the refractive index through the electro-optics coefficient. Microscopic phenomena for space charge buildup involves electrons charge diffusion or drift under an external applied electric field. There are several specific characteristics of the photorefractive effect which differ from other known non-linear mechanisms. First, there is no threshold effect and the material responds to the incident energy. In such conditions, the material response time can be easily controlled from a fraction of a microsecond to several seconds; second, the amplitude of the photoinduced index modulation is mainly determined by the value of the electro-optic coefficient and by the trapping center density. Moreover, photorefractive materials have a dark storage time constant, equivalent to memory effect[26, 7]. Photorefractive materials have already demonstrated their great importance in experiments based on the recording and erasure of holograms for optical information processing, high-gain wave mixing, and phase conjugation with low-power visible or near-infrared laser beams. Most of the experiments were performed with different types of EO materials such as $LiNbO_3$, $BaTiO_3$, and Bi_{12} (Si, Ge, Ti) O_{20}; semiconductors such as GaAs, InP, and CdTe; PLZT ceramics; and, more recently, doped EO polymers or liquid crystals[15].

12.2.6 Stimulated Brillouin scattering

The stimulated Brillouin scattering (SBS) effect originates from the electrostrictive effect in the transparent dielectric media such as liquids, gases or solids. The interference patterns due to the incident and spontaneous scattered optical fields generate a traveling acoustic wave that modulates the material refractive index through the elasto-optic effect. It thus results from the interaction, the equivalent of a coherent moving phase grating that propagates at sound velocity and that diffracts in the backward direction, a phase conjugate replica of the incident high intensity laser beam. SBS is a phenomenon that exhibits a threshold, and its typical response time ranges in the nanosecond due to the phonon lifetime. Moreover, because of the moving grating, the retro-reflected Stokes wave is frequency-shifted by the Doppler effect corresponding to the sound velocity by several gigahertz. SBS is a very well suited interaction for self-pumped phase conjugation and most efficient materials are high-pressure gases (SF_6, N_2, Xe, etc.), liquids (CS_2, CCl_4, $GeCl_4$, $SiCl_4$, $TiCl_4$, freon, acetone, etc.), or solids (silica fibers or bulk, quartz, organic crystals such as LAP, DLAP, etc.)[15].

In this chapter, PC wave generation in dye 4-[4-(Dimethylamino)styryl]-1- docosyl pyridinium bromide (DASPB) doped in PMMA-MA polymer matrix using low-power continuous wave laser excitation is presented. PC signal strength at different times for different dye doped concentration, PC reflectivity a function of angle between the probe beam and forward pump beam, dependence of PC reflectivity on the backward pump power and forward pump power, and transmittance as a function of time is considered.

12.3 Experimental Configuration

The ray diagram of the phase conjugate wave generation by degenerate four-wave mixing is shown in Fig. 12.2. The probe beam E_3 and the forward-pump beam E_1 interfere in the non-linear material and create a periodic interference pattern that modulates the physical properties of the non-linear medium. The resulting grating wave vector amplitude is $k = 2\pi/\Lambda$. The fringe period (Λ) can be determined by the well-known formula $\Lambda = \lambda/2\sin(\theta/2)$, where λ is the laser wavelength, and $\pm\theta$ is the forward-pump and probe beam incident angles with respect to the normal to the non-linear medium. The backward-pump beam E_2 is then diffracted under Bragg conditions by the dynamic volume hologram and generates a backward conjugate wavefront whose complex amplitude can be written as E_4.

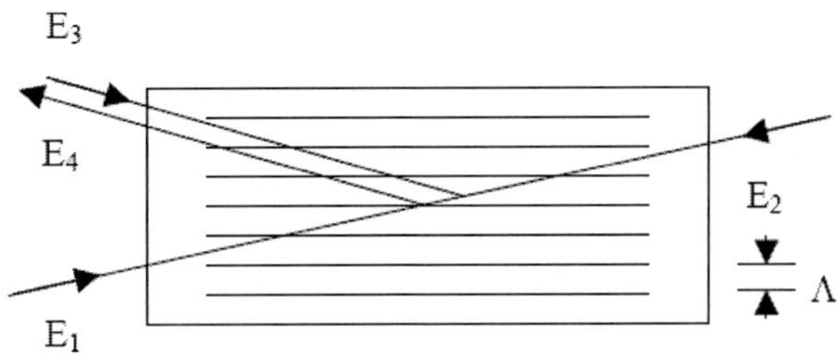

Figure 12.2 Ray diagram of phase conjugate wave generation by degenerate four-wave mixing

The schematic diagram of the phase conjugation experiment is shown in Fig. 12.3. A He-Ne laser (20 mW) beam at 633 nm was divided into three beams, two counter-propagating pump beams E_1 and E_2, namely forward-pump and backward-pump beams respectively and a probe beam E_3 form the DFWM configuration. The spot size of each of these three unfocused beams at the non-linear medium was 1.20 mm in diameter. The constant power ratio of the probe beam (E_3), forward-pump beam (E_1) and backward-pump beam (E_2) used in this work was $\approx 1:10:10$. The angle between the probe beam and the forward-pump beam was 8°. The sample was exposed simultaneously to all these three beams. The optical path lengths of all the three beams were made equal, so that they were coherent at the sample. The phase conjugate wave retraces the path in the opposite direction to that of the probe beam E_3 and was detected with the help of a photodetector and processed by a power meter and data recorder system. The experimental setup was mounted on a vibration isolation table to avoid the destruction of the laser-induced gratings formed in the DASPB dye-doped polymer matrix due to mechanical disturbances.

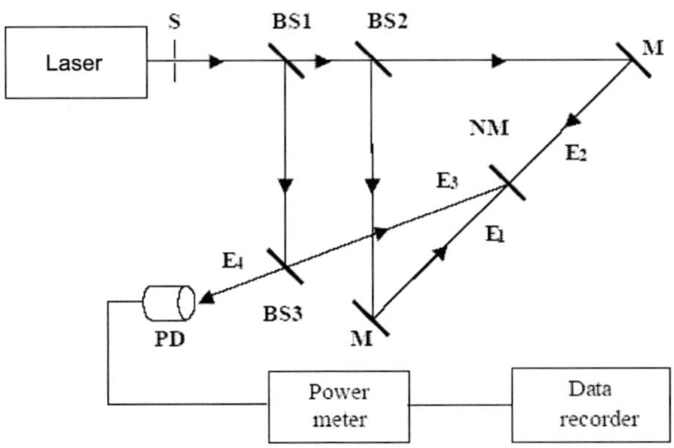

Figure 12.3 Experimental setup for observation of PC wave. S: Shutter; BS1, BS3: Beam splitters; M: Mirror; NM: Non-linear medium; PD: Photo-detector

12.4 Material

One design strategy proposed recently by Albota et al,[3] dealt with molecules based on benzene ring as π-center which is attached symmetrically by either electron–donor (D) or electron–acceptor (A) through various lengths of conjugated connectors; $D-\pi-D$ or $A-\pi-A$. They concluded that σ is increased by increasing the length of conjugation; change with the D/A strength and the extent of symmetric intramolecular charge-transfer (CT) from the D ends to the π-center or vice versa, meaning that symmetric charge redistribution effectively occurs upon excitation of such symmetric molecules.

A similar approach was made in designing molecules by Reinhardt and his coworkers[20] dealing with benzene ring as the π-center which is symmetrically coupled with two electron acceptors ($A-\pi-A$) or asymmetrically with D and A ($D-\pi-A$), respectively. There is no clear effect of structural symmetry on σ values, although increasing conjugation length of π-centers brings about a significant improvement of the value. This seems to suggest that there must be more crucial molecular factors other than structural symmetry involved. In this study, dye molecule as 4-[4-(Dimethylamino)styryl]-1- docosyl pyridinium bromide (DASPB) with π center is used.

12.5 Preparation of Dye-doped Film

Commercially available DASPB (Aldrich Chemical Co.) is purified by recrystallization twice with spectrograde ethanol and by vacuum sublimation. The purity is determined spectroscopically. Purified chloroform is used as the solvent. To prepare the film, polymethyl methacrylate – metacrylic acid was used as polymer matrix. The thin films of DASPB doped in PMMA-MA is prepared using hot press technique. Thin films of variable thickness are obtained between two glass slides. Film thickness of 0.3 mm was prepared with two different dopant concentrations of $d_o = 0.001$ M and 0.002 M respectively.

12.6 Results and Discussion

12.6.1 Linear optical properties of DASPB

The molecular structure of DASPB is shown in Fig. 12.4. A charge-transfer in between the aromatic moiety (electron donor) and bromine unit (electron acceptor) can be proposed to explain the large χ^3 value measured using Z-scan technique[2]. The linear absorption spectrum of DASPB in chloroform is measured on a VARIAN Cary UV-vis-IR recording spectrophotometer using quartz cuvette with one cm path length as well as doping it in polymethyl methacrylate methacrylic acid (PMMA-MA) film. Figure 12.5 shows the linear absorption spectrum of a DASPB in PMMA-MA with solute concentration of $d_0 = 0.001$ M. The spectral curve has shown that there is a strong absorption band with peak absorption located at 478 nm with a bandwidth of 100 nm, a medium absorption peaked at 270 nm with a bandwidth of 80 nm and no linear absorption is observed in the entire spectral range of 580 to 2000 nm except IR absorption between 1200 nm to 1600 nm. Styryl dyes are particularly preferred because they have a strong two-photon absorption with a cross-section that is significantly greater than commercial dyes, such as Rhodamine, stilbene and coumarin dyes. The dyes also exhibit intense emission at a wavelength ranging from about 300 to about 680 nm when excited by laser radiation having a wavelength of from about 660 nm to about 1300 nm. The observation of very strong frequency upconverted fluorescence emission and the subsequent quantitative measurement of the intensity dependence on the near infrared laser excitation proved it to be a two-photon absorption process.

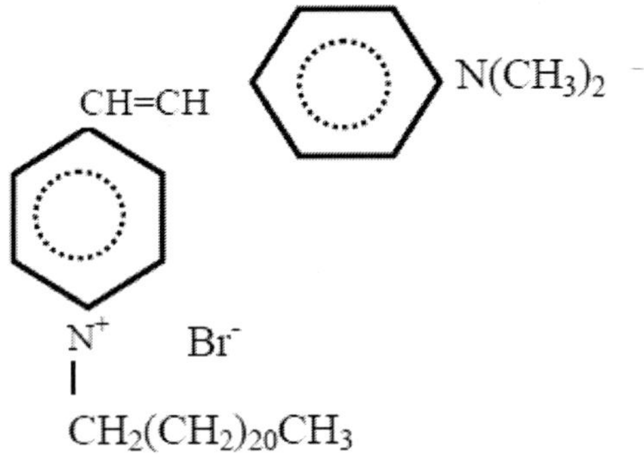

Figure 12.4 Molecular structure of DASPB

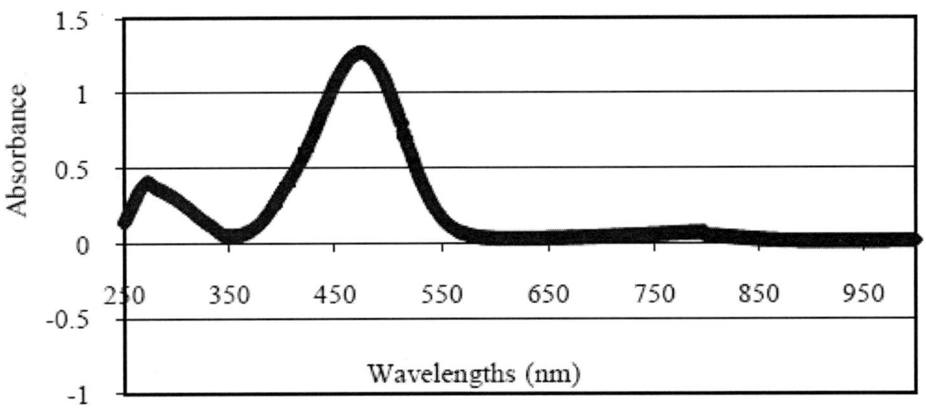

Figure 12.5 Linear absorption spectrum of DASPB in PMMA-MA matrix

12.6.2 Single and two photon florescence property of DASPB

The single photon fluorescence spectrum of the sample is measured for a 1 cm path DASPB in chloroform with the solute concentration of 0.001 mol/L using a spectral fluophotometer (Rf 50000U from Schmadza) with the spectral resolution of 1 nm. The peak wavelength of the single-photon induced fluorescence was 610 nm with a bandwidth of 60 nm (Fig. 12.6). Figure 12.7 corresponds to single photon fluorescence when DASPB is excited at 532 nm using second harmonic of an Nd:YAG laser beam.

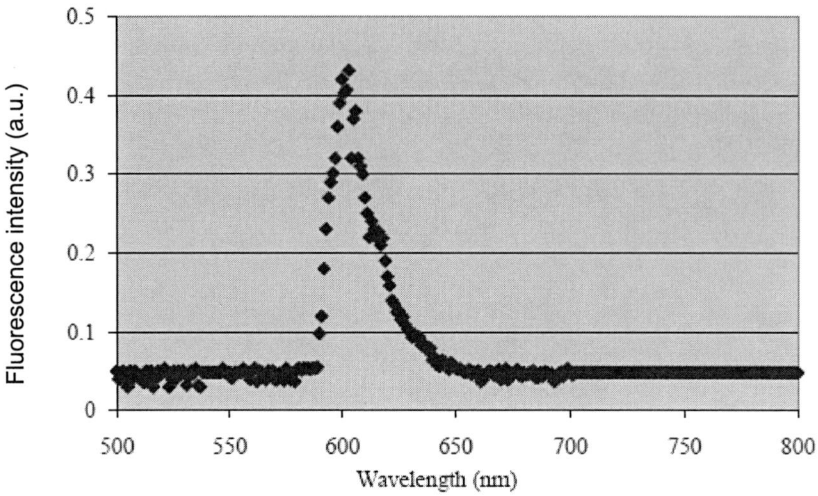

Figure 12.6 Ground state fluorescence spectrum

From the absorption spectrum of DASPB, we can see that there is no linear absorption in the entire spectral range from 580 nm to 1800 nm, except the fact that IR radiation between 1,200 nm to 1,600 nm is strongly absorbed by DSAPB solution. It has been observed that this dye shows quite strong frequency upconverted fluorescence when exposed to near IR and IR laser beam above 700 nm. This suggests that a very strong TPA process may occur inside the sample. The TPA induced emission spectrum of 0.001 M DASPB doped in PMMA-MA polymer matrix excited with 1064 nm laser beam is shown in Fig. 12.8. In the measurement of the upconversion efficiencies, VIS cutting filters were used to cut transmitted pump energy. Comparing Fig. 12.8 with Fig. 12.7, we can see that the TPA induced emission spectrum of DASPB with much higher concentration has a red-shift as compared to that in the much lower concentration single photon absorption study. This can be explained by re-absorption of dye material[28].

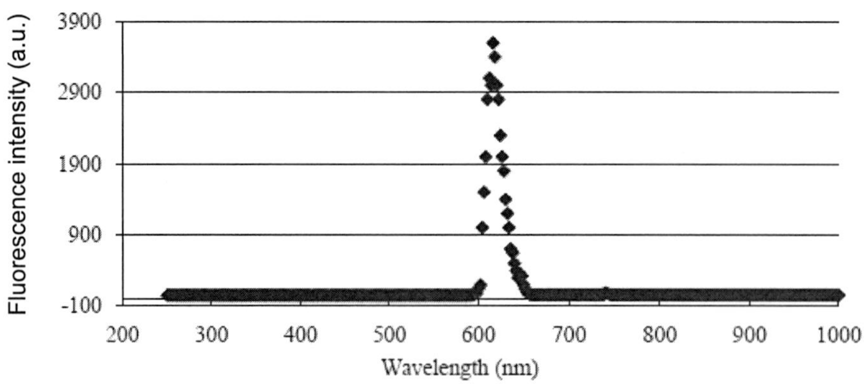

Figure 12.7 One-photon induced fluorescence at 532 nm irradiation

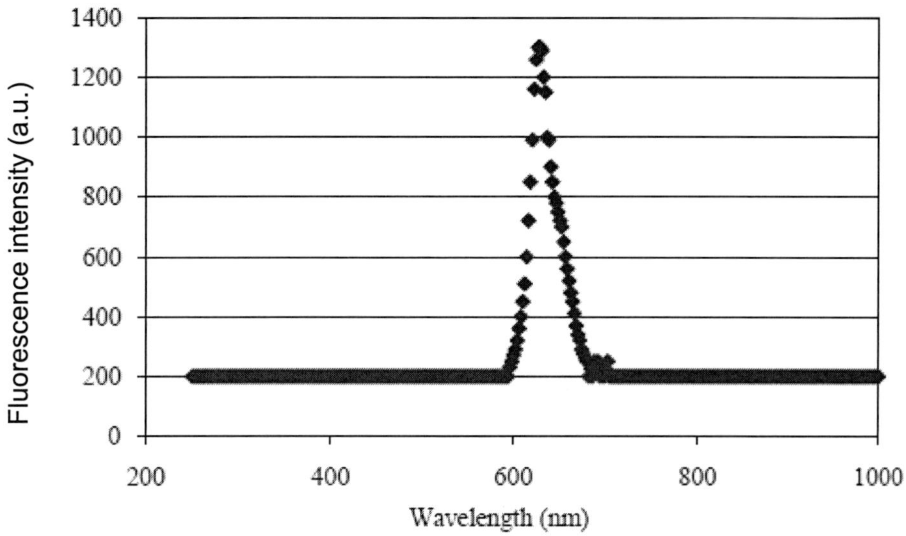

Figure 12.8 Two-photon induced emission spectrum of DASPB at 1.06 μm irradiation

12.6.3 Degenerate four-wave mixing property of DASPB

We consider a three-level system as an energy level diagram of the DASPB dye in PMMA–PA polymer matrix. In this system, S_0 and S_1 are the ground and excited singlet states, respectively, and level T_1 is the triplet state. Absorption of a photon by DASPB results in transition of the DASPB to the first excited singlet state ($S_0 \rightarrow S_1$). If the singlet-to-triplet cross-over is considerable, the dye molecules will switch over to the triplet state ($S_1 \rightarrow T_1$), where they will remain for a relatively longer time as the triplet-to-singlet transition is inhibited, and, consequently, the molecules will not be available for further absorption from the ground state. This will result in a saturation of absorption if the triplet lifetime is long enough. Thus, in the medium, the absorption becomes a function of intensity. Therefore, when the two write beams interfere, the intensity pattern modulates the complex refractive index, which results in the formation of a grating[21]. The probe beam E_3 and the forward-pump beam E_1 interfere in the non-linear material and create a periodic interference pattern that modulates the physical properties of the non-linear medium. The resulting grating wave vector amplitude is $k = 2\pi/\Lambda$. The resultant complex volume hologram is due to a spatial distribution of the refractive index (Kerr-like media), or of the absorption (saturable absorbers or two photon absorption), or of the gain when the conjugator is the laser medium itself. The second anti-parallel pump (backward-pump) beam E_2 is then diffracted under Bragg conditions by the dynamic volume hologram; following the classical formalism of holography, it generates a backward conjugate wavefront whose complex amplitude can be written as $E_4 = E_3^* E_1 E_2$. It is named as the conjugate image beam in holography. Since this geometry involves waves $E_1 - E_4$, which are simultaneously present in the medium, it is known as four-wave mixing. The phase conjugate reflectivity (R_{pc}) is defined as the ratio of the amplitude of the PC wave to the probe beam amplitude [Geethakrishnan T. and Palanisamy P. K., (2005b)].

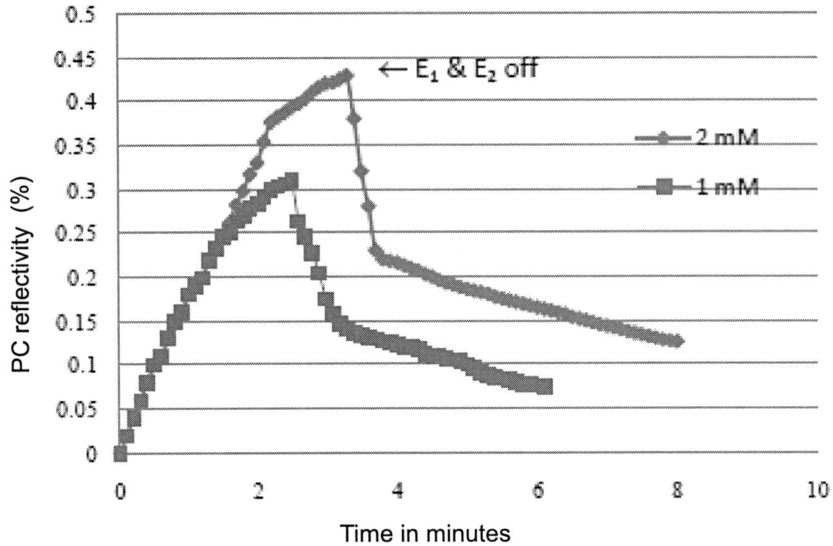

Figure 12.9 PC signal versus recording time for 0.001bM and 0.002 M concentrations

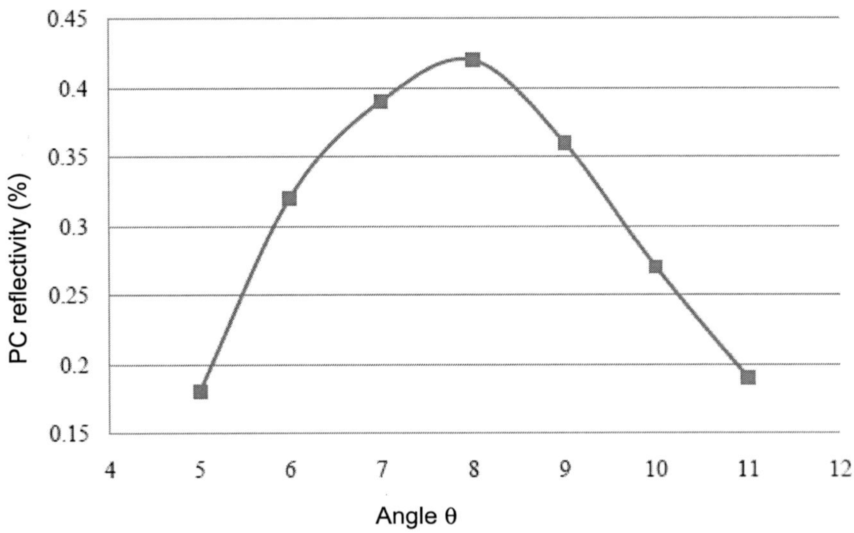

Figure 12.10 PC reflectivity as a function of angle between the probe and the forward pump beams

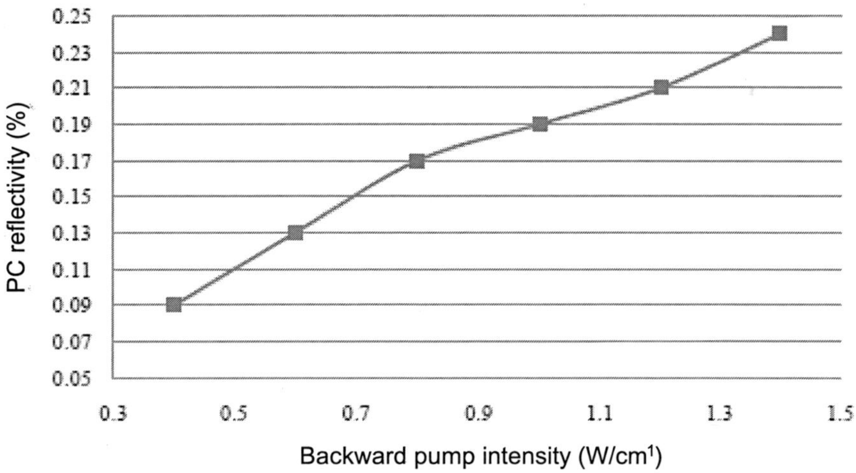

Figure 12.11 Dependence of PC reflectivity on backward pump intensity

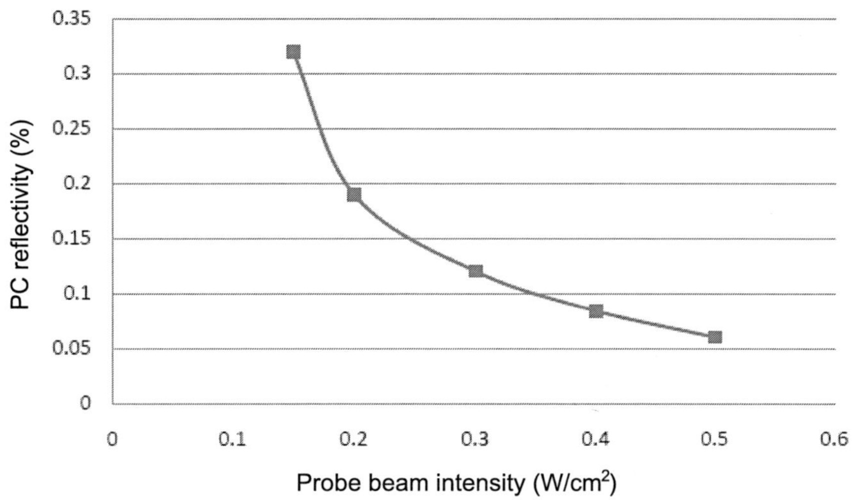

Figure 12.12 Conjugate reflectivity as a function of probe beam intensity

Organic dyes doped in polymer matrix have the capability of generating a phase conjugate wave by not only DFWM but also holographic process[9]. To distinguish the phase conjugate wave generated by DFWM from that by the holographic process, the transient behavior of the PC signal was studied.

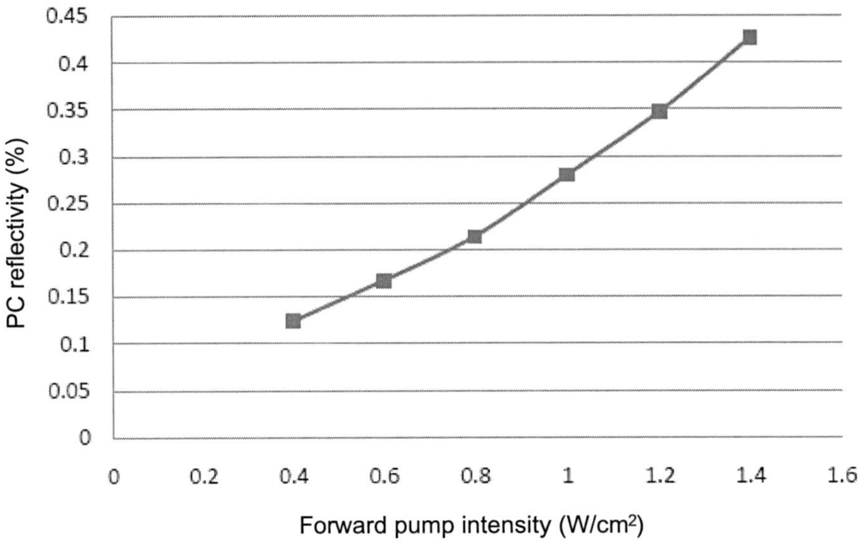

Figure 12.13 Dependence of PC reflectivity on forward pump power

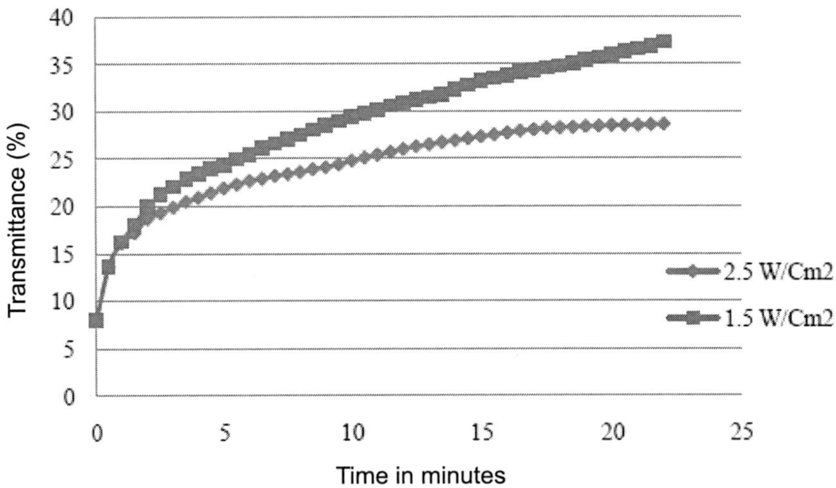

Figure 12.14 Transmittance as a function of time

For this, the DASPB dye doped in PMMA–PA polymer matrix was first illuminated with three waves E_1, E_2 and E_3 for a specified duration, and afterwards, E_1 and E_3 were successively turned off, so that only E_2 was incident on the dye film. Here, the duration for which all the three waves are incident on the dye film is called as the DFWM duration. Figure 12.9 shows the measured phase conjugate signal as a function of time. The initial rise to a peak within a few minutes is due to DFWM and holographic processes; the sudden drop in the intensity of the PC signal after shutting off both the write beams E_1 and E_3 indicates the contribution from the fast DFWM process. Due to the holographic process, the PC signal is present even after E_1 and E_3 are shut off, and it decays rather slowly. If the phase conjugate wave was generated only by DFWM, the lack of only one of the three beams E_1, E_2 and E_3 would have stopped generation of the phase-conjugate wave. Therefore, it is inferred that the rapidly decaying component corresponds to the phase conjugate wave which is generated by the DFWM. On the other hand, if spatially modulated information formed by E_1 and E_3 can be recorded in the DASPB dye in the PMMA–PA polymer film, the phase conjugate wave can still be generated when E_2 tries to read this stored information, during the lifetime of the holographic grating.

The PC signal measurements are taken by varying the parameters which influence the PC signal during the DFWM process. Figure 12.9 shows the PC signal versus the time for dye concentration of 0.001 M and 0.002 M doped polymer films. PC intensity rises linearly to a maximum and then starts decreasing. The phase grating formed is transient. To get maximum reflectivities, it is necessary that there be a perfect overlap of the probe and the pump beams in the non-linear medium. Figure 12.10 shows the PC reflectivity as a function of recording angle between the forward pump and probe beam. It seems from the figure that, as the angle between the probe beam and the forward pump beam increases, the PC reflectivity first increases and then decreases. This may be because as the angle increases, the probe beam becomes elliptical and only a fraction of its area falls within the interaction region. Because of two-wave coupling, the maximum PC reflectivity is achieved when the angle is 8 degrees. The effect of the backward pump beam power on the PC reflectivity by keeping the power of the forward pump and probe beams constant and varying the backward pump beam is shown in Fig. 12.11.

Figure 12.12 shows the influence of the input probe beam intensity on the conjugate beam reflectivity. A maximum reflectivity value of 0.42% is observed for probe beam intensity at 1.5 W/cm^2, and further increase in probe beam intensity resulted in decreasing in PC reflectivity. Similar observations have been reported in other kinds of material doped with organic dyes[19]. Figure 12.13 shows the variation of reflectivity for different power of forward pump beam. The PC reflectivity increases linearly with the power of the forward pump beam. There are two main processes which must be considered in the discussion of the origin of OPC in dye doped PMMA–PA films: (1) the formation of thermal grating and (2) third order non-linear optical processes. The DASPB dye doped film illuminated with 632 nm radiation of variable intensity and the transmittance of the sample is measured simultaneously by using a photodetector. If the effect observed in our experiments is of purely thermal nature, bleaching of the sample film should have been observed. The results obtained for the sample are shown in Fig. 12.14. It is clearly demonstrated that the transmission of the sample increases with time. The experiment described above indicates that third order non-linear processes like two photon absorption is mainly responsible for OPC in the sample under study.

12.7 Conclusion

To summarize, low-intensity optical phase-conjugation is observed in DASPB dye in PMMA–PA polymer matrix using a degenerate four-wave mixing setup, employing 633 nm light radiation from a He-Ne laser. The mechanism of phase conjugate wave generation associated with this dye-doped system is discussed. The phase conjugate signal is found to have contributions from the DFWM and the holographic processes. The maximum phase conjugate beam reflectivity observed in these dye films is about 0.42%. The maximum PC reflectivity is achieved when the angle between the probe and the forward pump beam is 8 degrees. The effects of dye concentration, intensity of backward, forward pump and interbeam angle between probe and forward pump beam on phase conjugation reflectivity are also studied. PC signal first increases with time and then decreases. PC reflectivity is also increased by increasing the intensity of the backward and forward pump beam.

The polarization and intensity profile are verified to be preserved in the conjugate signal. The predominant phase conjugation signal is attributed to the fact that saturable absorption and two photon induced the florescence property of the dye molecules. Since the DASPB dye in PMMA–PA polymer film is used at 633 nm, this may be suitable for low-power semiconductor lasers in the red wavelength region. Thus, DASPB dye in PMMA–PA polymer matrix may be a promising material for real-time double-exposure phase conjugate interferometry.

Bibliography

1. R. L. Abrams and R. C. Lind. 1978. 'Degenerate four-wave mixing in absorbing media.' *Optics Letters* 2, 4: 94–96.
2. S. Aithal, P. S. Aithal, and N. G. Bhat. 2011. 'Optical non-linearity of dye-doped polymer film using Z-scan technique.' *Proceedings of Second International Conference on Photonics IEEEXplore*, 1: 62–66.
3. M. Albota, D. Beljonne, J. W. Perry, G. Subramanium and C. Xu. 1998. 'Design of organic molecules with large two-photon absorption cross sections.' *Science* 281: 1653–1656.
4. R. W. Boyd. 1992. *Non-linear Optics*. New York, NY, USA: Academic Press.
5. R. Y. Choie, T. H. Barnes, W. J. Sandle, A. D. Woolhouse and I. T. Mc Kinnie. 2000. 'Observation of a thermal phase-grating contribution to diffraction in erythrosin-doped gelatin films.' *Opt. Commun*. 186, 13: 43–50.
6. H. Eichler, G. Salje and H. Stahl. 1973. 'Thermal diffusion measurements using spatially periodic temperature distributions induced by laser light.' *Journal of Applied Physics* 44, 12: 5383–5388.
7. H. J. Eichler and O. Mehl. 2001. 'Phase conjugate mirrors.' *J. Non-linear Optical Physics Mater*. 10: 43–52.
8. R. A. Fisher. 1983. *Optical Phase Conjugation*. New York, NY, USA: Academic Press.
9. H. Fujiwara and K. Nakagawa. 1985. 'Transient phase conjugation by degenerate four-wave mixing in saturable dyes.' *Opt. Commun*. 55: 386–390.
10. T. Geethakrishnan and P. K. Palanisamy. 2005. 'Generation of phase-conjugate wave in acid blue 7 dye-doped gelatin film.' *Current Science* 89, 11: 1894–1898.
11. T. Geethakrishnan and P. K. Palanisamy. 2006a. 'Optical phase-conjugation in erioglaucine dye-doped thin film.' *Pramana Journal of Physics* 66, 2: 473–478.
12. T. Geethakrishnan and P. K. Palanisamy. 2006b. 'Degenerate fourwave mixing experiments in methyl green dye-doped gelatin film.' *Optik*. 117, 6: 282–286.
13. G. S. He, J. D. Bhawalkar, C. F. Zhao and P. N. Prasad. 1995. 'Optical limiting effect in a two-photon absorption dye doped solid matrix.' *Appl. Phys. Lett*. 67: 2433–2435.
14. R. W. Hellwarth. 1977. 'Generation of time-reversed wave fronts by non-linear refraction.' *J. Opt. Soc. Am*. 67: 1–3.

15. J. P. Huignard and A. Brignon. 2004. *Overview of Phase Conjugation, Phase Conjugate Laser Optics*. Edited by A. Brignon and J. P. Huignard. 1–17 John Wiley & Sons, Inc. Hoboken, New Jersey.

16. M. A. Kramer, W. R. Tompkin and R. W. Boyd. 1986. 'Non-linear optical interactions in fluorescein doped boric acid glass.' *Physical Review A*. 34, 3: 2026–2031.

17. M. S. Malcuit, R. W. Boyd, L. W. Hillman, J. Krasinski and C. R. Stroud. 1984. 'Saturation and inverse-saturation absorption line shapes in alexandrite.' *J. Opt. Soc. Am. B*. 1: 73–75

18. A. Miniewicz, S. Bartkiewicz, J. Sworakowski, J. A. Giacometti and M. M. Costa. 1998. 'On optical phase conjugation in polystyrene films containing the azobenzene dye Disperse Red 1.' *Pure and Applied Optics* 7, 4: 709–721.

19. B. R. Reddy, P. Venkateswarlu and M. C. George. 1991. 'Laser induced gratings in a styryl due.' *Opt. Commun*. 84, 5–6: 334–338.

20. B. A. Reinhardt, L. L. Brott, S. J. Clarson, R. Kannan and A. G. Dillard. 1997. 'The design and synthesis of new organic molecules with large two photon absorption cross sections for optical limiting applications.' *Mater. Res. Soc. Sympo. Proc*. 479: 3–8.

21. A. T. Reghunath, C. K. Subramanian, P. S. Narayanan and M. R. Sajan. 1992. 'Optical phase conjugation in methylene blue films.' *Appl. Opt*. 31: 4905–4906.

22. M. Samoc, A. Samoc, B. L. Davies, H. Reish and U. Scherf. 1998. 'Saturable absorption in poly(indenofluorene): a picketfence polymer.' *Opt. Lett*. 23: 1295–1297.

23. A. E. Siegman. 1986. *Lasers* 207. University Science, California: Mill Valley.

24. Y. Silberberg and I. Bar-Joseph. 1981a. 'Transient effects in degenerate four-wave mixing in saturable absorbers.' *IEEE Journal of Quantum Electronics* 17, 9: 1967–1970.

25. Y. Silberberg and I. Bar-Joseph. 1981b. 'Low power phase conjugation in thin films of saturable absorbers.' *Opt. Commun*. 39, 4: 265–268.

26. L. Solymar, D. J. Webb and A. Grunnet-Jepsen. 1996. *The Physics and Applications of Photorefractive Materials*. Clarendon: Oxford.

27. H. Tanaka, A. Horikoshi, H. Fujiwara and K. Nakagawa. 2002. 'Phase conjugation in saturable absorbing dye films by degenerate four-wave mixing and holographic processes using nanosecond pulse and CW lasers.' *Optical Review* 9, 3: 106–111.

28. C. Wang, Y. Ren, Z. Shao, X. Zhao, G. Zhou, D. Wang, Q. Fang and M. Jiang. 2001. 'Optical properties of new two photon absorbing material HMASPS.' *Non-linear Optics* 28: 1–13.

29. A. Yariv. 1978. 'Phase conjugate optics and real-time holography.' *IEEE J Quantum Electron QE* 14, 9: 650–660.

13 Realization of a Tunable Optical Filter Using a Photonic Crystal Structure

Smruti Rath and S. K. Tripathy

13.1 Introduction

Photonic crystals (PhCs) are a novel class of optical media represented by natural or artificial structures with periodic modulation of the refractive index. Depending on the geometry of the structure, PhCs can be divided into three broad categories, namely, one-dimensional (1D), two-dimensional (2D) and three-dimensional (3D) structures. Here, the periodicity of the refractive index (i.e. dielectric constant) gives rise to photonic band gaps (PBG), forbidden energy bands for photons. Basically, in case of 1D PBG structure, the periodicity is within the structure along 1D only and is considered in terms of Bragg gratings[1].

Villar et al.[2] have proposed a multiple wavelength optical filter for 1D PhC with defects. Further, L. Scolari et al.[3] have proposed a tunable bandpass filter based on PhC fiber filled with multiple liquid crystals. It is also seen that[4], photonic crystal polarization filter is a type of filter which uses the band gap property of the photonic crystal to control the polarization state of the light signal. Xi-Hua Zou et al.[5] proposed a 1D PhC based on multichannel filters using binary phase-only sampling approach. Some studies on Fourier transform theory has also been done on PhC for filtering action[6]. Qiu and Jaskorzynska[7] have proposed a model design for a channel drop filter in a two-dimensional triangular PhC.

In this chapter, an optical filter based on the tunability of its excitation pulse for different materials such as glass, GaAs, ebonite and titanium dioxide is proposed using finite difference time domain (FDTD) method with varying layer thickness and mesh elements.

This chapter has been organized in following manner: Design and analysis, result and discussion and conclusion.

13.2 Design and Analysis

13.2.1 Photonic crystal structure

A photonic crystal is a microstructure that exhibits frequency intervals within which electromagnetic waves propagate irrespective of their propagation direction. The basic function of PhCs is that it controls the recombination of an electron–hole pair in a semiconductor during spontaneous emission, such that when optical transition occurs within this structure, band gap overlapping takes place.

13.2.2 FDTD method

Various types of computational methods have been adopted to analyze photonic crystal structure. They are: plane wave expansion (PWE) method, transfer matrix method, Green's function method and finite difference time domain method. However, in this study the FDTD method based on Yee's algorithm[8] is employed to observe the 1D-PCS, because the computational time and memory requirements are reduced.

A region of space is selected for field sampling in space and time. The solution is based on permittivity distribution function, initial conditions, boundary conditions. Once all these conditions are set, the field distribution is computed step-by-step starting from the radiation source. The space and time steps relate to the accuracy, numerical dispersion and stability of the FDTD method. The minimum cell size is given as

$$\min(\Delta x, \Delta y, \Delta z) \leq \frac{\lambda_{\min}}{10 n_{\max}} \tag{13.1}$$

where, $n_{\max}$ is the maximum refractive index value in the computational domain. After the cell size of computation mesh is selected, it is necessary to determine the time step. It is required to ensure that radiation should not propagate through more than one computation mesh cell during a single time step. Considering the light speed in vacuum as the radiation propagation speed, the time step allowed for stable computation is given by

$$c\Delta t \leq \left(\frac{1}{(\Delta x)^2} + \frac{1}{(\Delta y)^2} + \frac{1}{(\Delta z)^2} \right)^{\frac{-1}{2}}, \tag{13.2}$$

where Δx, Δy, Δz are linear dimensions of the computation mesh cell and are known as the Courant conditions[9].

A perfectly matched layer (PML) has been applied to ensure uniqueness and validity of the numerical solution of Maxwell's equations inside the computation domain. The reflection-less property is completely independent of the angle of incidence, polarization and frequency of incident wave. When the radiation passes through PML, it decays exponentially, and again when it passes the layer no radiation returns to the computation region. The dispersion relation is obtained with respect to the wave vector. This dispersion results in band gap.

13.3 Computation and Analysis

In order to compute PBG for the PhC under investigation, radiation has to be introduced to the structure. The Gaussian modulated high frequency is considered excitation. The expression for such a pulse is

$$E_z = A.e^{\frac{-(t-t_0)^2}{w^2}} \cdot \sin\omega t \tag{13.3}$$

where, to provides the time delay of the Gaussian envelope function, w is its width and ω is the carrier wave frequency. For the above source of excitation, we have varied the carrier frequency for wavelengths of 800 nm, 1310 nm and 1550 nm considering the refractive index of materials such as glass, GaAs, ebonite and titanium dioxide. The corresponding PBG for the above materials is calculated. Considering the different layer thickness (i.e., thickness ratio 2:1 and 3:1) and mesh elements, the PBG is calculated keeping the mesh elements to be 40 which was initially 50. In a similar manner, the dispersion diagram is obtained and the PBG computed by varying the parameters.

Table 13.1 *Band gap for titanium dioxide of layer thickness ratio 2:1*

Wavelength of source pulse (nm)	Upper frequency	Lower frequency	Frequency difference (Hz)	Energy (J)	Energy (eV)
800	1.9×10^{14}	1.4×10^{14}	0.5×10^{14}	3.31 e-20	0.207
1310	1.85×10^{14}	1.45×10^{14}	0.4×10^{14}	2.65 e-20	0.165
1550	1.95×10^{14}	1.4×10^{14}	0.55×10^{14}	3.64 e-20	0.228

Table 13.2 *Band gap for glass of layer thickness ratio 2:1*

Wavelength of source pulse (nm)	Upper frequency	Lower frequency	Frequency difference (Hz)	Energy (J)	Energy (eV)
800	2.25×10^{14}	2.05×10^{14}	0.2×10^{14}	1.32 e-20	0.082
1310	2.3×10^{14}	2.05×10^{14}	0.25×10^{14}	1.65 e-20	0.103
1550	2.25×10^{14}	2.05×10^{14}	0.2×10^{14}	1.32 e-20	0.082

Table 13.3 *Band gap for glass of layer thickness ratio 3:1*

Wavelength of source pulse (nm)	Upper frequency	Lower frequency	Frequency difference (Hz)	Energy (J)	Energy (eV)
800	2.95×10^{14}	2.45×10^{14}	0.5×10^{14}	3.31 e-20	0.206
1310	2.9×10^{14}	2.45×10^{14}	0.45×10^{14}	2.98 e-20	0.186
1550	2.9×10^{14}	2.45×10^{14}	0.45×10^{14}	2.98 e-20	0.186

Table 13.4 *Band gap for titanium dioxide of layer thickness ratio 3:1*

Wavelength of source pulse (nm)	Upper frequency	Lower frequency	Frequency difference (Hz)	Energy (J)	Energy (eV)
800	2.8×10^{14}	2.1×10^{14}	0.7×10^{14}	4.63 e-20	0.289
1310	2.75×10^{14}	2.1×10^{14}	0.65×10^{14}	2.30 e-20	0.268
1550	2.75×10^{14}	2.05×10^{14}	0.7×10^{14}	2.63 e-20	0.289

Tables 13.1–13.4 show the various calculation of band gap for different layer thickness of glass and titanium dioxide.

13.4 Result and Discussions

With the abovementioned wavelength of source of excitation and its corresponding energy, the graph is plotted.

Figure 13.1 denotes the layer thickness ratio of 2:1 for glass when the mesh elements are 40.

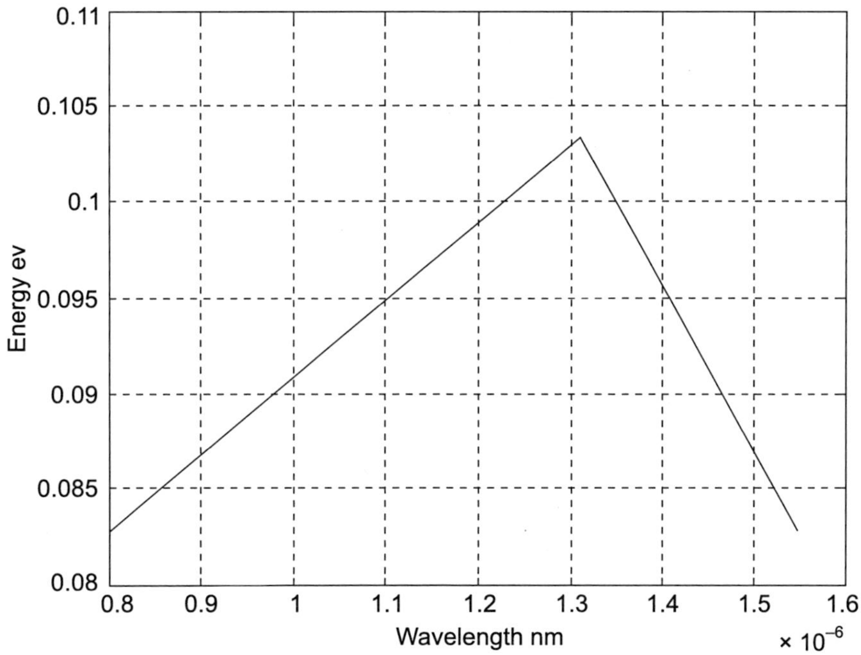

Figure 13.1 Layer thickness ratio 2:1 for glass

Figure 13.2 denotes the layer thickness ratio of 2:1 for titanium dioxide when the mesh elements are 40.

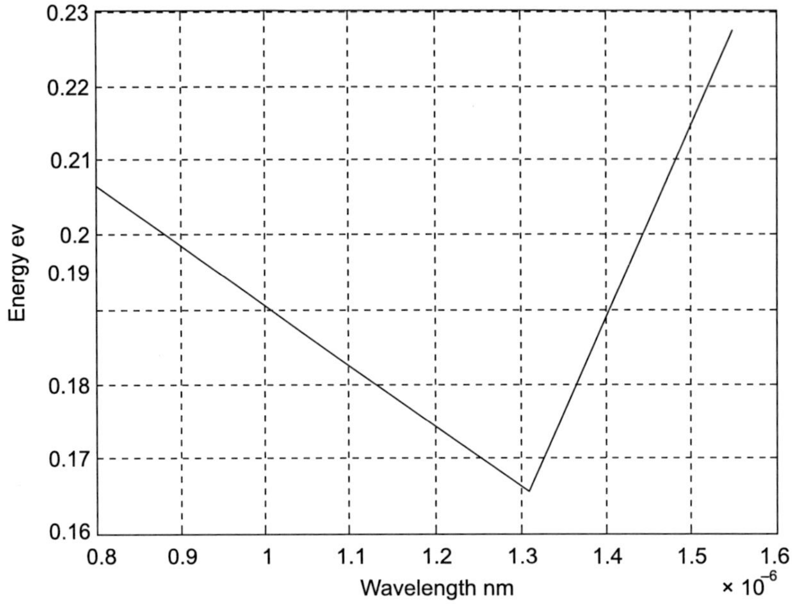

Figure 13.2 Layer thickness ratio 2:1 for titanium dioxide

Figures 13.3 and 13.4 depicts the layer thickness ratio of 3:1 for glass and titanium dioxide.

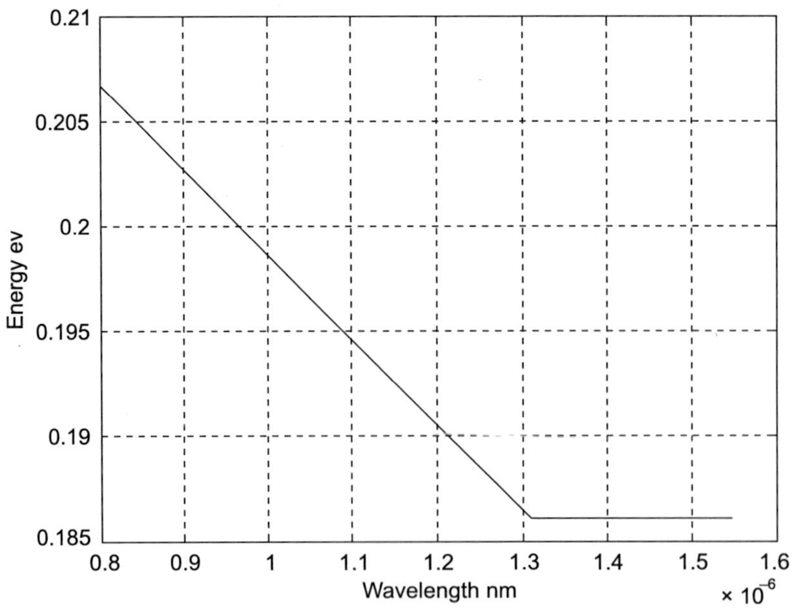

Figure 13.3 Layer thickness ratio 3:1 for glass

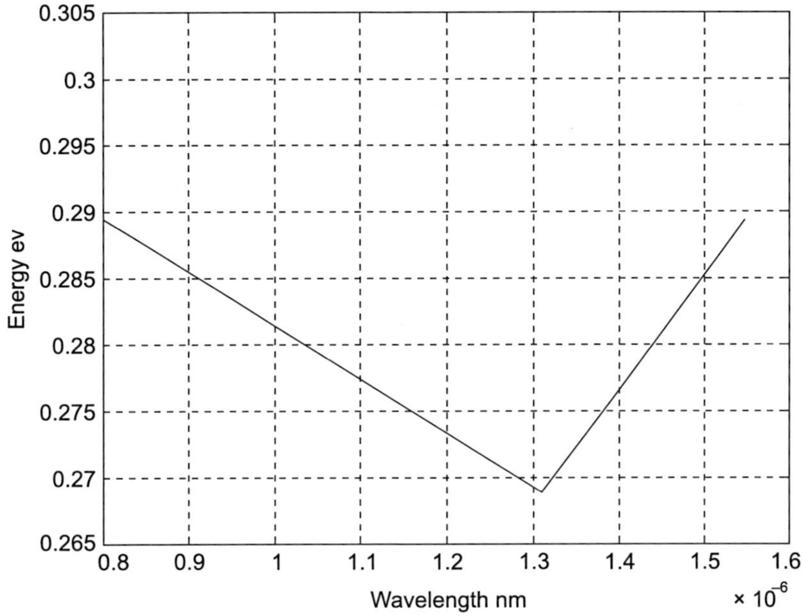

Figure 13.4 Layer thickness ratio 3:1 for titanium dioxide

From the above figures, it is inferred that the PCS can be used as a tunable optical filter. Since the different parameters such as mesh elements, the layer thickness and the pulse of excitation of the source for 800 nm, 1310 nm and 1550 nm are varied, based on the requirement of band gap or wavelength to be eliminated, the frequency of the excitation pulse can be tuned. Hence, this can be proposed for various optical filtering applications. Some interesting results are also found on this[10, 11].

Bibliography

1. H. Dey, and S. K. Varshney. 2009. 'Band gap engineering of one-dimensional photonic crystal structures.' *International Conference on Computers and Devices for Communication* 978-81-8465-152-2/09 CODEC.
2. I. D. Villar, I. R. Matas, and F. J. Arregui. June, 2004. 'Fiber-optic multiple-wavelength filter based on one-dimensional photonic band gap structures with defects.' *Journal of light wave technology* 22, 6: 1615-1621.
3. L. Scolari, G. Tartarini, E. Borelli, T. T. Alkeskjold, H. C. H. Mulvad, P. Bassi and A. Bjarklev. 2007. 'Tunable band pass filter based on photonic crystal fiber filled with multiple liquid crystals.' 1-4244-0925-X/07, *IEEE*, 319–320.
4. J. Shi, Y. Huang, K. Dong, S. Wen and X. Chen. 2009. 'One-dimensional photonic crystal polarization filter with multiple channels.' 978-1-4244-4412-0/09, *IEEE*, 978-1-4244-4412-0/09.

5. X. H. Zou, W. Pan, B. Luo, W. L. Zhang and M. Y. Wang. September, 2009. 'One-dimensional photonic crystal-based multichannel filters using binary phase-only sampling approach.' *Journal of light wave technology* 25, 9: Vol. 25, Issue 9, pp. 2482-2486 (2007).
6. M. C. Parker and S. D. Walker. September, 2002. 'A unified fourier transform theory for photonic crystal and FBG filters in the strong coupling regime.' *IEEE Photonics technology letters* 14, 9: 1041-1135/02.
7. M. Qiu and B. Jaskorzynska. 2003. 'Design of a channel drop filter in a two-dimensional triangular photonic crystal.' *Applied Physics Letter* 83: 1074–1076.
8. K. S. Yee. 1966. 'Numerical solution of initial boundary value problems involving Maxwell's equations in isotropic media.' *IEEE Trans. Antennas Propag.* 14: 302–07.
9. Igor A. Sukhoivanov and Igor V. Guryev. *Photonic Crystals*. ISBN: 978-3-642-02645-4 (Print) 978-3-642-02646-1 (Online)
10. Smruti Rath, Sonali Prava Dash and Sukanta Kumar Tripathy. 2011. 'Realization of logic gates using Y defect in a two dimensional photonic crystal structure.' *IEEE Xplore*. 978-1-4673-0074-2.
11. Smruti Rath, Sonali P. Dash, Mihir Hota, and Sukanta K. Tripathy. 2012. 'Realization of optical XOR and OR gates using asymmetric Y-structure in a two dimensional photonic crystal.' *AIP Conf. Proc.* 1461: 369.

14 Cavity Quantum Opto-mechanics: A Brief Review

Tarun Kumar, Sonam Mahajan, Aranya B. Bhattacherjee and Man Mohan

14.1 Introduction

The interaction between a movable mirror and the radiation field of an optical cavity has recently been the subject of extensive theoretical and experimental investigations. This field known as cavity opto-mechanics has played a vital role in the conceptual exploration of the boundaries between classical and quantum mechanical systems. These opto-mechanical systems couple the mechanical motion to an optical field directly via radiation pressure buildup in a cavity. The coupling of mechanical and optical degrees of freedom via radiation pressure has been a subject of early research in the context of laser cooling[1, 2, 3] and gravitational-wave detectors[4]. The recent improvements in nanofabrication techniques suggest that in the near future, these devices will reach the regime in which their sensitivity will be limited by the ultimate quantum limits set by the Heisenberg principle. The importance of the limits imposed by quantum mechanics on the resonator motion was first pointed out by Braginsky and coworkers[5] in the completely different context of massive resonators employed in the detection of gravitational waves[6]. However, in recent years, the quest for the experimental demonstration of genuine quantum states of macroscopic mechanical resonators has spread well beyond the gravitational wave physics community and has attracted wide interest.

Recently, there has been a great surge of interest in the application of radiation forces to manipulate the center-of-mass motion of mechanical oscillators covering a huge range of scales from macroscopic mirrors in the laser interferometer gravitational wave observatory (LIGO)

project[7, 8] to nano-mechanical cantilevers[9, 10, 11, 12], membranes[13] and Bose–Einstein condensates[14, 15, 16, 17]. The central accomplishment of the field of cavity opto-mechanics is the investigation of radiation pressure force which allows one to manipulate the motional state of micromechanical oscillators. In particular, it has become possible to substantially cool the thermal excitation of a single mechanical mode, down to a few tens of remaining phonons[18]. With these developments, micro- and nano-mechanical resonators now represent an important model system with the prospect of demonstrating quantum effects on a macroscopic scale.

The quantum optical properties of a mirror coupled via radiation pressure to a cavity field show interesting similarities to an intracavity Kerr-like interaction[19, 20]. Recently, in the context of classical investigations of non-linear regimes, the dynamical instability of a driven cavity having a movable mirror has been investigated[21]. Theoretical work has proposed to using the radiation pressure coupling for quantum non-demolition measurements of the light field[5]. A robust stationary entanglement between the mechanical resonator and the output fields of the cavity can be generated. This entanglement can be transferred to atomic ensembles placed within the cavity[22].

14.1.1 Jaynes–Cumming model

The system of a two level atom interacting with a multimode field is described by Dicke Hamiltonian[23] but the Eigenvalue equation derived from Dicke's Hamiltonian is not exactly solvable. Hence a new model was proposed by Jaynes and Cumming in 1963[24], which is exactly solvable under rotating wave and dipole approximations. The Jaynes–Cumming Hamiltonian describes the interaction of a single two level atom with a single mode radiation field. The model proposed by Jaynes and Cumming is not only used in optical physics but also successfully used in quantum field theory, laser physics and nuclear physics.

The Hamiltonian of a single electron atom interacting with a single mode radiation field $\vec{E}$ can be written as

$$H = H_{\text{field}} + H_{\text{atom}} + H_{\text{inter}}$$

where H_{field} is the energy of the single mode radiation field, H_{atom} is the energy of the free atom, and H_{inter} gives the interaction energy of the atom and the field. Under dipole and rotating wave approximation, Hamiltonian of the system can be written as

$$H = \hbar\omega a^{\dagger}a + \frac{1}{2}\hbar\omega_a\sigma_z + \hbar g(\sigma_+ a + a^{\dagger}\sigma_-)$$

This Hamiltonian represents the interaction of a single two level atom with a single mode radiation field and is known as a Jaynes–Cummings Hamiltonian. This Hamiltonian is the starting point of many calculations in quantum optics.

14.1.2 Micromirrors

Micromirrors are the mirrors that have dimensions in the range of the microscopic world. The use and fabrication of micromirrors have received a great deal of boost in the last decade because of its potential use in quantum optics to measure weak forces like gravitational force. They can also be used to confine light between two or more mirrors and are also known as optical cavity or Febry–Perot cavity. An optical cavity or optical resonator is an arrangement of mirrors that forms a standing wave for the light waves. They are extensively used in lasers, where they surround the gain medium and provide feedback of the laser light. The potential applications of optical cavities are in optical parametric oscillators and in interferometers. The standing wave patterns produced in a cavity are known as modes. Modes can be of two types: (i) longitudinal mode, which differ in frequency only and (ii) transverse mode which differ for different frequencies and have a different frequency pattern. The most common types of optical cavities consist of two facing plane or spherical mirrors. The simplest of these is the plane-parallel or Fabry–Perot cavity, consisting of two opposing flat mirrors. The end mirrors in the Fabry–Perot cavity can be movable or stationary depending on the system under consideration. In the subsequent sections both the configurations are used.

14.1.3 Radiation pressure

Radiation pressure i.e, force due to the momentum of light has been known as long as from the 17th century when scientists predicted that the inclination of comet tails is because of solar radiation pressure[25, 26]. Later in the 20th century, experimentalists Hull[27] and Lebedev[28] studied the strength of radiation pressure. The application of radiation pressure in the context of gravitational wave detection was first studied by Braginsky and co-workers. They predicted that the radiation pressure inside a cavity with finite decay time gives rise to dynamic backaction which is the underlying phenomena for cooling of mechanical oscillators and parametric instabilities. Braginsky et al. also studied the sensitivity limits due to the quantum nature of light inside a cavity[29, 30]. Later in 1980, other groups mainly of Caves[31] and Meystre[32] analyzed the radiation pressure noise in interferometers. After this, a great surge of interest creeped in to find out the potential application of radiation pressure. Many theoretical models were also predicted including the possible application of radiation pressure such as quantum non-demolition measurement of photon numbers[33, 34], generation of squeezed light[35, 36], feedback cooling of mechanical oscillators[37], optical and mechanical mode entanglement[38], quantum state transfer from light to mechanical oscillator[39] etc. Moreover, very recently some very interesting phenomena have been realized experimentally such as strongly coupled opto-mechanics[40], cavity cooling of mechanical oscillator[41, 42], parametric amplification of mechanical motion[43]. There have been a number of review articles discussing in great details, the possible application of optical cavities and radiation pressure[44, 45]. The radiation pressure interaction has also been used to study the dynamics of the mechanical oscillator such as cooling of mechanical oscillator in a resolved side band regime[46]. Cooling in a resolved sideband is possible because of radiation pressure. The optical mode (photons) and the mechanical mode (phonons) get entangled, which results in energy exchange between the two modes. As a result, if a photon is created in the side band then this results in the annihilation of a phonon of the mechanical mode and vice-versa.

14.1.4 Properties of Fabry–Perot cavity

Consider an optical cavity consisting of two mirrors M_1 and M_2. The mirror M_1 is partially transmitting whereas the mirror M_2 is nearly reflecting. Under such a condition, a coherent light enters the cavity, bounces off the mirror M_2 and stays inside the cavity. Let E_I be the incident light field; E_C the intracavity field, E_R the reflected light and E_T is transmitted light. Also let $r_1, t_1(r_2, t_2)$ be the reflection and transmission coefficient of the two mirrors M_1 and M_2 respectively. Let λ_0 be the wavelength of light used and L, the cavity length. We also assume that incident light is monochromatic and is along the positive x-direction of the form $e^{i(kz-\omega_0 t)}$; ω_0 is the frequency of the incident laser light.

The power stored inside the cavity can be defined as

$$W = |E_C|^2 = \frac{t_1^2 E_0^2}{1 + r_1^2 r_2^2 - 2r_1 r_2 \cos(\phi)}$$

where $E_C = \dfrac{E_0 t_1}{1 - r_1 r_2 e^{i\phi}}$,

Therefore the cavity is resonant for $\phi = 2\pi n$, n being an integer. Moreover, the resonant condition can be defined in terms of cavity length L. The cavity will also be resonant if L is equal to half integer multiple of the incident wavelength λ_0, which corresponds to the condition of constructive interference. Therefore, we can write the resonance condition as $L = \frac{n\lambda_o}{2}$. Thus, the cavity has two resonant conditions. Cavity resonance condition can also be estimated in terms of frequency difference between two adjacent power resonances of the cavity as $\Delta\omega = \frac{\pi c}{L}$. For a perfectly reflecting mirror M_2, it can be assumed that $r_2 = 1$, $r_1^2 = 1 - t_1^2$. This gives the line width as $\gamma = \frac{ct_1^2}{4L}$. Also the finesse of a Fabry–Perot cavity is defined as the ratio of frequency power response to the line width.

Therefore, $$f = \frac{\Delta\omega}{\gamma} = \frac{2\pi}{t_1^2}.$$

The finesse of an optical cavity is analogous to the quality factor Q of a mechanical oscillator.

14.1.5 Input–output formalism

In this part, the formalism of the quantum damping theory is reviewed in which the explicit nature of the inputs from a heat bath and of the outputs into it, is taken. These can be described by standard master equations, where the external field is supposed to be from an external bath source. The Quantum Langevin equations of the system will be discussed[47]. The theory has also been described in[48].

14.1.6 One sided cavity

Consider a single cavity mode interacting with an external multi-mode field. To start with, let us assume that the cavity has one partially transmitting mirror, say M_1, and one completely reflecting mirror, say M_2 (Fig. 14.1) Mirror M_1 couples the external field to the intracavity mode. Coupling between the intracavity mode and the external modes is determined by the geometry of the cavity and nature of the dielectric interface. Here it is assumed that the external field is essentially one-dimensional.

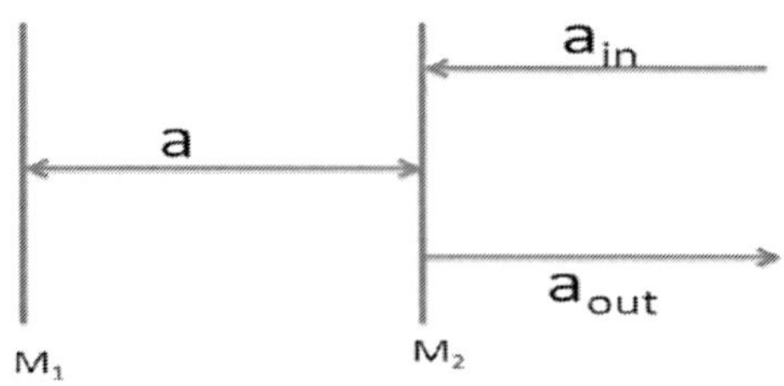

Figure 14.1 Single cavity mode interacting with an external multi-mode field

The Hamiltonian of the system can be written as

$$H = H_S + H_B + H_I$$

where H_S is the Hamiltonian of the intracavity field, H_B is the free Hamiltonian of the bath and H_I describes the interaction between the bath and the cavity field. Also the H_B and H_I can be written as

$$H_I = \hbar \int_{-\infty}^{\infty} d\omega \omega b^{\dagger}(\omega) b(\omega) \text{ and}$$

$$H_B = i\hbar \int_{-\infty}^{\infty} d\omega g(\omega)[b^{\dagger}(\omega)a - a^{\dagger} b(\omega)].$$

Here $b(\omega)$ is the annihilation operator of the bath and c is one of the several possible system operators, which finally gives

$$b(t,\, \omega) = e^{-i\omega(t-t_o)} b_0(\omega) + g(\omega) \int_{t_0}^{t} e^{-i\omega(t-t')} a(t')dt'.$$

Here $b_0(\omega)$ is the value of $b(\omega)$ at $t = t_0$.

Moreover, the equation of motion for the intracavity field can be written as

$$\dot{a}(t) = \frac{i}{\hbar}[H_S,\, a] - \int_{-\infty}^{\infty} d\omega g(\omega) b(t, \omega),$$

Using $b(t,\, \omega)$,

$$\dot{a}(t) = \frac{i}{\hbar}[H_S,\, a] - \int_{-\infty}^{\infty} d\omega g(\omega) e^{-i\omega(t-t_0)} b_0(\omega) - \int_{-\infty}^{\infty} d\omega g(\omega)^2 \int_{t_0}^{t} e^{-i\omega(t-t')} a(t') \tag{14.1}$$

Here $g(\omega)$ is independent of frequency and we take $g(\omega) = \sqrt{\gamma/2\pi}$ and let the input field operator be written as

$$a_{in}(t) = -\frac{1}{2\pi} \int_{-\infty}^{\infty} d\omega e^{-i\omega(t-t_0)} b_0(\omega).$$

Using the relations $\int_{-\infty}^{\infty} d\omega e^{-i\omega(t-t')} = 2\pi\delta(t-t')$ and $\int_{t_0}^{t} \delta(t-t')a(t') = a(t)/2$ in Eq. (14.1), we get

$$\dot{a}(t) = \frac{i}{\hbar}[H_S,\, a] - \frac{\gamma}{2} a(t) + \sqrt{\gamma} a_{\text{in}}(t). \tag{14.2}$$

This is the quantum differential equation for the intracavity fields. Similarly, in terms of output field, $\dot{a}(t)$ can also be written as

$$\dot{a}(t) = \frac{i}{\hbar}[H_S,\, a] + \frac{\gamma}{2} a(t) - \sqrt{\gamma} a_{\text{out}}(t), \tag{14.3}$$

where $a_{\text{out}}(t) = -\frac{1}{\sqrt{2\pi}} \int_{-\infty}^{\infty} d\omega e^{-i\omega(t-t_1)} b_1(\omega)$.

Subtraction of Eq. (14.3) from Eq. (14.2) gives

$$a_{\text{in}}(t) + a_{\text{out}}(t) = \sqrt{\gamma}a(t).$$

This is the boundary condition which relates the intracavity field with the external field.

Similarly for a two-sided cavity, boundary conditions can be written as

$$a_{\text{in}}(t) + a_{\text{out}}(t) = \sqrt{\gamma_1}a(t),$$

$$b_{\text{in}}(t) + b_{\text{out}}(t) = \sqrt{\gamma_2}a(t),$$

where γ_1 and γ_2 are the loss coefficients of the two mirrors.

Cooling of micro mechanical mirrors

There has been considerable interest in using micromirrors for producing superpositions of macroscopic quantum states but such micromirrors has to be cooled to their quantum ground states. Two ways of cooling the micro mechanical mirrors have been proposed. One is the active feedback scheme, where a viscous force is fed back to the movable mirror which reduces its Brownian motion. The other is the passive feedback scheme, where the Brownian motion of the movable mirror is damped by the radiation pressure force exerted by photons in an appropriately detuned optical cavity. Constant work is going on to cool the micromirror toward its ground state. The great success of cavities with parametric amplifiers in the production of nonclassical light. The movable mirror can reach a minimum temperature of about a few hundred mK, a factor of 500 below room temperature 300 K. The lowering of the temperature is achieved by changes in photon statistics due to parametric. Since radiation pressure depends on the number of photons, the cooling of the micromirror can be manipulated by using effects of the photon statistics. Huang et al.[49] has shown that an optical parametric amplifier inside a cavity considerably improves the cooling of a micro-mechanical mirror by radiation pressure. They propose and analyze a method to achieve cooling of a movable mirror to sub-Kelvin temperatures by using a type I optical parametric amplifier inside a cavity. They showed that if the mirror is already precooled to say about 1 K, then by using an optical parametric amplifier the mirror can be cooled to about milliKelvin temperatures or less.

It has also been shown that ground state cooling of micro-mechanical mirrors is possible only in the resolved side band regime (RSB) where the mechanical resonance frequency exceeds the bandwidth of the driving resonator[5]. The cooling of mechanical oscillators in the RSB regime at high driving power can entail the appearance of normal mode splitting (NMS)[50].

Opto-mechanically coupled mirrors has been investigated earlier[51, 52, 53]. A continuous variable entanglement between the two mirrors was maintained by the light bouncing between the mirrors and was found to be robust against thermal noise[51, 52]. Entanglement between two mechanical oscillators coupled to a non-equilibrium environment showed that there is an optimal dissipation strength for which the entanglement between two coupled oscillators is maximized[53]. A new cooling method which involves the two-sided irradiation of the vibrating mirror inside an optical cavity has been proposed recently[54]. This method provides a stiffer trap for cooling the mirror and has several advantages over conventional methods of opto-mechanical cooling. Recent experimental advances achieved promising results via cryogenic cooling, backaction cooling, in which the off-resonant operation of the cavity results in a retarded backaction on the mechanical system[10], or by cold-damping quantum feedback where the oscillator position is measured through a phase-sensitive detection of the cavity output and the resulting photocurrent is used for a real-time correction of the dynamics. Recent theoretical results[55, 56] have shown that these cooling schemes are capable of bringing the micro resonator down to its ground state. Both mechanisms achieve cooling by increasing the damping of the mechanical resonator so that it becomes insensitive to thermal noise. It is important to analyze all the possible limitations of ground state cooling, because this would open up the experimental realization of a number of genuinely quantum phenomena, such as the possibility of entangling an acoustic mode to a cavity quantum field, or to another mechanical oscillator, and even continuous variable quantum information protocols such as quantum teleportation[57], and entanglement swapping[58]. In the next section, we discuss the effect of Kerr medium, placed inside a Fabry–Perot cavity. Giant optical Kerr non-linearities are obtained by placing an $\chi^{(3)}$ medium inside a cavity[59]. This gives rise to a strong non-linear interaction between photons. A single photon in a cavity can block the injection of a second photon due to a photon blockade effect[60]. This dispersive interaction could be useful in realizing a single-photon turnstile device for possible use in quantum computation[59]. Greentree et al.[61] have shown that by adding photons to a two-dimensional array of coupled optical cavities each containing a single two-level atom in the photon blockade regime, polaritons are formed that are both long-lived and strongly interacting. They proposed that this system could be extremely beneficial for quantum information processing.

14.2 Fabry–Perot Cavity Containing Kerr Medium

Consider an optical Kerr medium with an $\chi^{(3)}$ non-linearity inside a Fabry–Perot cavity with one fixed partially transmitting mirror and one movable totally reflecting mirror in contact with a thermal bath in equilibrium at temperature T, as shown in Fig. 14.2. The movable mirror is treated as a quantum mechanical harmonic oscillator with effective mass m, frequency Ω_m and energy decay rate Γ_m. The system is also coherently driven by a laser field with frequency ω_L through the cavity mirror with amplitude ε. It is known that high-Q optical cavities significantly isolate the system from its environment, thus reducing decoherence and ensuring that the light field remains quantum-mechanical for the duration of the experiment. We also consider that the induced resonance frequency shift of the cavity and the non-linear interaction coefficient η are much smaller than the longitudinal mode spacing, so that the system is restricted to a single longitudinal mode ω_c. Moreover, $\Omega_m \ll \pi c/L$ (adiabatic limit); c is the speed of light in vacuum and L the cavity length in the absence of the cavity field. The system in a frame rotating at the laser frequency ω_L can be modeled by the following Hamiltonian

$$H = \hbar(\omega_c - \omega_L)a^\dagger a - \hbar g_m a^\dagger a q + \frac{1}{2}\left(\frac{p^2}{m} + m\Omega_m^2 q^2\right) + i\hbar\varepsilon(a^\dagger - a) + \hbar\eta a^{\dagger 2} a^2$$

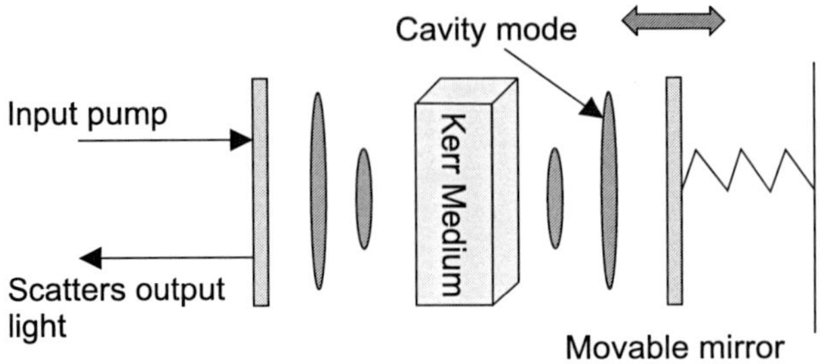

Figure 14.2 Opto-mechanical realization of parametric coupling of a mechanical oscillator to an optical mode of a non-linear cavity

Here a and $a^\dagger$ are the annihilation and creation operators for the cavity field respectively. Also q and p are the position and momentum operators for the movable mirror. The parameter $g_m = \omega_c/L$ is the coupling parameter between the cavity field and the movable mirror and η is the anharmonicity parameter and is proportional to the third-order non-linear susceptibility $\chi^{(3)}$ of the Kerr medium: $\eta = 3\hbar\omega_c^2 Re[\chi^{(3)}]/2\varepsilon_0 V_c$, ε_0 is the dielectric constant of the medium and V_c is the volume of the cavity. The input laser field populates the intracavity mode which couples to the movable mirror through the radiation pressure. The field in turn is modified by the backaction of the cantilever. Here the system is intrinsically open as the cavity field is damped by the photon-leakage through the massive coupling mirror and the mirror is connected to a bath at finite temperature. In the absence of the radiation pressure coupling, the cantilever would undergo a pure Brownian motion driven by its contact with the thermal environment. To study the dynamics of the system, the linearized Heisenberg equation of motion for position, momentum, and intracavity field operators are calculated as

$$\delta\dot{q} = \frac{\delta p}{m}, \tag{14.4}$$

$$\delta\dot{p} = -m\Omega_m^2\delta q + \hbar g_m(a_s\delta a^\dagger + a_s^*\delta a) - \Gamma_m\delta p + \xi, \tag{14.5}$$

$$\delta\dot{a} = -i\Delta\delta a + ig_m\delta q a_s - 2i\eta(2|a_s|^2\delta a + a_s^2\delta a^\dagger) - \kappa\delta a + \sqrt{2\kappa}\delta a_{in}, \tag{14.6}$$

$$\delta\dot{a}^\dagger = i\Delta\delta a^\dagger - ig_m\delta q a_s^* + 2i\eta(2|a_s|^2\delta a^\dagger + a_s^{*2}\delta a) - \kappa\delta a^\dagger + \sqrt{2\kappa}\delta a_{in}, \tag{14.7}$$

where

$$a \to a_s + \delta a,\ p \to p_s + \delta p,\ q \to q_s + \delta q.$$

In the transformed quadratures, $\delta x = (\delta a^\dagger + \delta a)$, $\delta y = i(\delta a_{\text{in}}^\dagger - \delta a)$, $\delta x_{\text{in}} = \delta a_{\text{in}} + \delta a_{\text{in}}$, $\delta y_{\text{in}} = i(\delta a_{\text{in}} - \delta a_{\text{in}})$.

The above equations reduces to

$$\delta\dot{x} = ig_m\delta q(a_s - a_s^*) + \delta x[-\kappa - i\eta(a_s^2 - a_s^{*2})] + \delta y[\Delta + 4\eta|a_s|^2 - \eta(a_s^2 + a_s^{*2})] + \sqrt{2\kappa}\delta x_{\text{in}}. \tag{14.8}$$

$$\delta\dot{y} = \delta x(-\Delta - 4\eta|a_s|^2 - \eta(a_s^2 + a_s^{*2})) + \delta y(-\kappa - i\eta(a_s^2 - a_s^{*2})) + g_m\delta q(a_s + a_s^*) + \sqrt{2\kappa}\delta y_{\text{in}}. \tag{14.9}$$

Equations (14.4), (14.5), (14.8) and (14.9) represent a set of linearized coupled equations for the fluctuation operators. These equations in general can be written as

$$\dot{f}(t) = Af(t) + \eta(t), \tag{14.10}$$

where $f(t)$ is the column vector of the fluctuations and $\eta(t)$ is the column vector of the noise sources which can be written as:

$$f(t) = \begin{pmatrix} \delta q \\ \delta p \\ \delta x \\ \delta y \end{pmatrix}. \tag{14.11}$$

$$\eta(t) = \begin{pmatrix} 0 \\ \xi \\ \sqrt{2\kappa}\delta x_{\text{in}} \\ \sqrt{2\kappa}\delta y_{\text{in}} \end{pmatrix}. \tag{14.12}$$

$$A = \begin{pmatrix} 0 & 1/m & 0 & 0 \\ -m\Omega_m^2 & -\Gamma_m & \hbar g_m \frac{a_s + a_s^*}{2} & \hbar g_m \frac{a_s - a_s^*}{2i} \\ ig_m(a_s - a_s^*) & 0 & -\left[\kappa + i\eta\left(a_s^2 - a_s^{*2}\right)\right] & \left[\Delta + 4\eta|a_s|^2 - \eta\left(a_s^2 + a_s^{*2}\right)\right] \\ g_m(a_s + a_s^*) & 0 & -\left[\Delta + 4\eta|a_s|^2 + \eta\left(a_s^2 + a_s^{*2}\right)\right] & -\left[\kappa + i\eta\left(a_s^2 - a_s^{*2}\right)\right] \end{pmatrix}.$$

The solution of Eqs (14.4), (14.5), (14.8) and (14.9) are stable only if all the eigenvalues of the matrix A have negative real parts. Here it is always assumed that $\Gamma_m \ll \kappa$. Equations (14.4), (14.5), (14.8) and (14.9) constitute a system of four first order coupled operator equations, for which the Routh–Hurwitz criterion implies that the system is stable for the following conditions:

$$(\Delta + 4\eta|a_s|^2)^2 + \kappa^2 + \Omega_m^2 + 2\Gamma_m\kappa + 2i\eta(a_s^2 - a_s^{*2})(\kappa + \Gamma_m) - 2\eta^2(a_s^4 + a_s^{*4}) > 0. \tag{14.13}$$

$$\Gamma_m(\Delta + 4\eta|a_s|^2)^2 + 2\Omega_m^2\{\kappa + i\eta(a_s^2 - a_s^{*2})\} - \Gamma_m\eta^2(a_s^2 + a_s^{*2})^2 > 0. \tag{14.14}$$

$$\Omega_m^2\{[\kappa + i\eta(a_s^2 - a_s^{*2})]^2 + [\Delta + 4\eta|a_s|^2]^2\} + (\frac{\hbar g_m^2\eta}{m} - \Omega_m^2\eta^2)(a_s^2 + a_s^{*2})^2 - \frac{2\hbar g_m^2|a_s|^2}{m}(\Delta + 4\eta|a_s|^2) > 0.$$

14.2.1 Displacement spectrum and photon blockade mechanism

The displacement spectrum of the movable mirror is defined as

$$S_q(\omega) = \frac{1}{4\pi}\int d\Omega e^{-i(\omega+\Omega)t}\langle\delta q(\omega)\delta q(\Omega) + \delta q(\Omega)\delta q(\omega)\rangle.$$

Using the above results, the displacement spectrum is obtained as

$$S_q(\omega) = \hbar|\chi|^2\left\{m\Gamma_m\omega\coth\left(\frac{\hbar\omega}{2k_BT}\right) + \frac{4\hbar\kappa g_m'^2\left(\omega^2 + \kappa^2 + \delta^2\right)}{(\kappa^2 + \delta'^2 - \omega^2)^2 + 4\kappa^2\omega^2}\right\}, \tag{14.15}$$

where

$$\chi^{-1}(\omega) = m\left[\Omega_{\text{eff}}^2 - \omega^2\right] - i\omega\Gamma_{\text{eff}}. \tag{14.16}$$

$$\Omega_{\text{eff}}^2 = \Omega_m^2 - \frac{2\hbar g_m'^2 \delta'' \left(\kappa^2 + \delta'^2 - \omega^2\right)}{m\left[(\kappa^2 + \delta'^2 - \omega^2)^2 + 4\kappa^2\omega^2\right]}. \tag{14.17}$$

$$\Gamma_{\text{eff}} = m\Gamma_m + \frac{4\hbar\kappa g_m'^2 \delta''}{\left[(\kappa^2 + \delta'^2 - \omega^2)^2 + 4\kappa^2\omega^2\right]}. \tag{14.18}$$

Displacement spectrum $S_q(\omega)$ gives the information about the modified motion of the mirror. An immediate observation reveals that $S_q(\omega)$ is peaked at a frequency Ω_{eff}. In the expression for $S_q(\omega)$ [Eq. (10.15)], the first term represents the contribution due to the thermal contact with the bath while the second term is attributed to the contribution due to radiation pressure. In the absence of radiation pressure, any dependence from the effective detuning Δ vanishes and the resulting spectrum is simply that of a harmonic oscillator undergoing Brownian motion at temperature T.

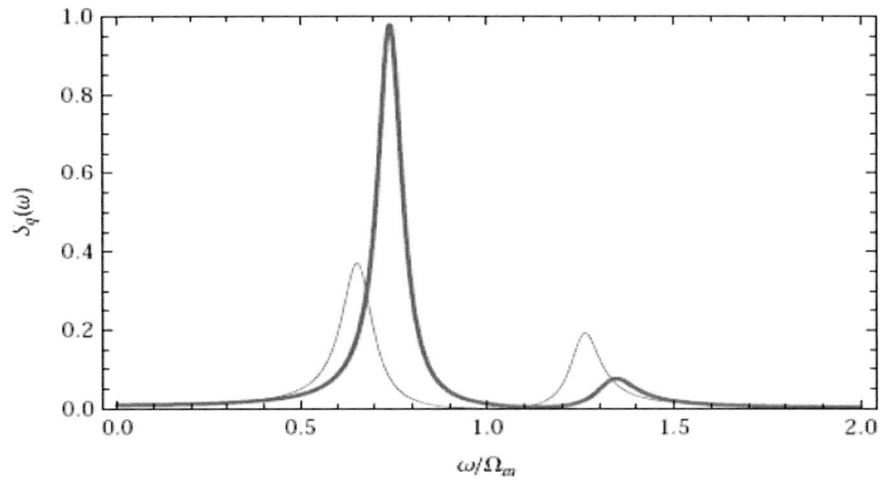

Figure 14.3 Plot of the displacement spectrum $S_q(\omega)$ with non-linear coefficients, $\eta'/\Omega_m = 0$, 0.04. Parameters used are: $\Gamma_m/\Omega_m = 0.01$, $\kappa/\Omega_m = 0.1$, $g_m' = 0.1$ and $\Delta/\Omega_m = 1$

In Fig. 14.2, $S_q(\omega)$ is plotted as a function of ω/Ω_m for $\eta/\Omega_m = 0$, 0.04 with successively increasing thickness of plot. When Kerr is non-linear, an NMS(normal mode splitting) is observed in the displacement spectrum with small amplitude. The NMS shows a mixing between the mechanical mode and the fluctuation around the steady state of the cavity field. As Kerr non-linearity increases, that one of the peak increases while the other peak decreases. The increase in the amplitude of one peak and decrease in the amplitude of the other peak with increasing Kerr non-linearity is because of the following reason: If $\eta \gg \kappa$, ε, the applied field will couple the vacuum state to the Fock state with a single photon resonantly. The higher lying photon-number states may be neglected since they are out of resonance.

Now if initially a photon is injected in the cavity with a probability determined by the drive strength, injection of a second photon will be blocked. This is because the presence of two photons in the cavity will require an additional $\hbar\eta$ energy, which cannot be provided by the pump laser. Only after the first photon leaves the cavity, can a second photon be injected. The strong interaction between the photons therefore causes a photon (Kerr) blockade of cavity transmission and this drastically reduces the photon number fluctuation. A decrease in photon number fluctuation implies a reduced mixing between the mechanical mode and the cavity field fluctuation. In the limit of high non-linearity, NMS would be negligibly small. In any case if $\eta < \kappa$, ε, the cavity would contain more than one photon and the above argument is still valid due to the photon–photon repulsion.

14.2.2 Optically coupled Fabry–Perot cavities

In the previous sections, we considered the dynamics of a Fabry–Perot cavity containing a Kerr medium and discussed in detail the dynamics of an NMS and the photon blockade mechanism. In this section, we consider another system having two Fabry–Perot cavities connected with each other through their movable mirrors (as shown in Fig. 14.4) and study the occurrence of NMS and ground state cooling in the system. Here mirror M_1 and mirror M_4 of the two optically coupled cavities are fixed and are partially transmitting whereas mirrors M_2 and M_3 are movable and totally reflecting. The two mirrors M_2 and M_3 can both oscillate under the effect of the radiation pressure. The motion of each mirror is the result of the excitation of many oscillation modes which can be either external or internal.

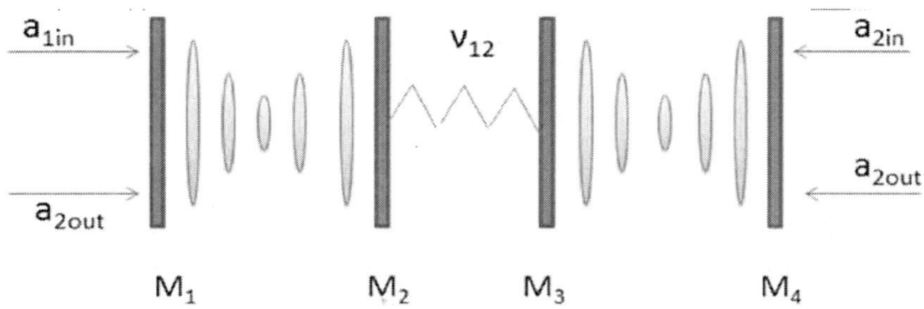

Figure 14.4 Schematic representation of two mechanically coupled Fabry–Perot cavities

The external modes correspond to pendulum modes which leads to global displacements of the mirror while the internal modes correspond to deformations of the mirror surface due to excitation of the internal acoustic modes of the mirror surface. These various degrees of freedom have different resonance frequencies and experimentally, it is possible to select the mechanical response of a single mode by using a bandpass filter in the detection circuit. Consequently, a single mechanical mode is considered for each movable mirror, which will be therefore described as a simple harmonic oscillator. The system under consideration is in contact with a thermal bath, in equilibrium with the thermal bath at temperature T. The movable mirrors are treated as quantum mechanical oscillators with masses m_1 and m_2, frequencies ν_1 and ν_2, and energy decay rates γ_{m_1} and γ_{m_2}, respectively of the mirrors M_1 and M_4. The system is coherently driven by two laser fields ($a_{1,\text{ in}}$ and $a_{2,\text{ in}}$) with frequencies ω_1 and ω_2 as shown in Fig. 14.4. Here, it is also assumed that the induced cavity resonance frequency shift of each cavity is much smaller than the longitudinal spacing, so the model is restricted to a single longitudinal mode for each cavity. Let ε_1 and ε_2 be the amplitudes of the two laser fields. As we know in a Fabry–Perot cavity, when a photon collides with the surface of the movable mirror, it exerts radiation pressure on the mirror and the force that the mirror will experience is proportional to the photon number inside the cavity. However, in present system, the force experienced by one of the movable mirrors, say M_2 not only depends on the number of photons of the corresponding cavity but also depends on the number of photons of the second cavity. This is because the two mirrors are coupled; therefore position of one mirror is influenced by the position of the other mirror.

The Hamiltonian of the system can be written as

$$H = \hbar\omega_1 a_1^\dagger a_1 + \hbar\omega_2 a_2^\dagger a_2 + \hbar\nu_1\left(b_1^\dagger b_1 + 1/2\right) + \hbar\nu_1\left(b_2^\dagger b_2 + 1/2\right) - \hbar g_1 a_1^\dagger a_1\left(b_1^\dagger e^{-i\theta_1} + b_1 e^{i\theta_1}\right)$$
$$-\hbar g_2 a_2^\dagger a_2\left(b_2^\dagger e^{-i\theta_2} + b_2^\dagger e^{i\theta_2}\right) + \hbar\nu_{12}\left(b_1^\dagger e^{-i\theta_1} + b_1 e^{i\theta_1}\right)\left(b_2^\dagger e^{-i\theta_2} + b_2 e^{i\theta_2}\right)$$
$$+ i\hbar\varepsilon_1\left(a_1^\dagger - a_1\right) + i\hbar\varepsilon_2\left(a_2^\dagger - a_2\right). \tag{14.19}$$

Here a_1 ($a_1^\dagger$) and a_2 ($a_2^\dagger$) are the annihilation (creation) operators of the two cavity fields, b_1 ($b_1^\dagger$) and b_2 ($b_2^\dagger$) are the phonon annihilation (creation) operators of the two movable mirrors M_2 and M_3, respectively. The parameters g_1 and g_2 are the coupling parameters between the cavity fields and fixed mirrors M_1 and M_4 respectively, ν_{12} is the coupling frequency of the two movable mirrors and θ_1 and θ_2 are the phases of the two movable mirrors. The phases θ_1 and θ_2 can be thought of as arising from the complex mirror–photon coupling strengths $g_i(i = 1, 2)$.

14.2.3 Small fluctuations dynamics: Normal mode splitting and cooling of micromirrors

In this subsection, we show that the coupling of the two mechanical oscillators and the two cavity field fluctuations leads to the splitting of the normal mode into two modes (normal mode splitting (NMS)) for each cavity depending on the system parameters. The opto-mechanical NMS however involves driving four parametrically coupled non-degenerate modes out of equilibrium. The NMS does not appear in the steady state spectra but rather manifests itself in the fluctuation spectra of the mirror displacement. In order to study the dynamics of the coupled mirror, the fluctuations in the mirror's need to be determined position. Here, the problem involved is non-linear. It is assumed that the non-linearity is small. Therefore, we study the dynamics of fluctuations around the steady state of the system. Each canonical operator of the system is written as a sum of its steady state mean value and a small fluctuation with zero mean value,

$$q_1 \to q_{1s} + \delta q_1,\ q_2 \to q_{2s} + \delta q_2.$$
$$p_1 \to p_{1s} + \delta p_1,\ p_2 \to p_{2s} + \delta p_2. \tag{14.20}$$
$$a_1 \to a_{1s} + \delta a_1,\ a_2 \to a_{2s} + \delta a_2.$$

In order to achieve ground state cooling, $\gamma_{m_i} \ll \kappa_i$, $g_i < \nu_i$ and $\nu_i > \kappa_i$ (with $i = 1, 2$). The last condition is the resolved side band regime necessary for ground state cooling. Note that these conditions necessary for cooling also implies that the system is stable. Linearized Heisenberg–Langevin equations for the fluctuation operators are obtained as:

$$\delta\dot{q}_1 = \nu_1\delta p_1 + A\delta p_2 + B\delta q_2 + 2D_2 a_{1s}\delta a_1^\dagger + 2D_2 a_{1s}^*\delta a_1, \tag{14.21}$$

$$\delta\dot{q}_2 = \nu_2\delta p_2 + A\delta p_1 - B\delta q_1 + 2D_4 a_{2s}\delta a_2^\dagger + 2D_4 a_{2s}^*\delta a_2, \tag{14.22}$$

$$\delta\dot{p}_1 = -\nu_1\delta q_1 - A\delta q_2 + B\delta p_2 + 2D_1 a_{1s}\delta a_1^\dagger + 2D_1 a_{1s}^*\delta a_1 - \gamma_{m_1}\delta p_1 + \sqrt{2\gamma_{m_1}}\delta p_{\text{in}}^1, \tag{14.23}$$

$$\delta\dot{p}_2 = -\nu_2\delta q_2 - A\delta q_1 - B\delta p_1 + 2D_3 a_{2s}\delta a_2^\dagger + 2D_3 a_{2s}^*\delta a_2^\dagger - \gamma_{m_2}\delta p_2 + \sqrt{2\gamma_{m_2}}\delta p_{\text{in}}^2, \tag{14.24}$$

$$\delta\dot{a}_1 = -i\omega_1\delta a_1 + iD_1(a_{1s}\delta q_1 + q_{1s}\delta a_1) - iD_2(a_{1s}\delta p_1 + p_{1s}\delta a_1) - \kappa_1\delta a_1 + \sqrt{2\kappa_1}\delta C_{\text{in}}^1 \tag{14.25}$$

$$\delta\dot{a}_2 = -i\omega_2\delta a_2 + iD_3(a_{2s}\delta q_2 + q_{2s}\delta a_2) - iD_4(a_{2s}\delta p_2 + p_{2s}\delta a_2) - \kappa_2\delta a_2 + \sqrt{2\kappa_2}\delta C_{\text{in}}^2, \tag{14.26}$$

where $\left(b_i + b_i^\dagger\right) = q_i$ and $i\left(b_i^\dagger - b_i\right) = p_i$; $i = 1, 2$.

Also $D_1 = g_1\cos(\theta_1)$, $D_2 = g_1\sin(\theta_1)$, $D_3 = g_2\cos(\theta_2)$, $D_4 = g_2\sin(\theta_2)$, $A = \nu_{12}\cos(\theta_2 - \theta_1)$, $B = \nu_{12}\sin(\theta_2 - \theta_1)$.

C_{in}^1 and C_{in}^2 are input noise operators with zero mean value and who obey the following commutation relations:

$$\left\langle \delta C_{\text{in}}^i(t)\delta C_{\text{in}}^{j\dagger}(t')\right\rangle = \delta_{ij}(t-t'), \left\langle \delta C_{in}^i \delta C_{\text{in}}^j\right\rangle = \left\langle \delta C_{\text{in}}^{i\dagger}\delta C_{\text{in}}^{j\dagger}\right\rangle = 0. \tag{14.27}$$

Also $p_{\text{in}}^i = i\left(\xi^{i\dagger} - \xi^i\right)$, ξ is the Brownian noise operator arising due to the thermal bath. The Brownian noise operator has zero mean value and obeys the following correlation at temperature T:

$$\left\langle \xi^i(t)\xi^{j\dagger}(t')\right\rangle = 2\gamma_{m_i}(1 + 2n_T)\delta_{ij}(t-t'), \tag{14.28}$$

$$\left\langle \xi^i(t)\xi^j(t')\right\rangle = \left\langle \xi^{i\dagger}(t)\xi^{j\dagger}(t')\right\rangle = \left\langle \xi^{i\dagger}(t)\xi^j(t')\right\rangle = 0, \tag{14.29}$$

and $n_T = \coth\left(\dfrac{\hbar\omega}{2k_B T}\right)$,

where k_B is the Boltzmann constant and T is the temperature of the thermal bath. Now the spectrum of fluctuation of a mirror can be defined as

$$S_q(\omega) = \frac{1}{4\pi}\int d\Omega e^{-i(\omega+\Omega)t}\,\langle \delta q(\omega)\delta q(\Omega) + \delta q(\Omega)\delta q(\omega)\rangle.$$

The displacement spectrum of mirror M_2, i.e, $S_{q_1}(\omega)$ is finally obtained as

$$\begin{aligned} S_{q_1}(\omega) = \frac{1}{d(\omega)d(-\omega)}\Bigg[&\left(\frac{4|a_{1s}|^2\kappa_1}{\kappa_1^2 + (\omega - \Delta_1)^2} + \frac{4|a_{1s}|^2\kappa_1}{\kappa_1^2 + (\omega + \Delta_1)^2}\right)X(\omega)\,X(-\omega) \\ &+ \left(\frac{4|a_{2s}|^2\kappa_1}{\kappa_2^2 + (\omega - \Delta_2)^2} + \frac{4|a_{2s}|^2\kappa_1}{\kappa_2^2 + (\omega + \Delta_2)^2}\right)Y(\omega)\,Y(-\omega) \\ &+ (1 + 2n_T)\left(\gamma_{m_1}Z(\omega)Z(-\omega) + \gamma_{m_2}T(\omega)T(-\omega)\right)\Bigg]. \end{aligned} \tag{14.30}$$

Here the commutation relation for ξ defined in Eqs (14.28) and (14.29) is used. In Eq. (14.30), the first two terms corresponding to $X(\omega)X(-\omega)$ and $Y(\omega)Y(-\omega)$ are due to the radiation pressure contribution of the optical modes in the two cavities, whereas the last term is due to the thermal noise contribution from the two cavities. In Fig. 14.5, the contour plot of the displacement spectrum $S_{q_1}(\omega)$ is shown as a function of dimensionless effective detuning $\frac{\Delta_1}{\nu}$ for $\theta_1 = 0$, $\theta_2 = 2\pi/3$, $\frac{\Delta_2}{\nu} = 0$, $\frac{\gamma_1}{\nu} = \frac{\gamma_2}{\nu} = 0.005$, $\frac{g_1}{\nu} = \frac{g_2}{\nu} = 0.25$, $\frac{\nu_{12}}{\nu} = 0.3$, $\frac{\kappa_1}{\nu} = \frac{\kappa_2}{\nu} = 0.15$ for different values of the photon numbers in the two cavities, (A): $|a_{1s}|^2 = 0.15$ and $|a_{2s}|^2 = 0.15$, (B):$|a_{1s}|^2 = 0.15$ and $|a_{2s}|^2 = 0.30$, (C): $|a_{1s}|^2 = 0.30$ and $|a_{2s}|^2 = 0.15$, (D): $|a_{1s}|^2 = 0.30$ and $|a_{2s}|^2 = 0.30$. Clearly, four modes are visible corresponding to the two mechanical and two optical modes. The coupling of the cavity field fluctuations and the mirror fluctuations leads to splitting of the normal mode of each cavity into two modes (NMS). The NMS is associated with the mixing between the fluctuation of the cavity field around the steady state and the fluctuations of the mirror mode around the mean field. The origin of the fluctuations of the cavity field is the beat of the pump photons with the photons scattered from the mirrors. It is observed from the displacement spectra that NMS is observed only in plots (A) and (B) where the photon number in the first cavity is $|a_{1s}|^2 = 0.15$. Increasing the photon number in the first cavity destroys the NMS. Comparing plots (B) and (C), it can be noted that decreasing the photon number in the second cavity to $|a_{2s}|^2 = 0.15$ do not restore the NMS.

Radiation pressure couples the mirror to the optical cavity mode, which behaves as an effective additional reservoir for the mechanical oscillator. As a consequence, the effective temperature of the mirror mode will be intermediate between the initial thermal reservoir temperature and that of the optical reservoir, which is practically zero due to the condition that the mean number of photons is extremely small. Therefore one can approach the mechanical ground state of the mirror when the number of photons is small. In the present case, the mechanical mode of mirror M_2 not only couples to the optical mode of the first cavity but also to the optical mode of the second cavity via the mechanical mode of the mirror M_3. This explains why significant mechanical cooling of the mechanical mode is obtained when the number of photons in both the cavities is low. Figure 14.6 shows the plot of T_{eff} of the mirror M_2 as a function of detuning $\frac{\Delta_1}{\nu}$ corresponding to the cases shown in Fig. 14.5.

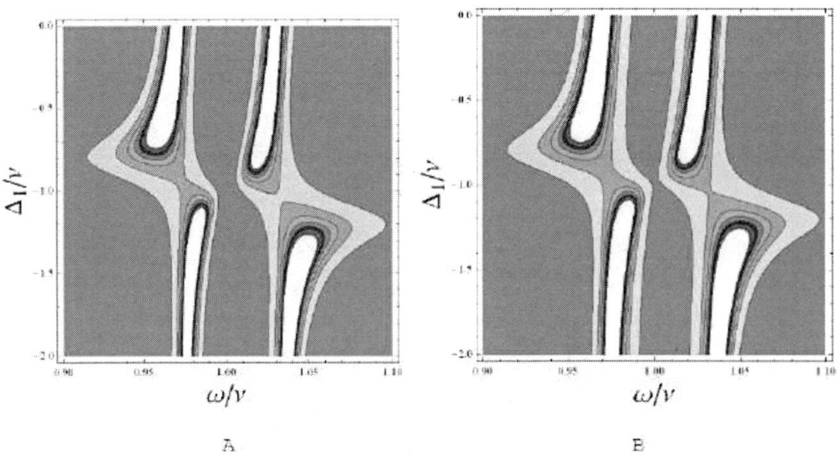

Figure 14.5 Contour plot of the displacement spectrum $S_{q1}(\omega)$ as a function of normalized effective detuning $\frac{\Delta_1}{\nu}$ for the following parameters: $\theta_1 = 0$, $\theta_2 = 2\pi/3$, $\frac{\Delta_2}{\nu} = 0$, $\frac{\gamma_1}{\nu} = \frac{\gamma_2}{\nu} = 0.005$, $\frac{g_1}{\nu} = \frac{g_2}{\nu} = 0.25$, $\frac{\nu_{12}}{\nu} = 0.03$, $\frac{\kappa_1}{\nu} = \frac{\kappa_2}{\nu} = 0.15$. (A): $|a_{1s}|^2 = 0.15$ and $|a_{2s}|^2 = 0.15$, (B): $|a_{1s}|^2 = 0.15$ and $|a_{2s}|^2 = 0.30$, (C): $|a_{1s}|^2 = 0.30$ and $|a_{2s}|^2 = 0.15$, (D): $|a_{1s}|^2 = 0.30$ and $|a_{1s}|^2 = 0.30$

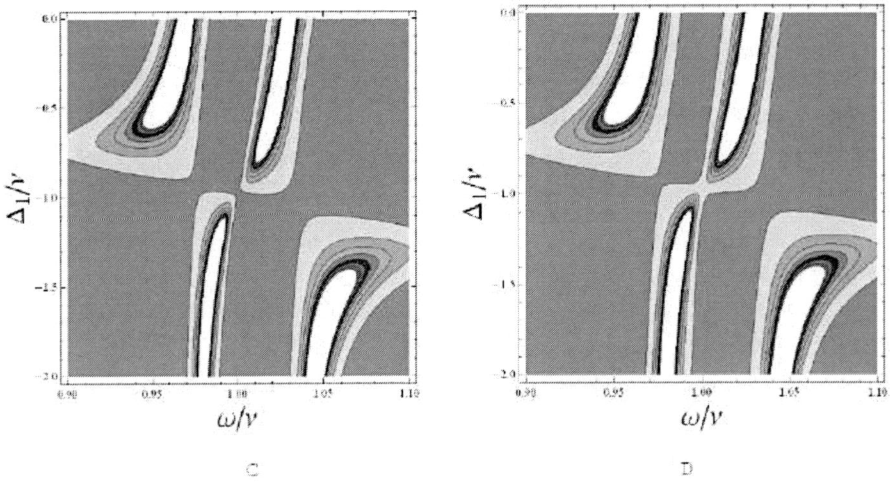

Figure 14.5 Contd...

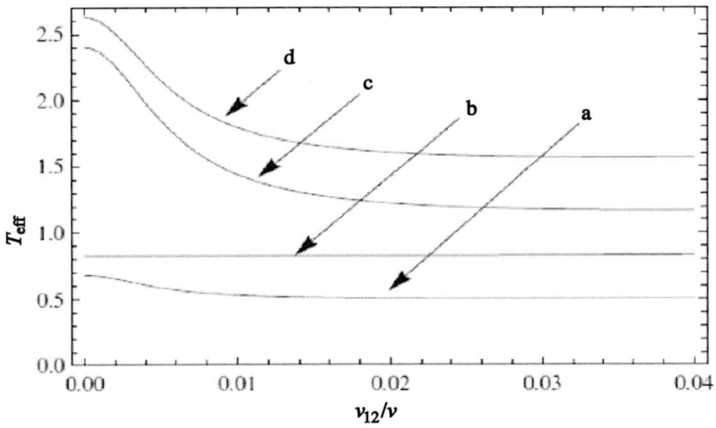

Figure 14.6 Plot of effective temperature vs coupling strength. Parameters used are the same as in Fig. 14.5

14.3 Conclusion

In conclusion, this chapter provides a brief review of the quantum mechanical system in quantum regime. The various properties of a Fabry–Perot cavity and their input–output formalism are discussed. It is shown that a non-linear Kerr medium present inside a cavity can be used as a tuneable quantum mechanical device by a photon blockade mechanism via photon blockade mechanism. The various aspects of cooling of a micromirror in two optically coupled micromirrors are discussed and the effect of photon numbers present inside the cavities in the process of normal mode splitting as well as in the cooling of a micro-mrchanical mirror is shown.

Bibliography

1. T. W. Hansch and A. L. Schawlow. 1975. *Optics Comm.* 13: 68.
2. D. J. Wineland, R. E. Drullinger and F. L. Walls. 1978. *Phy. Rev. Letts.* 40: 1639.
3. S. Chu, L. Hollberg, J. E. Bjorkholm, A. Cable and A. Ashkin. 1985. *Phy. Rev. Letts.* 55: 48.
4. C. M. Caves. 1980. Phys. Rev. Letts. 45: 75 R. Loudon 1981. *Phys. Rev. Letts.* 47: 815.
5. V. B. Braginsky and F. Ya Khalili. 1992. *Quantum Measurements*, Cambridge: Cambridge University Press.
6. B. Abbott et al. 2004. Phys. Rev. D. 69: 122004. F. Acernese et al. 2005. *Class. Quantum Grav.* 22 S869.
7. T. Corbitt and N. Mavalvala. 2004. *J. Opt. B: Quantum Semi-class. Opt.* 6: S675.
8. T. Corbitt, Y. Chen, E. Innerhofer, H. Mller-Ebhardt, D. Ottaway, H. Rehbein, D. Sigg, S. Whitcomb, C. Wipf and N. Mavalvala. 2007. *Phy. Rev. Letts.* 98: 150802.
9. C. Hohberger-Metzger and K. Karrai. 2004. *Nature* 432: 1002.
10. S. Gigan, H. R. Bhm, M. Paternostro, F. Blaser, G. Langer, J. B. Hertzberg, K. Schwab, D. Buerle, M. Aspelmeyer and A. Zeilinger. 2006. *Nature* 444: 67.
11. O. Arcizet, P. F. Cohadon, T. Briant, M. Pinard and A. Heidmann. 2006. *Nature* 444: 71.
12. D. Kleckner and D. Bouwmeester. 2006. *Nature* 444: 75.
13. J. D. Thompson, B. M. Zwickl, A. M. Jayich, Florian Marquardt, S. M. Girvin and J. G. E. Harris. 2008. *Nature* 452: 72.
14. F. Brennecke, Stephan Ritter*, T. Donner and T. Esslinger. 2008. *Science* 322: 235.
15. K. W. Murch, K. L. Moore, Subhadeep Gupta and D. M. Stamper-Kurn. 2008. *Nature Physics* 4: 561. G. De. Chiara et al. 2011. *Phys. Rev. A.* 83: 052324. S. K. Steinke et al. 2011. *Phys. Rev. A.* 84: 023834. D. Hunger et al. 2011. *Comptes Rendus Physique* 12: 871. B. Chen et al. *J. Opt. Soc. Am.* 28: 2007.
16. A. B. Bhattacherjee. 2009. *Phys. Rev. A.* 80: 043607.
17. A. B. Bhattacherjee. 2010. *J. Phys. B.* 43: 205–301.
18. A. Schliesser, O. Arcizet, R. Riviere, G. Anetsberger and T. J. Kippenberg. 2009. *Nature Physics* 5: 509.
19. C. Fabre, M. Pinard, S. Bourzeix, A. Heidmann, E. Giacobino and S. Reynaud. 1994. *Phys. Rev. A.* 49: 1337.
20. S. Mancini and P. Tombesi. 1994. *Phys. Rev. A.* 49: 4055.
21. F. Marquardt, J. G. E. Harris and S. Girvin. 2006. *Appl. Phys. Lett.* 96: 103901.
22. C. Genes, A. Mari, D. Vitali and P. Tombesi. *arXiv*: 0901.2726.
23. Jin-Sheng Peng and Gao-Xiang Li. 1998. *Introduction to modern quantum optics*. Singapore: World Scientific.
24. E. T. Jaynes and F. W. Cummings. 1963. *Proc. IEEE.* 51: 89.
25. J. Kepler, Letter to Galileo Galilei (1610).
26. J. Kepler, De Cometis Libelli Tres (1619).

27. E. F. Nichols and G. F. Hull. 1903. 'About radiation pressure.' *Ann. Phys.* 12: 225.
28. P. Lebedev. 1901. 'Untersuchungen uber die druckkrafte des lichtes.' *Ann. Phys.* 6: 433.
29. V. B. Braginsky. 1967. *Eksp. Teor. Fiz.* 53: 1434.
30. V. B. Braginsky and Y. I. Vorontsov. 1975. *Sov. Phys. Usp.* 17: 644.
31. C. M. Caves. 1980. *Phys. Rev. Lett.* 45: 75.
32. P. Meystre, E. M. Wright, J. D. McCullen and E. Vignes. 1985. *J. Opt. Soc. Am. B.* 2: 1830.
33. K. Jacobs, P. Tombesi, M. J. Collett and D. F. Walls. 1994. *Phys. Rev. A.* 49: 1961.
34. M. Pinard, C. Fabre and A. Heidmann. 1995. *Phys. Rev. A.* 51: 2443.
35. C. Fabre, M. Pinard, S. Bourzeix, A. Heidmann, E. Giacobino and S. Reynaud. 1994. *Phys. Rev. A.* 49: 1337.
36. S. Mancini and P. Tombesi. 1994. *Phys. Rev. A.* 49: 4055.
37. S. Mancini, D. Vitali and P. Tombesi. 1998. *Phys. Rev. Lett.* 80: 688.
38. S. Bose, K. Jacobs and P. L. Knight. 1997. *Phys. Rev. A.* 56: 4175.
39. J. Zhang, K. Peng and S. L. Braunstein. 2003. *Phys. Rev. A.* 68: 013808.
40. S. Groblacher, K. Hammerer, M. R. Vanner and M. Aspelmeyer. 2009. *Nature* 460: 724.
41. S. Gigan, H. R. Bohm, M. Paternostro, F. Blaser, G. Langer, J. B. Hertzberg, K. C. Schwab, D. Bauerle, M. Aspelmeyer and A. Zeilinger. 2006. *Nature* 444: 67.
42. O. Arcizet, P. F. Cohadon, T. Briant, M. Pinard and A. Heidmann. 2006. *Nature* 444: 71.
43. T. Kippenberg, H. Rokhsari, T. Carmon, A. Scherer and K. Vahala. 2005. *Phys. Rev. Lett.* 95: 033901.
44. I. Favero and K. Karrai. 2009. *Nature Photonics* 3: 201.
45. F. Marquardt and S. M. Girvin. 2009. *Physics* 2: 40.
46. D. J. Wineland and W. M. Itano. 1979. *Phys. Rev. A.* 20: 1521.
47. M. J. Collett and C. W. Gardiner. 1984. *Phy. Rev. A.* 30: 1386.
48. D. F. Walls and G. J. Milburn. *Quantum Optics*, 2*nd* edition, Verlag, Heidelberg: Berlin, Springer.
49. S. Huang and G. S. Agarwal. 2009. *Phys. Rev. A.* 79: 013821.
50. J. M. Dobrindt, I. Wilson-Rae and T. J. Kippenberg. 2008. *Phys. Rev. Letts.* 101: 263602.
51. S. Mancini, D. Vitali, V. Giovannetti and P. Tombesi. 2003. *Eur. Phys. J. D.* 22: 417.
52. S. Mancini, V. Giovannetti, D. Vitali and P. Tombesi. 2002. *Phys. Rev. Letts.* 88: 120401.
53. M. Ludwig, K. Hammerer and F. Marquardt. 2010. *Phys. Rev. A.* 82: 012333.
54. Bhattacharya and P. Meystre. 2007. *Phys. Rev. Letts.* 99: 073601.
55. C. Genes, D. Vitali, P. Tombesi, S. Gigan and M. Aspelmeyer. 2008. *Phys. Rev. A.* 77: 033804.
56. A. Dantan, C. Genes, D. Vitali and M. Pinard. 2008. *Phys. Rev. A.* 77: 011804(R).
57. S. Mancini, D. Vitali and P. Tombesi. 2003. *Phys. Rev. Lett.* 90: 137901.
58. S. Pirandola, D. Vitali, P. Tombesi and S. Lloyd. 2006. *Phys. Rev. Lett.* 97: 150403.

59. A. Imamoglu, H. Schmidt, G. Woods and M. Deutsch. 1997. *Phys. Rev. Letts*. 79: 1467.
60. Tarun Kumar, Aranya B. Bhattacherjee and Man Mohan. 2010. *Phys. Rev. A*. 81: 013835.
61. A. D. Greentree, C. Tahan, J. H. Cole and Lloyd C. L. Hollenberg. 2006. *Nature Physics* 2: 856.

V
Laser Applications

15 Generation of Mid-infrared and Terahertz Radiation for Defence Applications

A. K. Gupta, Ajay Mishra, Maria Farooqui, Nimish Dixit and A. N. Kaul

15.1 Introduction

Broadband optical sources find a wide range of inter-disciplinary application areas encircling science, medicine, defence, engineering etc. They have long been used owing to their spectroscopic applications. Recently, these sources have found their place in defence applications also, the most notable application being the directed infrared countermeasure (DIRCM) system and light detection and ranging (LIDAR) for detection of chemical and biological agents (an application that falls in the domain of spectroscopy). Tunability in the wavelength can be achieved by a number of techniques. Common tunable lasers include dye laser, tunable Excimer lasers, tunable solid state lasers, semiconductor lasers, free electron lasers (FEL) etc. However, for wide tunability, researchers have started exploiting the second order optical non-linearity associated with certain materials by means of optical parametric processes. The arrival of new non-linear optical materials as alternatives to the traditional oxides and silver-containing compounds has allowed development of compact and efficient optical parametric oscillators (OPO) with tunability extending from mid-infrared (mid-IR) to far-infrared wavelengths (also known as terahertz waves).

Optical parametric generation (OPG) essentially relies on phase matching. Phase matching can either be achieved by exploiting inherent birefringence of the material, in which case it is termed as birefringent phase matching (BPM), or by quasi phase matching (QPM), which is based on periodically reversing the non-linear coefficient of the material[1]. In QPM, the accumulated phase mismatch is offset by modulating the second order non-linear coefficient (d_{33}) at a period that is twice the coherence length[1]. The distinct advantages of QPM over BPM have encouraged researchers to develop periodically poled materials. One of the most popular materials for accompli-shing QPM is lithium niobate (LN). A most efficient and commonly used technique to realize QPM structure is electric field poling (E-field poling) in which a high electric field greater than the material's coercive field is applied across the thickness of the crystal[1–2]. For example, periodically poled lithium niobate (PPLN) can be fabricated with this technique and it has been widely used for broadband mid-IR generation using OPO[2].

Recently, there has been a lot of interest in THz generation using PPLN. Although conventional PPLN may be used for THz generation using optical rectification of femtosecond pulses[3], the geometries of PPLN grating for THz generation using nanosecond laser is significantly different. At the same time, build-up of the THz power is also different from that during mid-IR generation from PPLN[4–5]. Apart from PPLN, a collinearly phase matched (CPM) material, DAST (4-dimethylamino-N-methyl-4-stilbavolium tosylate) has also been widely used for THz generation by exploiting second order optical non-linearity[6].

In this chapter, non-linear optical methods and material for mid-IR and THz generation have been discussed. In Section 15.2, the material and sources for mid-IR generation will be elaborated with more emphasis on QPM materials. Section 15.3 discusses the defence applications of mid-IR laser. Subsequently, the development of PPLN based mid-IR tunable laser with PPLN fabrication facility at IRDE is discussed in Section 15.4. Different optical methods for THz generation along with materials have been elucidated in Section 15.5 followed by application of THz sources in Section 15.6 THz generation in PPLN as well as how the PPLN used for the generation of THz waves differs from the PPLN used for mid-IR generation has been discussed in Section 15.7. This would include the difference in the grating period as well as the grating structure. In Section 15.8, THz generation using difference frequency generation (DFG) in DAST is explained.

15.2 Sources and Material for Mid-IR Generation

There are a number of portions for mid-IR laser sources. In most of the cases, the required category of the laser is driven by a particular application. There are a wide variety of criterion that mid-IR laser has to satisfy for its applications, which include sufficient optical power, suitable linewidth for selectivity, required modal characteristics, wavelength tunability, excellent beam quality, small beam divergence, beam pointing stability etc. In addition to all these, for defence applications, mid-IR lasers must have a compact size, lesser weight and high level of robustness. Figure 15.1 depicts the spectral coverage of the most frequently employed tunable mid-IR sources[7].

It can be observed from this figure, that there are a variety of mid-IR sources which include gas lasers, all solid state lasers, quantum cascade laser (QCL)[8] etc. Recently, QCLs have been widely researched upon and commercial QCLs have been developed for various applications in the mid-IR region. However, mid-IR tunable laser based on non-linear parametric processes such as OPO remains an active area of research. These kinds of lasers possess high powers, excellent beam quality, smaller beam divergence, variable repetition rates etc. In addition, these lasers have a high degree of robustness due to their all solid state nature. In this chapter, we shall focus our attention to these lasers which are based on non-linear parametric conversion only.

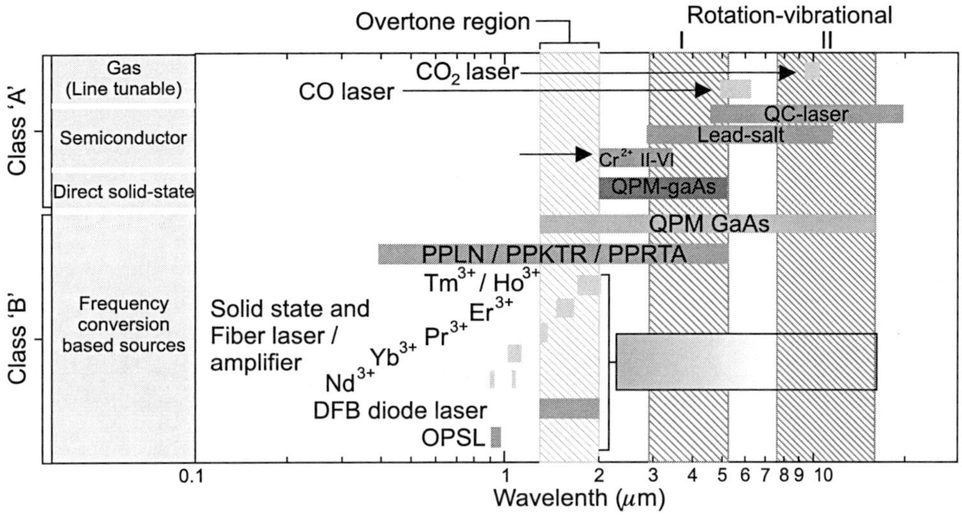

Figure 15.1 Tunable laser sources and their typical wavelength coverage. Atmospheric window (3–5 μm and 8 – 16 μm) are also given[7]

The choice of a suitable material for non-linear parametric conversion is governed by several requirements that the material has to fulfill. The material has to possess the relevant second order non-linear susceptibility. This excludes all materials with inversion symmetry. The material should be transparent at all the frequencies involved. If this is not the case, the effective interaction length is limited by absorption. Another necessity is that the damage threshold should be high enough so that high intensities can be tolerated. Another important factor which needs attention is its thermal conductivity. Materials, which have good thermal conductivity, show less thermal lensing and photorefractive effects. Generally, non-linear materials are classified into two main categories, viz., oxide crystals (such as $LiNbO_3$, $KNbO_3$, $LiTaO_3$, $KTiOPO_4$ etc.) and semiconductors (such as GaAs, ZnSe etc.)[8]. $LiNbO_3$ (LN), $KNbO_3$, $LiTaO_3$ (LT) and $KTiOPO_4$ (KTP) are ferroelectric and can be periodically poled using E-field poling technique. A comparison of three widely used materials, namely, LN, LT and KTP is shown in Table 15.1. Although all these materials have been widely used for mid-IR generation, $LiNbO_3$ has emerged as the most interesting material due to its highest value of figure of merit (FOM).

Table 15.1 *Comparison of non-linear materials*

Crystal	$d_{eff}^{1)}$ (pm/V)	Transparency (μm)	$E_{th}^{2)}$ (J/cm^2)	FOM ($E_{th} \times d_{eff}^2$)
$LiNbO_3$	17.2	0.4 – 5.0	30	8875
$LiTaO_3$	8.8	0.28 – 5.0	41	3213
$KTiOPO_4$	8.7	0.35 – 4.5	40	3041

$^{1)}$Non-linear coefficient; $^{2)}$Damage threshold

LN is chosen for periodic poling and wavelength generation in mid-IR region. The performance of PPLN, however, is limited due to photorefractive damage that occurs at room temperature. This effect can be alleviated either by heating the PPLN chip at elevated temperatures or by doping it with MgO, which is popularly known as MgO doped PPLN (MgO:PPLN)[9].

15.3 Defence Applications of Mid-IR Laser

As discussed earlier, although mid-IR sources have a number of defence and civilian applications, our main focus is to elaborate on the defence applications of these sources. Two major applications of these lasers are in LIDAR and DIRCM systems. In the former case, these lasers have been used to detect chemical and biological agents, which have spectroscopic signatures in this band. Therefore, application in LIDAR is inclined towards spectroscopy in which these lasers should have a very precise linewidth for high degree of selectivity and sensitivity[10].

DIRCM systems are used for defeating/deceiving heat seeking missiles. The seeker head of heat seeking IR missiles use I, II and IV band of mid-IR spectrum, which can be covered by these lasers[11]. The application of these lasers in DIRCM system is owing to the fact that these lasers have very high power, high repetition rates, excellent directionality and tunability in these bands[11]. Although there are different sources for mid-IR wavelength generation (Fig. 15.1), mid-IR OPO based on non-linear frequency conversion in PPLN have been a source of choice for applications in DIRCM systems. Northrop Grumman has produced the first commercial DIRCM system using PPLN based mid-IR laser that they call as VIPER laser[12]. PPLN has been a material of choice due to the reasons stated in Section 15.2.

15.4 PPLN Based Mid-IR Tunable OPO

In the preceding sections, we have explained the applications of mid-IR tunable laser and the advantages of PPLN as a non-linear parametric converter. In this section, we will elucidate the detailed mechanism involved in the development of PPLN based mid-IR OPO at IRDE along with the PPLN fabrication facility that has been done for the first time in the country[13].

In OPG, a pump photon breaks into two photons of higher and lower frequency due to second order non-linear interaction in the material. By convention, the higher frequency photon is called signal and that with the lower frequency is called idler photon. For efficient non-linear parametric conversion, these photons must satisfy the conditions of energy and momentum conservation, popularly known as phase matching. In the case of QPM these two conditions can be written as

$$\frac{1}{\lambda_p} = \frac{1}{\lambda_s} + \frac{1}{\lambda_i}, \tag{15.1}$$

$$\Delta k = 2\pi\left(\frac{n_p}{\lambda_p} - \frac{n_s}{\lambda_s} - \frac{n_i}{\lambda_i} - \frac{1}{\Lambda}\right) = 0, \tag{15.2}$$

where the first order QPM has been considered. λ_p, λ_s and λ_i are the pump, signal and idler wavelengths; n_p, n_s and n_i are extraordinary refractive indices at pump, signal and idler wavelengths respectively; Λ is the period of QPM domain grating period. It can be seen from Eq. (15.1) that any wavelength within the transparency range of the material can be generated by suitably tailoring the grating period for a particular pump wavelength. These two equations can easily be solved by using temperature dependent Sellmeier equations[14] and the grating period required for a particular wavelength generation can be estimated. A typical wavelength tuning curve with respect to grating period for PPLN is shown in Fig. 15.2.

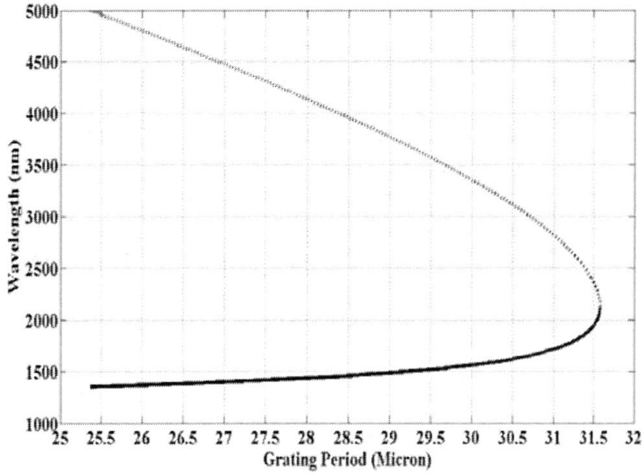

Figure 15.2 PPLN wavelength tuning curve

15.4.1 PPLN fabrication

As discussed before, IRDE has established a complete setup for periodic poling of lithium niobate (LN) using E-field poling technique[13]. PPLN fabrication is achieved at room temperature by the electric field poling process. A positive voltage pulse slightly exceeding the coercive field of LN (~21 kV/mm) is applied on the $z+$ patterned crystal face by using a LiCl liquid electrolyte. The LN wafers of 0.5 mm thickness have been poled using this setup. Fanned grating and multi-grating PPLN chips of 50 mm length have been fabricated with very uniform period of almost 50% duty cycle and are shown in Fig. 15.3. In fanned grating, PPLN period continuously varies from 25.5–31.5 μm while, in multi-grating PPLN chip, the same variation is accomplished in the steps of 0.5 μm. As can be understood from Fig. 15.2, this period variation of 25.5–31.5 μm is required for the generation of 2–5 μm tunable radiation.

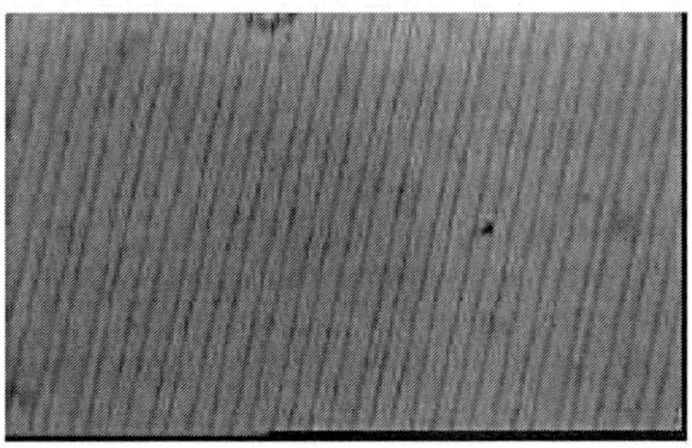

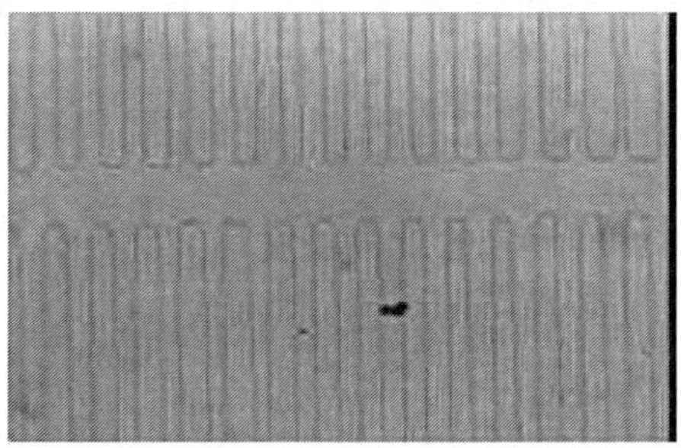

Figure 15.3 Fanned grating PPLN chip (left) multi grating PPLN chip (right)

15.4.2 OPO based on PPLN

IRDE has setup a bench-top experiment for PPLN based 2.8–4.8 μm tunable nanosecond OPO[2] for counter-measure applications. The modified experimental results are presented here. The pump beam from a Q-switched 1064 nm diode pumped solid state (DPSS). Nd:YVO_4 laser was focused onto a PPLN chip. This laser generates a pulse of nearly 10 ns at 25 KHz repetition rate. MgO:PPLN was kept inside an oven integrated with a temperature controller at 50°C with a stability of ±0.1°C. OPO cavity mirrors with appropriate broadband coatings are used so as to resonate the signal. Therefore, a singly resonant linear OPO cavity configuration has been setup in which only the signal beam is resonated and the idler beam is automatically generated due to energy conservation requirements. Signal beam is resonated because firstly, it has a better overlap with the pump beam as compared to the idler beam and secondly, it has a smaller wavelength variation which lead to simpler mirror coatings.

The multi-grating MgO:PPLN chip used in the experiment has QPM gratings with periods varying from 25.5 μm to 31.5 μm in steps of 0.5 μm. The dimensions of the chip were 50.0 mm (L) × 16.0 mm (W) × 0.5 mm (T). This specific grating period corresponds to the generation of the signal beam from 1.37 to 1.64 μm and idler beam from 2.8 to 4.8 μm respectively. It is to be noted that the wavelength tuning curve and grating period for MgO:PPLN will be different from that shown in Fig. 15.2 because of the difference in Sellmeier equations[14]. Power variation with the change in the input pump power is shown in Fig. 15.4 for a grating period of 30.5 μm which gives rise to a signal generation of 1.53 μm and idler generation of 3.47 μm. The optic-optic (signal + idler) and idler conversion efficiencies as a function of input pump power are depicted in Fig. 15.5. It can be observed from this figure that maximum idler conversion efficiency is obtained to be 12.5%.

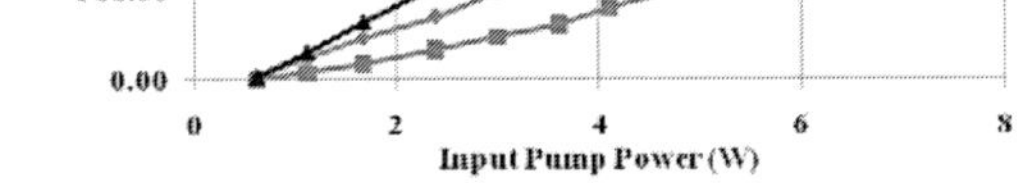

Figure 15.4 Output power variation as a function of input pump power for a grating period of 30.5 μm

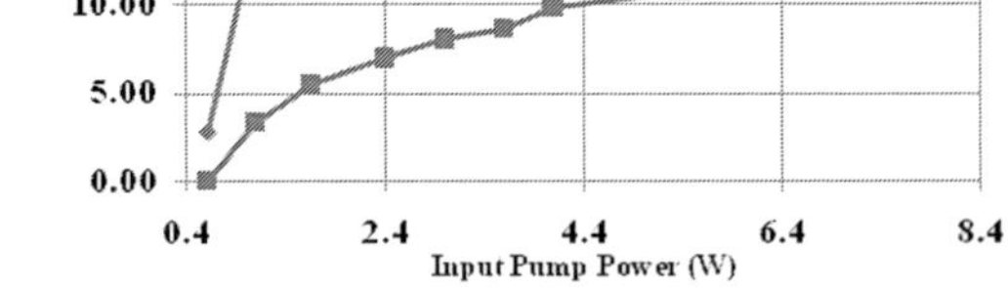

Figure 15.5 Conversion efficiency variation as a function of input pump power for grating period of 30.5 μm

Figure 15.6 depicts the experimentally observed output signal and idler wavelengths from OPO. It can be seen from this figure that they are in good agreement with theoretical values of the wavelengths. A prototype has also been developed for PPLN based tunable mid-IR OPO, which is shown in Fig. 15.7. Final specifications of the prototype are shown in Table 15.2.

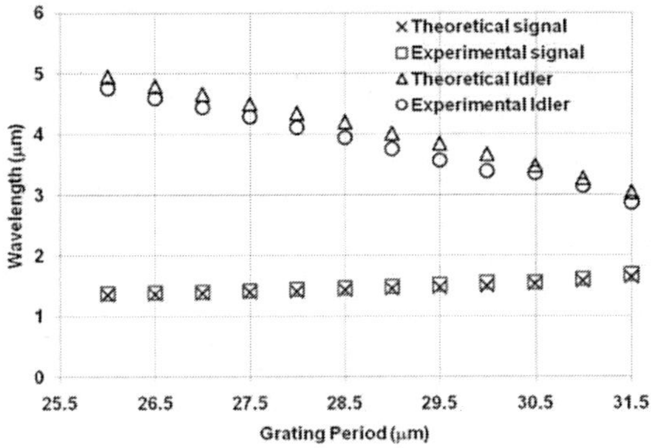

Figure 15.6 Experimentally measured signal/idler wavelength and comparison with numerically obtained results

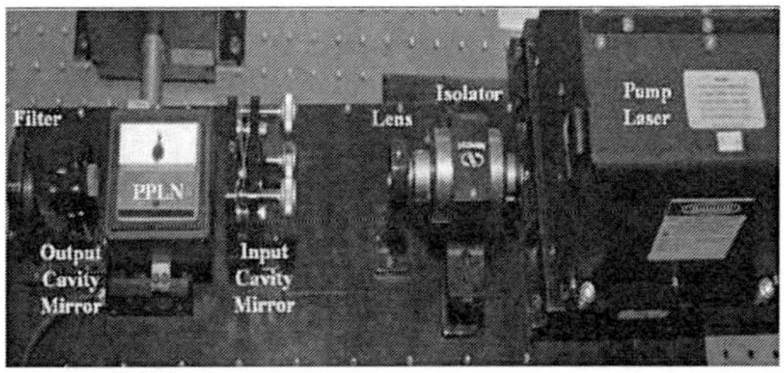

Figure 15.7 Prototype for tunable OPO developed at IRDE

Table 15.2 *Specifications of OPO prototype*

Pump source	DPSS Nd:YVO_4 Laser@1.064 μm
Pump power	7.3 W
Pulse width	~ 10 ns
Pulse repetition rate	25 KHz
PPLN grating period	25.5 to 31.5 μm
PPLN dimensions (length × width × thickness)	$50 \times 15.7 \times 1$ mm^3
Output wavelength	2.8 to 4.7 μm tunable
Output power	~ 1 W Average
Dimension (L×W×H) and weight	570 mm × 161 mm × 137 mm, ~ 9 kg

15.5 Methods and Materials for THz Generation

There are a number of methods available for THz generation approaching either from the microwave side or from the photonic side. The optical methods for generation of THz include free electron lasers (FEL)[15], photoconductive switching[16], non-linear optical methods[17], THz-QCL[18] etc. THz-QCL is a semiconductor device that operates on intersubband transitions. When an electric current flows through the multiple layer structure of a QCL, electrons cascade down the energy staircase of the intersubband[18]. This has been a recent area of research for providing THz radiation in a very compact package. However, room temperature operation of THz-QCL has been a challenge to researchers[19]. Non-linear optical techniques for generating THz radiation, on the other hand, are very attractive for generating high power THz radiation at room temperature in a compact manner. They can very well take the advantage of progress in solid state and fiber laser technology. Therefore, the scope of this chapter is limited to generation of THz via non-linear optical methods. Non-linear methods can further be classified into two categories, viz., optical rectification of ultrashort pulsed lasers[20] and DFG/OPG in nanosecond/CW lasers[21–22].

Optical rectification of wide bandwidth ultrashort pulsed lasers is very attractive; however, they suffer from disadvantages like limited output power, difficult phase matching, larger linewidths and higher costs. On the other hand, DFG enables the use of nanosecond/CW lasers, which are comparatively much cheaper than the ultrashort laser. Besides, it provides rapid tuning with narrower linewidth. Hence, we focus our attention to the generation of THz radiation via DFG/Terahertz PG (TPG). This will be explained in the subsequent sections.

With the advancement of material research, a wide variety of non-linear materials for THz generation have been proposed, which include electro-optics crystals[21], semiconductors[23], organic compounds[24–26] and most recently metamaterials[27]. Advances in poling technology favors electro-optic crystals while minimum absorption and less difference between the refractive indices of optical and THz waves are attributes of semiconductors. Organic compounds have the advantage of having high non-linear coefficients. There are a number of parameters that a material has to possess for being able to be used for THz generation. These include sufficient transparency in the THz region, large non-linear coefficient, high optical damage threshold, less difference between the refraction of optical and THz waves etc. Table 15.3 shows the properties of some important materials for THz generation. Emphasis is laid on $LiNbO_3$ and DAST due to their distinct advantages compared to other materials.

Table 15.3 *Properties of most widely used crystals (transparent in the* $0-4$ *THz range)*

Crystals	GaAs	GaP	DAST	$LiNbO_3$
Optical index	3.33	3.11	3.38 @800 nm	2.16
THz index	3.6	3.31	3.00	5.2
THz absorption at $1-2$ THz (cm^{-1})	1	3.3	100	21.7
Non-linear coefficient (pm/V)	$d_{14} = 46$	$d_{14} = 22$	$\chi_{111} = 123$	$d_{33} = 152$
FOM_1 at long pulses	0.83	0.06	–	1*
FOM_2 at short pulses	0.64	1.67	–	1*

15.6 Applications of THz Sources

THz technology is finding use in an increasingly wide variety of applications—information and communication technology[28] (ICT), medical sciences and pharmaceutical sciences[29–31], non-destructive evaluation in semiconductor industry, homeland security[32], defence[33], quality control of food and agricultural products, global environmental monitoring and ultrafast computing are some examples. The three major factors/attributes which contribute primarily to the above-mentioned applications are listed below[34–35]:

1. Terahertz radiation is readily transmitted through most non-metallic and non-polar mediums, thus enabling THz systems to 'see through' concealing barriers.
2. Many materials of interest for security applications including explosives, chemicals and biological agents have characteristic THz spectra.
3. Terahertz radiation possess either no or minimal health risk to either a suspect being scanned by a THz system or the system's operator[34].

15.7 THz Generation Using PPLN

Generation of THz in PPLN using nanosecond and CW lasers can be accomplished by using non-linear optical techniques such as DFG. This technique involves the mixing of two closely spaced wavelengths such that their difference lies in the THz regime. In conventional PPLN, where grating vector $k_\Lambda$is parallel to the optical beams, it can be used to satisfy the momentum conservation by choosing suitable domain periods. However, one of the major drawbacks of the QPM collinear phase matching in conventional PPLN is the long path inside the crystal, which is prohibited due to the strong absorption of THz in $LiNbO_3$($\alpha_{THz} \approx$ 12 cm^{-1} at 1 THz to α_{THz} ≈170 cm^{-1} at 2.5 THz)[7].

A solution to the above problem is the emission of generated THz radiation perpendicular to the surface of the PPLN crystal in which case it is termed as surface emitted THz. In this way, the path length within the PPLN can be sufficiently reduced and the area of THz emission can be enlarged, resulting in a smaller divergence and higher output powers. The phase matching for surface emitted THz in conventional PPLN requires very hard focusing of the input beams, limiting the effective interaction length and inducing the crystal damage.

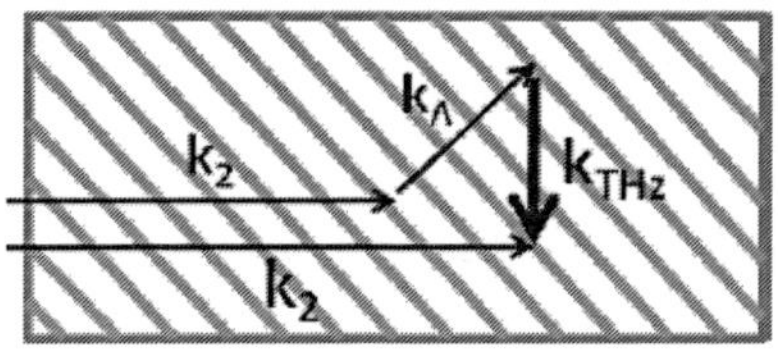

Figure 15.8 Phase matching scheme for slant-stripe PPLN

The surface emission of THz radiation without requiring tight focusing can be achieved in $LiNbO_3$ by poling it in a slant direction and hence called slant-stripe PPLN[4] as shown in Fig. 15.8. The phase matching equation for the geometry in this figure can be given as

$$k_1 = k_2 + \sqrt{k_\Lambda^2 - k_{THz}^2}, \tag{15.3}$$

where k_1, k_2 and k_{THz} are the wave vectors corresponding to λ_1, λ_2 and λ_{THz} respectively. k_Λ is the grating vector.

Equation (15.3) can be used to calculate the PPLN period required for THz generation. The THz wavelength as a function of grating is plotted in Fig. 15.9. The grating period required for the generation of THz wavelength from 50–300 μm varies from 54–9 μm for an optical pump wavelength of 1.064 μm. In this case, although two optical frequencies are collinear resulting in high interaction between them, the THz wave that propagates perpendicular to the optical waves have only a limited interaction with them, which limits the conversion efficiency.

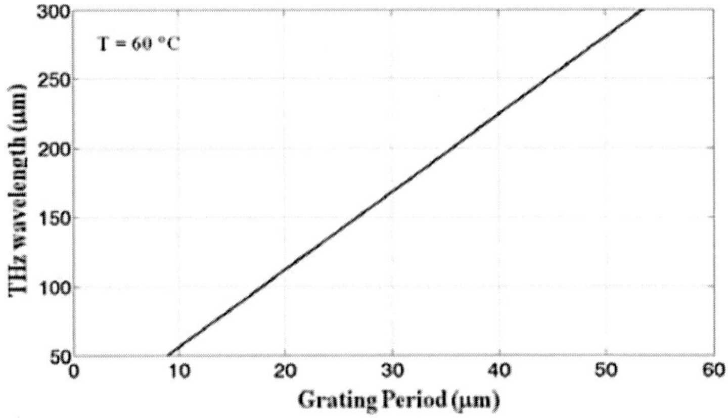

Figure 15.9 Variation of THz wavelength with change in PPLN grating period for slant stripe PPLN

In order to enhance the interaction between THz waves and the optical beams, another kind of phase matching scheme[5] can be used in which the THz wave is emitted at an angle close to the Cerenkov angle. In this scheme, the grating vector is made normal to the propagation vectors (novel phase matching) of the input optical beams as shown in Fig. 15.10.

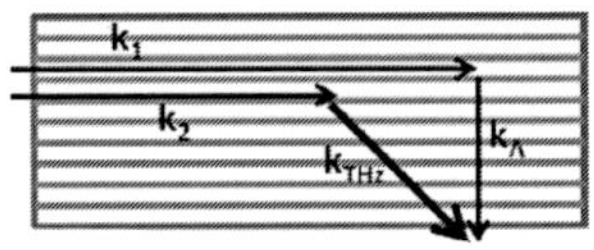

Figure 15.10 Phase matching scheme for novel PPLN

The grating period required for the generation of THz wavelength from 50–300 μm has been calculated to be 64–11 μm and in plotted in Fig. 15.11.

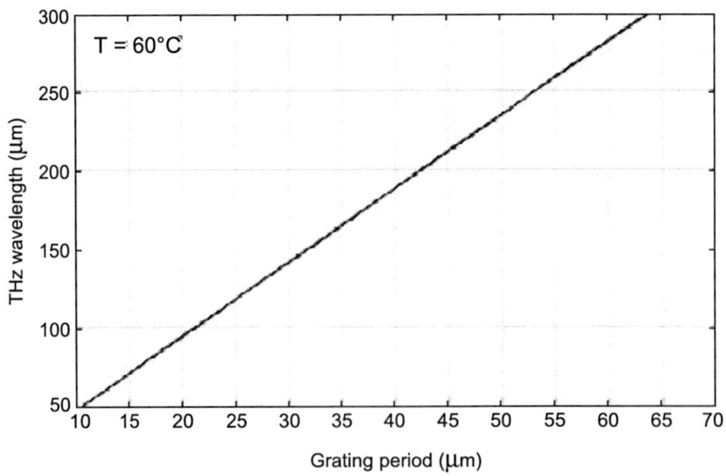

Figure 15.11 Variation of THz wavelength with change in PPLN grating period for novel PPLN

15.8 THz Generation Using DAST

DAST is an organic crystal that provides excellent velocity matching as the refractive index of the DAST in optical and THz region are nearly the same ($n_{opt} \approx n_{THz} \approx 2.2$)[24–26], i.e., it can provide CPM without the need for special techniques. In addition, DAST has very large second order susceptibility ($\approx$ 230 pm/V). Thus DFG of properly selected wavelengths in DAST crystal can provide a wide range of tunable THz radiation.

The phase matching equations for collinear DFG in DAST will be similar to Eqs (15.1) and (15.2) excluding the term with grating period. The growth of THz power after propagating the thickness 'L' is obtained by solving the non-linear coupled wave equation to be

$$P_3 = \frac{2\omega_3^2 d^2 L^2}{\varepsilon_0 n_1 n_2 n_3 C^3}\left(\frac{P_1 P_2}{\pi r^2}\right) \times S \tag{15.4}$$

where

$$S = \frac{\left\{1 + e^{-\alpha_3 L} - 2\cos(\Delta k L) e^{-\frac{\alpha_3 L}{2}}\right\}}{(\Delta k L)^2 + \left(\frac{\alpha_3 L}{2}\right)^2}, \text{ and } \Delta k = |k_1 - k_2 - k_3|$$

where d is the non-linear coefficient and L is the thickness of the DAST crystal. P_1, P_2 are the power of two input beams at λ_1 and λ_2 respectively; T_1, T_2 and T_3 are the Fresnel transmission coefficients; r is the focal spot radius and has been assumed equal for both the input beams and Δk is the momentum mismatch between optical beams and THz wavelength. n_1, n_2 and n_3 are the refractive indices whereas α_1, α_2 and α_3 are the absorption coefficients at wavelengths λ_1, λ_2 and λ_3 respectively. ω_3 is the angular frequency corresponding to λ_3. The refractive indices of DAST crystal at optical frequencies are given by Sellmeier equations[25] as shown in Fig. 15.12

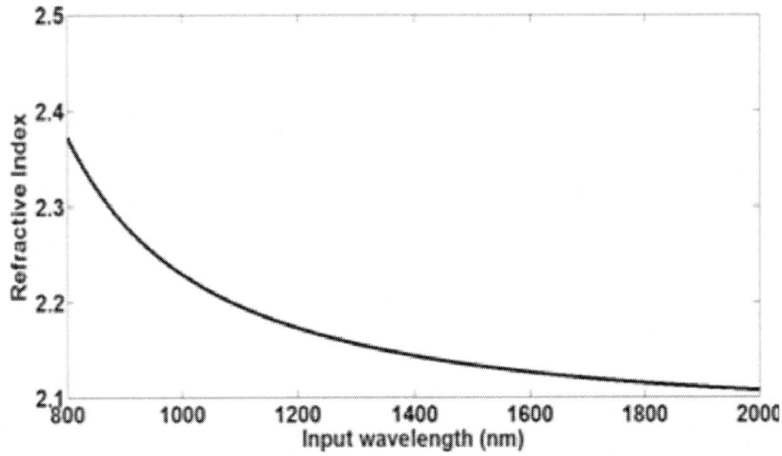

Figure 15.12 Variation refractive index of DAST at infrared wavelengths

Two Lorentzian line shape functions obtained on the basis of harmonic oscillator model for the evaluation of refractive index and absorption for DAST in THz wavelength regime have been summed up[26] (see Figs 15.13 and 15.14).

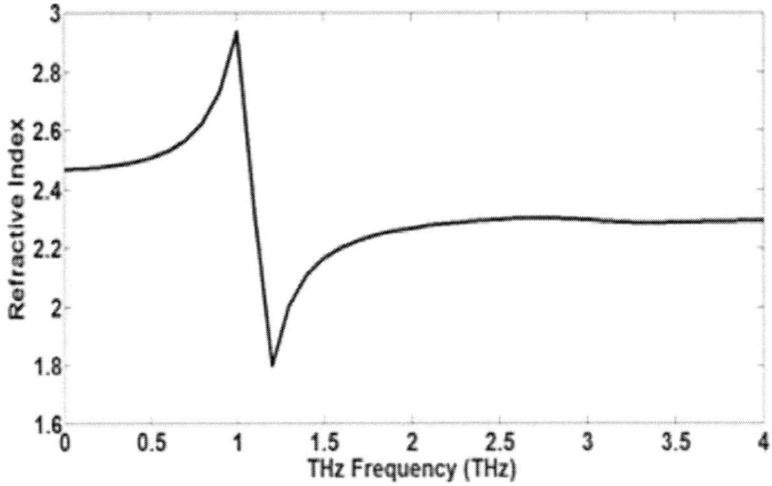

Figure 15.13 Refractive index of DAST at THz frequency

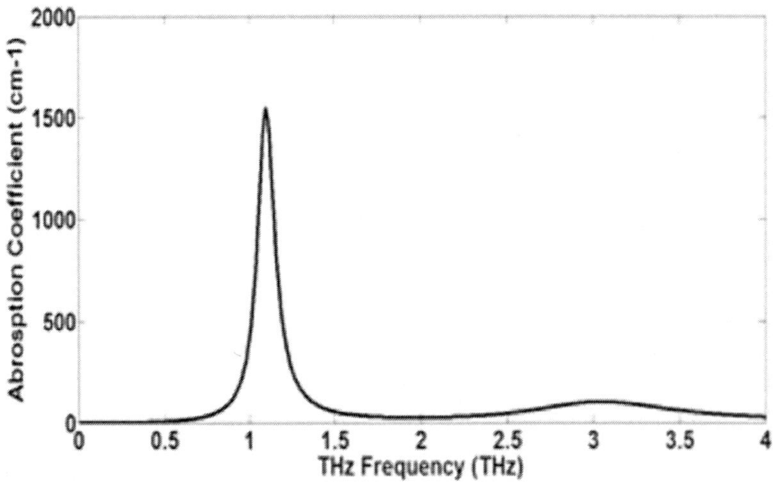

Figure 15.14 Absorption coefficient of DAST at THz frequency

The coherence length is first evaluated with the variation of one of the input wavelengths for obtaining the range of wavelengths suitable for DFG as shown in Fig. 15.15. For 100 μm (3 THz), 140 μm (2.14 THz) and 180 μm (1.67 THz) coherence length was maximum (or momentum mismatch was minimum) at input wavelengths of 1409 nm, 1452 nm and 1685 nm respectively. The estimated THz power as a function of input wavelength is shown in Fig. 15.16, where it is evident that the THz power attains the maximum value at the points where phase matching is satisfied.

Figure 15.16 also shows that phase matching is not very specific and hence the wavelength ranging from 1300–1700 nm can be used for efficient generation above 1 THz. Thus, one of the input wavelengths was fixed to be 1450 nm and the other input varied in its vicinity, i.e., from 1300–1450 nm, which theoretically corresponds to the frequency tuning from 0.1–23 THz.

From Eq. (15.4), it is clear that THz power increases as the square of the crystal thickness. However, even for the completely phase-matched condition it does not grow indefinitely. Rather, the power saturates beyond a certain value ($\approx$ 1 mm) of thickness due to the absorption of THz waves in DAST (Fig. 15.17). Besides, THz frequency has continuous tunability over the range 1–3 THz (Fig. 15.18). However, the actual tunability extends from 0.1–23 THz as the input wavelengths are tuned from 1300–1450 nm.

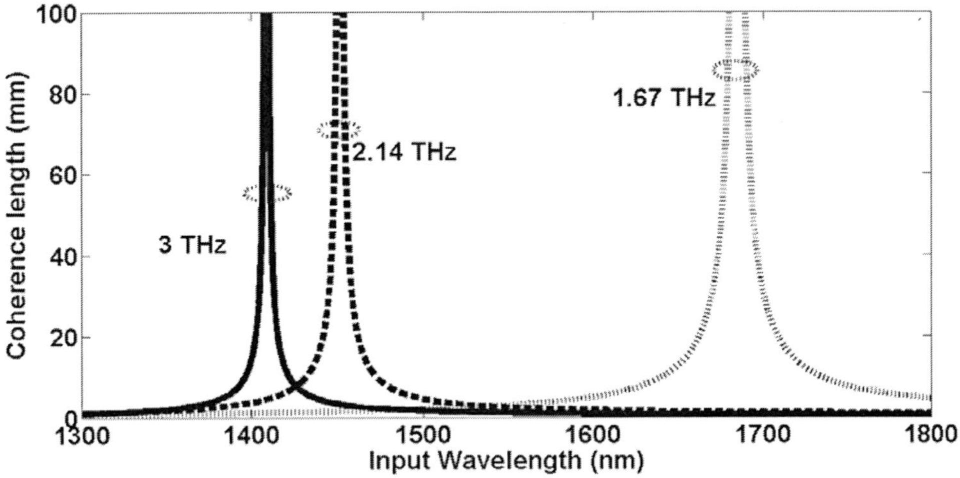

Figure 15.15 Coherence length as a function of input wavelengths

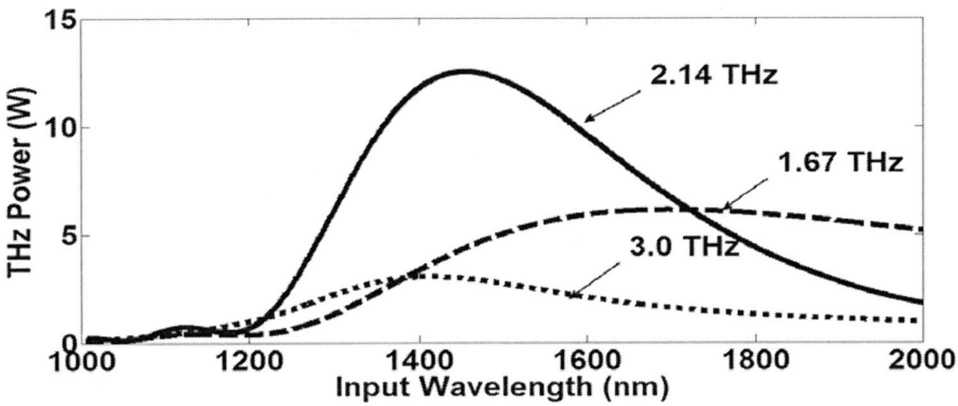

Figure 15.16 THz power as a function of input wavelengths

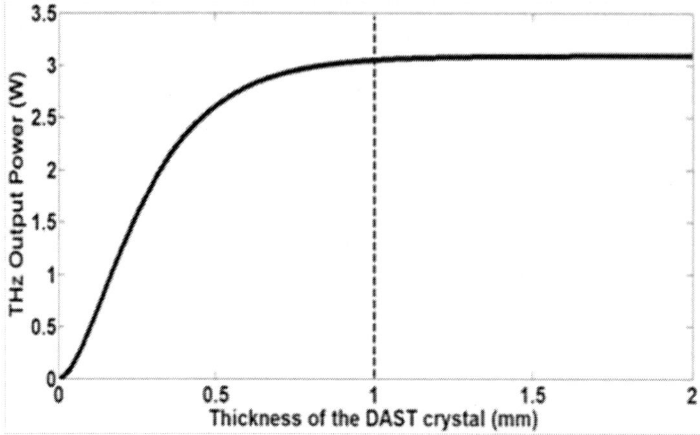

Figure 15.17 THz output power as a function of thickness of DAST crystal

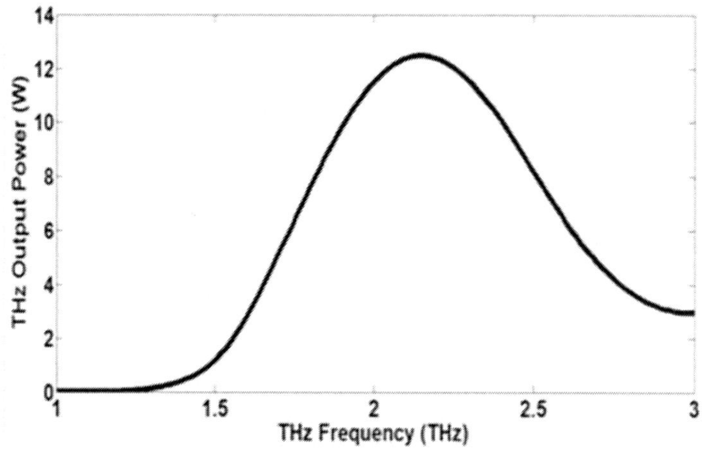

Figure 15.18 THz power as a function of the THz frequency

15.9 Conclusion

In this chapter, the design and development of mid-IR tunable OPO using PPLN was discussed. The PPLN fabrication facility at IRDE was explained. Moreover, the various non-linear optical materials available for the generation of mid-IR and THz radiation also discussed. Finally, THz generation using PPLN and DAST was also elaborated.

Bibliography

1. M. M. Fejer, G. A. Magel, D. H. Jundt and R. L. Byer. 1992. 'Quasi-phase matched second harmonic generation: Tuning and tolerances.' *IEEE J. Quantum. Electron* 28: 2631.
2. Nimish Dixit, Rouchin Mahendra, O. P. Naraniya, A. N. Kaul and A. K. Gupta. 2010. 'High repetition rate mid-infrared generation with singly resonant optical parametric oscillator using multi-grating periodically poled MgO:$LiNbO_3$.' *Opt. Laser Technol.* 42: 18.
3. C. Weiss, G. Torosyan, Y. Avetisyan and R. Beigang. 2001. 'Generation of tunable narrow-band surface-emitted terahertz radiation in periodically poled lithium niobate.' *Opt. Lett.* 26: 563.
4. Y. Sasaki, Y. Avetisyan, K. Kawase and H. Ito. 2002. 'Terahertz-wave surface-emitted difference frequency generation in slant-stripe-type periodically poled $LiNbO_3$ crystal.' *Appl. Phys. Lett.* 81: 3323
5. D. Molter, M. Theuer and R. Beigang. 2009. 'Nanosecond terahertz optical parametric oscillator with a novel quasi phase matching scheme in lithium niobate.' *Optics Express* 17: 6623.
6. T. Taniuchi, S. Okada and H. Nakanishi. 2004. 'Widely tunable THz wave generation in an organic crystal and its spectroscopic application.' *J. Appl. Phys.* 95: 5984.
7. Frank K. Tittel, Dirk Richter and Alan Fried. 2003. 'Mid-Infrared laser applications in spectroscopy.' *Appl. Phys.* 89: 445.
8. Paul D. Mason and Laurent F. Michaille. 2008. 'Review of the development of non-linear materials for mid-IR generation.' *Proc. SPIE.* 7115: 71150N–1.
9. O. Gayer, Z. Sacks, E. Galun and A. Arie. 2008. 'Temperature and wavelength dependent refractive index equations for MgO-doped congruent and stoichiometric $LiNbO_3$.' *Appl. Phys. B.* 91: 343.
10. V. Vaicikauskasa, Z. Kuprionisb, M. Kaucikasa, V. Svedasa and V. Kabelkaa. 2006. 'Mid-infrared all solid state DIAL for remote sensing of hazardous chemical agents.' *Proc. SPIE.* 6214: 62140E.
11. H. H. P. T. Bekman, J. C. van den Heuvel, F. J. M. van Putten and H. M. A. Schleijpen. 2004. 'Development of Mid-IR laser for study of infrared counter measures techniques.' *Proc. SPIE.* 5615: 27.
12. '*Viper Mid IR Laser*'. Accessed on 11th June 2009. http://www.es.northropgrumman.com/countermanpads/tested proven/multi band.html.
13. Ashok. Kaul and Ajay Mishra. 2010. 'Fabrication of periodically poled lithium niobate for optical parametric oscillators.' *Pramana – J. Phys.* 75: 817.
14. D. H. Jundt. 1997. 'Temperature-dependent Sellmeier equation for the index of refraction, n_e, in congruent lithium niobate.' *Opt. Lett.* 22: 1553.
15. T. Ping, H. Jiang, L. Kaifeng, X. Y. Qian and F. M. Wu. 2012. 'Terahertz radiation sources based on free electron lasers and their applications.' *Sci China Inf Sci.* 55, 1: 1.
16. Klaus Reimann. 2007. 'Table-top sources of ultra short THz pulses.' *Rep. Prog. Phys.* 70: 1597.

17. J. A. 'Huillier, G. Torosyan, M. Theuer, C. Rau, Y. Avetisyan and R. Beigang. 2007. 'Generation of THz radiation using bulk, periodically and aperiodically poled lithium niobate – Part 1: Theory.' *Appl. Phys. B.* 86: 197

18. Benjamin S. Williams. 2007. 'Terahertz quantum-cascade lasers.' *Nature Photonics* 1: 517.

19. B. S. Williams, S. Kumar, Q. Hu and J. L. Reno. 2005. 'Operation of terahertz quantum-cascade lasers at 164 K in pulsed mode and at 117 K in continuous-wavemode.' *Optics. Express* 13: 3331.

20. T. Meade, Y. S. Lee, V. Perlin, H. Winful, T.B. Norris and A. Galvanauskas. 'Narrow band THz waveform generation in PPLN.' *Laser Focus World.* "http://www.nrao.edu/meetings/isstt/papers/2000/2000326335.pdf" accessed on 13th August 2014.

21. J. A.' Huillier, G. Torosyan, M. Theuer, C. Rau, Y. Avetisyan and R. Beigang. 2007. 'Generation of THz radiation using bulk, periodically and aperiodically poled lithium niobate – Part 2: Experiments.' *Appl. Phys. B.* 86: 197.

22. J. A.' Huillier, G. Torosyan, M. Theuer, C. Rau, Y. Avetisyan and R. Beigang. 2007. 'Generation of THz radiation using bulk, periodically and aperiodically poled lithium niobate – Part 1: Theory.' *Appl. Phys. B.* 86: 197.

23. G. Imeshev, M. E. Fermann, K. L. Vodopyanov and M. M. Fejer. 2006. 'High-power source of THz radiation based on orientation-patterned GaAs pumped by a fiber laser.' *Optics Express* 14, 10: 4439.

24. T. Taniuchi, S. Okada and H. Nakanishi. 2004. 'Widely tunable THz wave generation in an organic crystal and its spectroscopic application.' *J. Appl. Phys.* 95: 5984.

25. M. Jazbinsek, L. Mutter and P. Gunter. 2008. 'Photonics application with the organic non-linear optical crystal DAST.' *IEEE J. Sel. Top. Quantum Electron* 14: 1298.

26. A. Schneider, M. Neis, M. Stillhart, B. Ruiz, R. U. A. Khan and P. Gunter. 2006. 'Generation of THz pulse through optical rectification in organic DAST crystals: theory and experiment.' *J. Opt. Soc. Am. B.* 23: 1822.

27. Withawat Withayachumnankul. 2009. 'Metamaterials in the terahertz regime.' *IEEE Photonics Journal* 1, 2: 99.

28. John Hopkins. 2004. 'THz wave for communication and sensing.' *APL Technical Digest* 25, 4.

29. Manoj P. Tarachandani, Nalini S. Kurup. March 2008. 'Terahertz technology pharmaceutical and medical applications.' *Pharma Times* 40, 3: 11.

30. M. Pepper, D. D. Arnone, C. M. Ciesla, A. Corchia and S. Egusa. 2009. 'Application of THz technology to medical imaging.' *Invited Paper, Proc. SPIE.* 3828: 209.

31. Ruth M. Woodward. 2004. 'Terahertz technology in the medical and pharmaceutical industry.' *Preclinica* 2, 5: 328.

32. Torsten May and Hans Georg Meyer. 'How to perform a security check without undressing people.' '*Optik & Photonik* 3, 4: 31–34.

33. R. M. Langdon, V. Handerek, P. Harrison, H. Eisele, M. Stringer, C. F. Rae and M. H. Dunn. 2004. *Military Applications of Terahertz Imaging.* Edinburgh: 1st EMRS DTC Technical Conference.

34. Adrian Dobroiu, Chiko Otani and Kodo Kawase. 2006. 'Terahertz-wave sources and imaging applications.' *Meas. Sci. Technol.* 17: R161–R174.

35. Vincent P. Wallace, Emma MacPherson, J. Axel Zeitler and Caroline Reid. 2008. 'Three-dimensional imaging of optically opaque materials using non-ionizing terahertz radiation.' *J. Opt. Soc. Am. A*. 25, 12: 3120.

16 A New Test System Design for Comprehensive Characterization of Non-imaging IR Guided Missiles

Anil Kumar Maini, Varsha Agrawal, Manoj Sharma, R. S. Solanki and Gagan Preet

16.1 Introduction

Non-imaging types of IR guided missiles make use of infrared emission corresponding to the thermal signatures of the exhaust and the mainframe of the target aircraft to home on to it. Emission in 3–5 and 8–12 micron bands is characteristic of electromagnetic emission from jet exhausts and mainframe of the aircraft. Spectral content of infrared emission as received by an IR seeker head is the superposition of the spectral emission from the aircraft on the transmission characteristics of the atmosphere and is shown in Fig. 16.1[5]. This spectral signature is judiciously used in guidance of air-to-air and surface-to-air IR guided missiles. Both air-to-air and surface-to-air IR guided missiles receive the target's infrared signatures in the presence of background radiation from the sky and also infrared signatures of flares, if any, deployed by target aircraft platform. The seeker head should be able to discriminate between infrared signatures of the background and flares from those of the target.

16.1.1 Infrared signatures of background, target and flare

Signature prediction and analysis constitutes a very important component in the design of test systems that simulate true battlefield like scenario. This topic has been extensively researched and reported in literature.

Major sources of background infrared noise include the earth, the sky, thermal emissions from clouds and solar reflections[6]. The background sources are generally modelled as a black body whose radiance is dependent upon its temperature and emissivity. The total radiance incident on the missile seeker is estimated after taking into consideration contributions from all these sources and also the transmission characteristics of the atmosphere. Though background signatures are dynamic in nature and vary with weather condition, path length and altitude; for the purpose of IR guided missile guidance, static background noise spread over 3–12 micron band is a reasonably good approximation.

Different sources of infrared radiation in an aircraft can be classified as internal and external sources[7]. Internal sources include hot engine parts, airframe skin heated by the engine and aerodynamic heating, combustion of the hot gases in the plume and the skin of the airframe. External sources include reflected ambient radiation from the sun, the sky and the ground. The hotter surfaces like the hot engine parts, the tail pipe region and the plume mainly contribute in the lower wavelength band whereas the fuselage region, the airframe and the reflected radiation are the dominant factors in the longer wavelength region. The target can be modelled as an object with different temperature zones or sub-targets. The infrared signature model of a target can be determined from the known values of emissivity, temperature, range from the guided weapon and area of the sub-target[8, 9]. The infrared emission from a given sub-target is then multiplied by a corresponding atmospheric transmission coefficient at that temperature to compute the signal reaching the IR seeker head. The total signal is a summation of the signals received from various sub-targets. The target signatures reaching the IR seeker head are typically considered in 3–5 micron and 8–12 micron bands (Fig. 16.1)[10].

Flares are chemically heated sources that radiate in accordance with blackbody or gray body characteristics. They are much smaller, lighter and cheaper than the protected target[11]. The smaller physical size of the flare is overcome by heating it to a much higher temperature so as to match its spectral characteristics with that of the target. The infrared signatures of a flare depend upon the spectral size of the flare, the dispenzation sequence, ballistic trajectory and the irradiance properties of the flare material[12]. Ideally, the infrared signatures of a flare should be spectrally and spatially matched with that of the platform to be protected but with a higher radiant intensity. However, conventional flares act like a hot point source with temperature as high as 1400–1800 K. For generation of infrared signatures of flare, one needs to know the important parameters of a flare, which includes flare rise time, burn time, burn time constant, rise time constant, flare temperature and emissivity of the flare. Typical infrared flare signature is in the 1.5–5 micron band.

16.1.2 Requirement specifications

Based on infrared signature analysis reported in literature, if the test systems for IR guided missile seekers were to simulate realistic battlefield like situation; they should have the following features.

1. Target signatures in 3–5 and 8–12 micron bands
2. Static background simulation in 3–12 micron band
3. Spectral output modes
    a. 3–5 micron output with or without static background
    b. 8–12 micron output with or without static background
    c. Simultaneous 3–5 and 8–12 micron outputs with or without static background

4. Power density (3–5 micron and 8–12 micron bands) across seeker cross-section in the range of 1 $\mu W/m^2$–10 mW/m^2 to simulate both maximum and minimum of operational range.
5. Power density to simulate static background noise in the range of 10–100 nW/m^2 in 3–12 micron band.
6. Power density in the range of 2–1000 $\mu W/m^2$ to simulate typical IR flare signature in the 1.5–5 micron band.

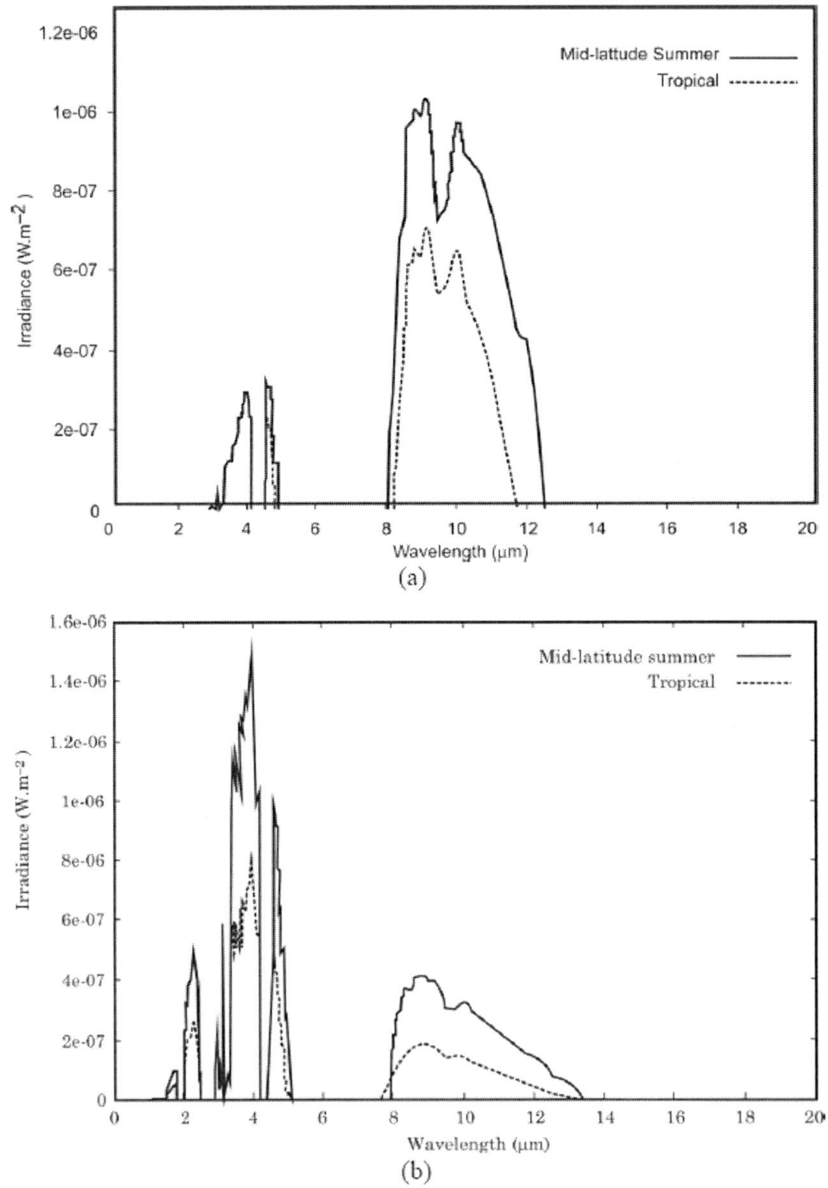

Figure 16.1 Infrared signatures as seen by an IR seeker head: (a) Target at zenith (b) Target at 120° from horizon

Since spectral matching is the basis of testing an IR guided missile, it is necessary to generate infrared signatures of the target in the 3–5 micron band for testing single colour missiles and both the 3–5 micron and 8–12 micron bands to test dual colour missiles. In addition to target signatures, one would also need to generate background and flare signatures when carrying out comprehensive testing of the missile. As the missile approaches the target, target signatures grow in magnitude while the flare signatures become weaker. Therefore, it is important that the test system generates the dynamics of target and flare signatures.

16.1.3 Design concept

The proposed design concept is configured as three independent channels namely *Target channel* for generation of target signatures, *background channel* for generation of background signatures and *flare channel* for generation of flare signatures. The target channel generates target signatures in the 3–5 micron and 8–12 micron bands according to the operational mode selected by the user. The background channel generates static background radiation in the 3–12 micron band at a power density level selected by the user. The flare channel generates flare signatures in the 3–5 micron band with desired temporal and spatial characteristics as per the operational mode selected by the user. Desired spectral profiles are obtained by using suitable band pass filters in the paths of the three beams; desired temporal and spatial profiles for the target channel is obtained by using motorized iris whereas that for the flare channel is obtained by using a combination of a motorized iris mounted on linear X-Y stages. Each of the three channels is driven by a dedicated drive circuitry. All drive and control circuitries are in turn controlled by an embedded circuit based master controller. The optical outputs from the three channels are combined in a set of suitable beam combiners to derive the final output.

Figure 16.2 shows the conceptual block diagram of the proposed design. The test system has five major components, namely, the target channel, the flare channel, the background channel, control and drive electronics and the beam combination optics. As is evident from the block schematic of Fig. 16.2, the drive and control electronics module contains drive and control circuits for the target, the background and the flare channels. These drive and control circuits are in turn controlled by an embedded controller. The IR emitter block comprises IR emitters for all the three channels. The beam shaping optical assembly contains beam shaping optics of all the three channels and also the beam combining optics used to generate the composite beam at the output. The user interface with the test system is through panel display.

The *target channel* generates the signatures of the target in the 3–5 micron and 8–12 micron bands as seen by the infrared seeker unit of the missile. Figure 16.3(a) shows the proposed layout of the target channel. The channel comprises two infrared emitters for generating the spectrum in the 3–5 and 8–12 micron bands. The two infrared emitters respectively have a mid-IR band pass filter (3–5 micron band) and a far infrared band pass filter (8–12 micron band) placed at the output. Each output has a suitable beam expander placed in front of the band pass filter to get a beam of desired size and divergence. The beams from the two expanders are combined in a suitable beam combiner. The combined beam is passed through a motorized iris. The iris is used to control the size of the target spot as seen by the missile seeker unit. The target spot size varies inversely with the missile-to-target distance.

The *background channel* comprises an IR emitter, a band pass filter transmitting in the 3–12 micron band and a beam expander operating in the 3–12 micron band. The optical layout of the background channel is shown in Fig. 16.3(b). The beam expander is used to generate the beam of desired size.

The *flare channel* as shown in Fig. 16.3(c) simulates the flare signatures with desired intensity, spectral, temporal and spatial profiles. The flare channel comprises of an infrared emitter, a band pass filter transmitting in the 3–5 micron band and a beam expander. The beam at the output of the beam expander is passed through a motorized iris, which is mounted on the X-Y linear stages. The size and the X-Y position of the iris are controlled through the electronics module to simulate temporal and spatial profile of the flare signatures.

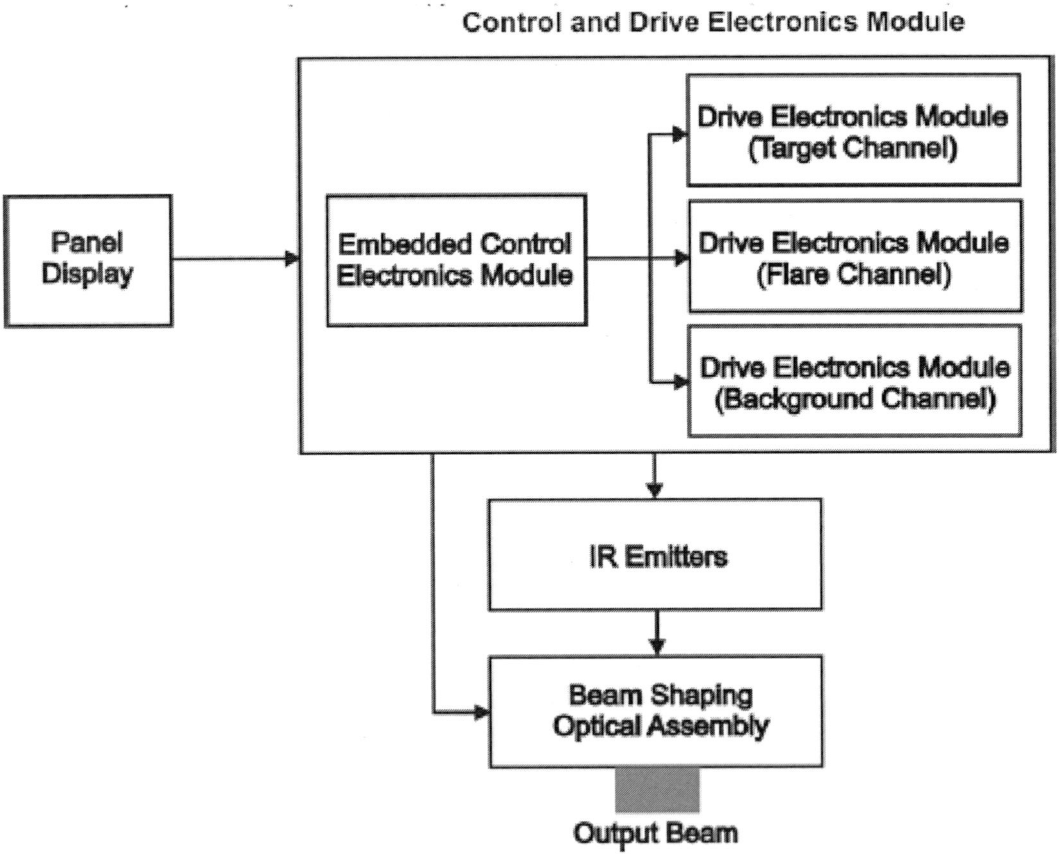

Figure 16.2 Conceptual block diagram of the proposed design

The *control and drive electronics* is used to feed the drive voltages and waveforms to the infrared emitters of the three channels, motorized iris and the linear stages used for X-Y movement depending upon the profile selected by the user. Figure 16.4 shows the block diagram of the control and drive electronics module of the system.

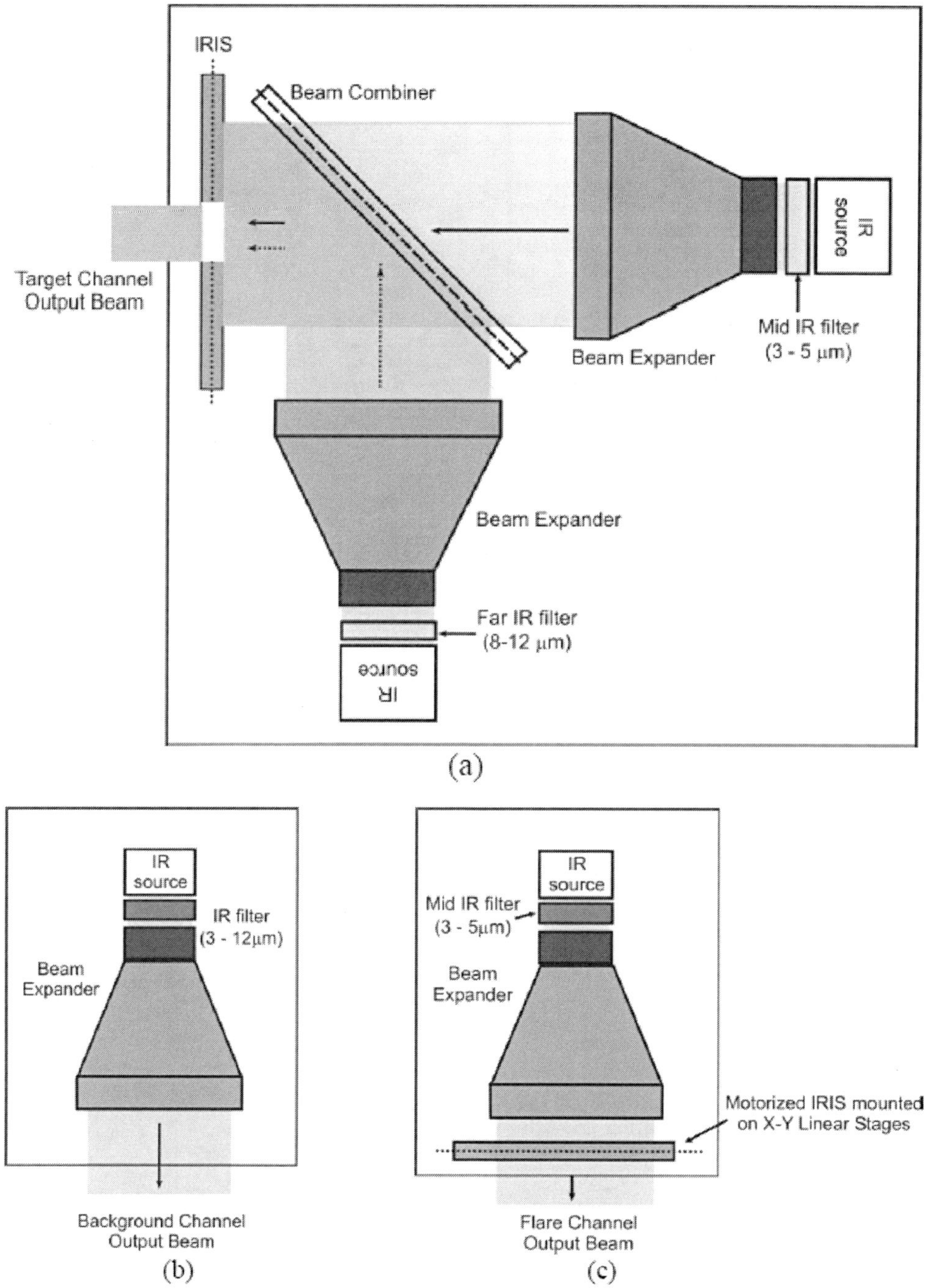

Figure 16.3 (a) Target channel (b) Background channel (c) Flare channel

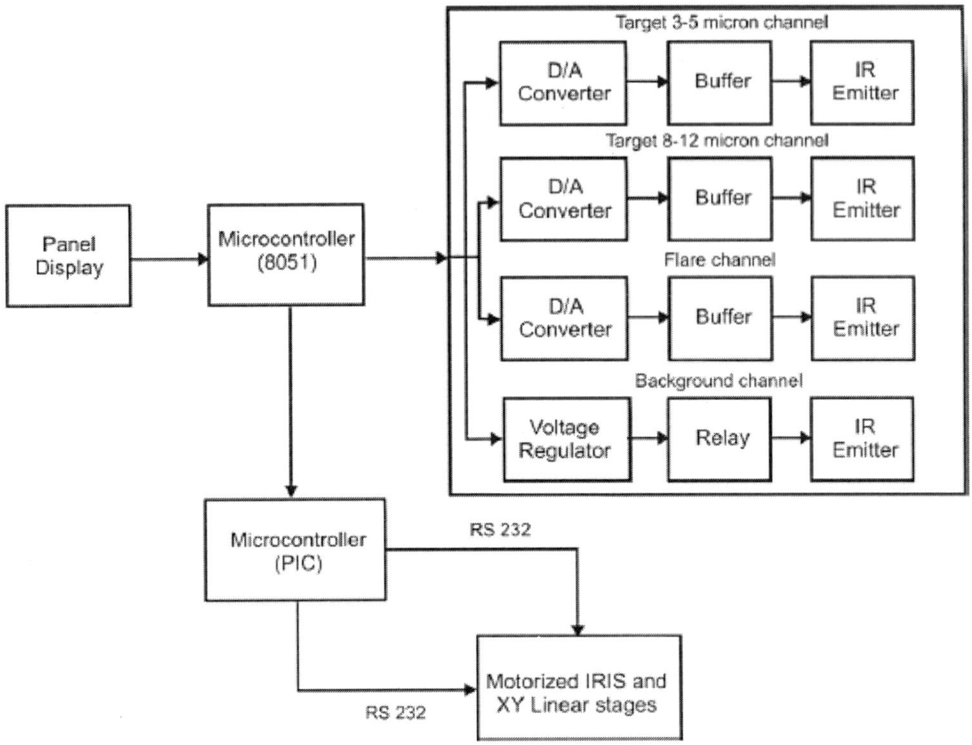

Figure 16.4 Control and drive electronics module

Figure 16.5 shows the layout for the *beam combination optics* used in the proposed design concept. In this, outputs from the flare and background channels are first combined in a beam combiner (Beam combiner 1). The combined output is further combined with the output of the target channel in another beam combiner (Beam combiner 3). The target channel output is a combination of 3–5 micron and 8–12 micron band outputs combined in a beam combiner (Beam combiner 2). The proposed layout configuration has the least optical losses contributed by reflection and transmission.

16.1.4 Hardware implementation

Figure 16.6 shows the CAD view of the experimental hardware with the target channel, the flare channel and the background channel integrated on the breadboard as per the beam combining optics layout shown earlier in Fig. 16.5.

The *target channel* comprises two infrared emitters with corresponding band pass filters for generating 3–5 micron and 8–12 micron band infrared spectra. The two emitters are heated in such a manner that their temperature increases from 350°C to 820°C in approximately 40 seconds. This is done in order to increase the intensity of the target signatures to simulate an approaching missile. The DC voltages to be applied to the infrared emitters to heat them to the desired temperatures are determined from the manufacturer's data sheets of the emitters chosen for the purpose. Beam expanders are used in front of the filters to get beams of desired size and divergence. The output beams from the two expanders are combined and the combined beam is passed through a motorized iris. The iris is used to control the size of the target spot as seen by the missile. The size of the target increases from approximately 1.0 mm to 12 mm in 40 seconds. The spot size varies inversely with the missile-to-target range.

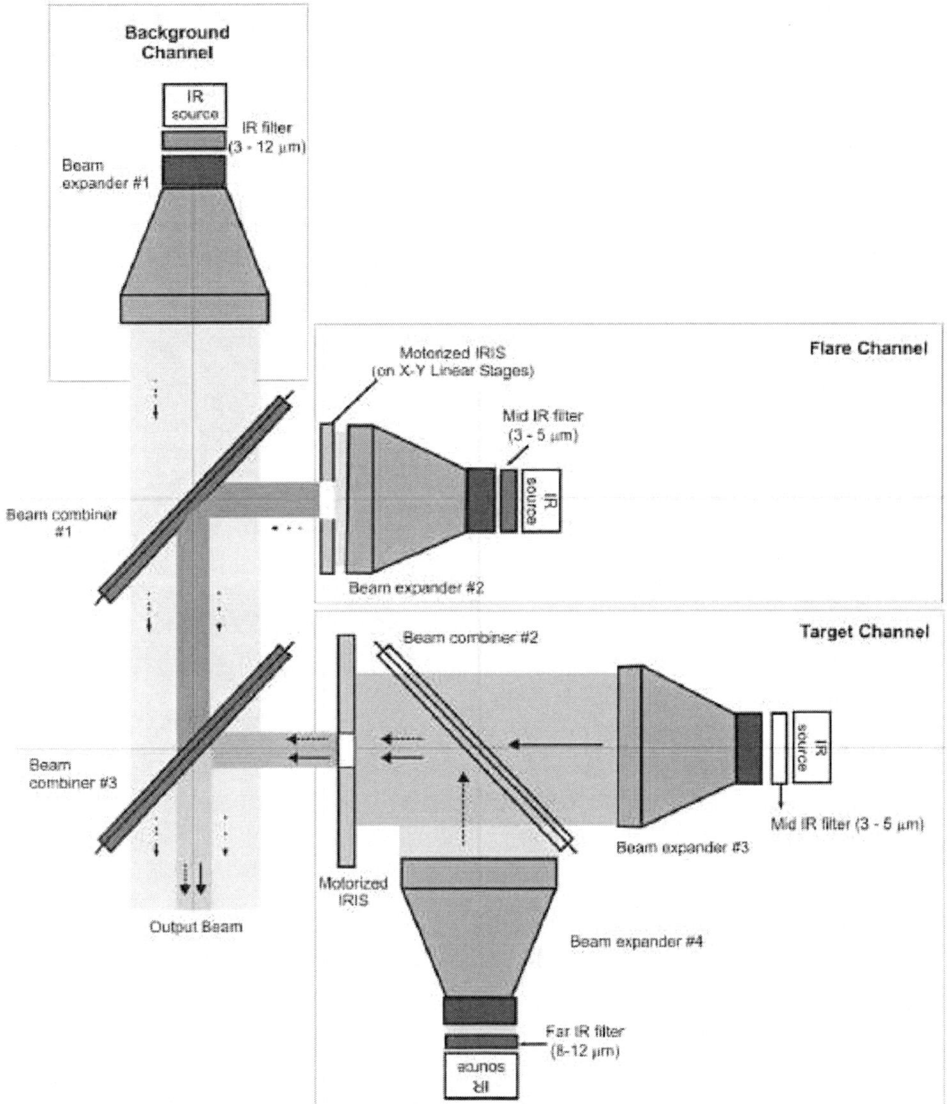

Figure 16.5 Proposed layout configuration of the beam combining optics

The IR emitters used for generating the target signatures both in the 3–5 micron and 8–12 micron bands are solid-state emitters with parabolic reflectors to reduce the divergence of the IR beam emitted by it. The FWHM values of the bare element and emitter with parabolic reflectors are ±50° and ±15° respectively. The band pass filter chosen for the 3–5 micron band has 85% transmission in the 3.2–4.8 micron band and the one chosen for 8–12 micron band has 85% transmission in

8.3–12.3 micron band. The two filters offer no transmission outside their respective pass bands. The output spectrum in the two bands of the target channel is obtained by superimposing the output spectra of the two emitters on the transmission characteristics of the corresponding band pass filters. The two chosen beam expanders expand the respective beams from an input diameter of 25 mm to an output diameter of approximately 70 mm. Although the FWHM of the radiation from the two emitters is ±15°, the design of the beam expanders is such that it expands only the portion of the beam lying within the ±2° cone. As a result of this, the full angle beam divergence at the output of beam expanders is approximately 25 milli-radians.

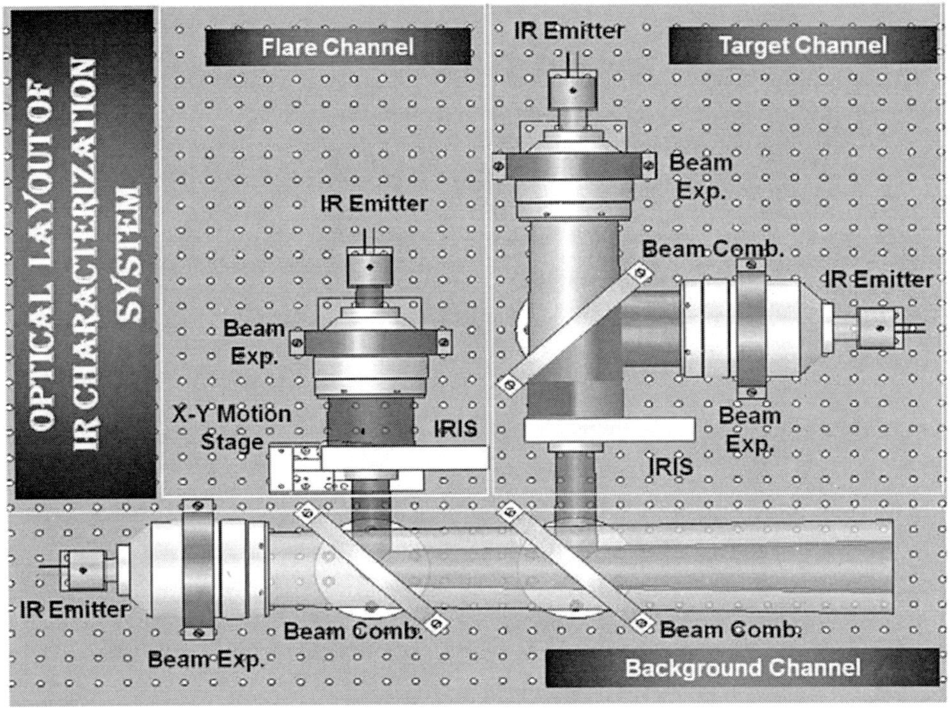

Figure 16.6 CAD view of experimental hardware

The beam combiner 2 combines the target channel beams in the 3–5 micron and 8–12 micron bands. The chosen beam combiner has a clear aperture of 115 mm and the two beams are incident at an angle of 45°. The coating on surface 1 is an anti-reflection coating with reflection of less than 5% in the 3–5 micron band. The coating on surface#2 is an antireflection coating ($R \leq 5\%$) for 3–5 micron band and a high reflection coating ($R \geq 90\%$) for the 8–12 micron band. The motorized iris is used to provide a variable aperture in the path of the infrared beam of the target channel. This is used to vary the target spot size to simulate the condition of an IR guided missile approaching the target. It may be mentioned here that the target spot size grows in size and intensity with reduction in missile-to-target distance. The chosen iris provides a continuous control of the aperture radius from 1 to 25 mm. The iris is driven by a fine resolution stepping motor, which is part of the assembly. The motor is interfaced with the embedded controller or a PC through RS-232 or IEEE-488 communication protocol.

The *background channel* is used for generating the background spectrum in the 3–12 micron band. It comprises an infrared emitter, a band pass filter, a neutral density filter, a beam expander and drive and control electronics to produce beams of desired spot size and divergence. The infrared emitter used in the background channel is the same as that used in the target channel. In this case, it is heated to a temperature of 80°C in order to generate the background signatures. The chosen band pass filter has transmission in the 3–12 micron band. A neutral density filter of OD of 4 is used to further attenuate the signal. The beam expander used in the background channel operates in the 3–12 micron band and expands the 25 mm diameter input beam to an expanded 70 mm beam. In this case too, the beam expander takes into account that portion of the beam falling within ±2° cone of the output beam profile. Therefore, the full angle divergence of the beam is of the order of 25 milli-radians.

The *flare channel* simulates the flare signatures not only in terms of its intensity and spectral profile specifications but also in terms of its temporal and spatial profile. Flare signatures are mainly present in the 3–5 micron band. While the intensity of the flare decreases with time; its size and distance from the aircraft increase with time. The flare channel comprises an infrared emitter, a band pass filter with transmission in the 3–5 micron band, a beam expander to generate beams of desired spot size and divergence and a motorized iris mounted on an X-Y linear stage. Motorized iris on X-Y linear stages is used here to generate the desired temporal and spatial profiles of the flare.

The infrared emitter is heated such that its temperature increases exponentially from the ambient value to 820°C in 3 s and then decreases exponentially to the ambient value in the next 9 s. It may be mentioned here that the flare is initiated 10 s after the target profile is initiated. The band pass filter has 85% transmission in 3–5 micron band and no transmission outside this band. The design of the beam expander is the same as that of the one used in the target channel. It produces a 70 mm spot diameter with a full angle divergence of 25 milli-radians. The motorized iris used here is the same as the one used for the target channel. The flare grows in size with time and moves away from the target. The movement of the flare in the X-Y plane is simulated using X-Y linear stages. The proposed design employs two single-axis linear stages. The movement of the stages is controlled using a PIC microcontroller through a motion controller provided with the stage. The function of the drive and control electronics is to generate the desired drive voltages for the infrared emitter and the drive waveform for the motorized iris and the X-Y linear stages.

The final output beam is produced by using two beam combiners. One of the beam combiners (beam combiner 1) is used for combining the background and flare channels. The combined output is then further added to the output of the target channel in another beam combiner (beam combiner 3). The two beam combiners are given appropriate coatings on their respective surfaces to get the intended beam combination. Figure 16.7 shows the final output in one of the operational modes. The design allows generation of other possible modes also. The estimated power density of the target channel increases from 3.8 $\mu W/m^2$ to 7.3 mW/m^2 in the 3–5 micron band and from 1.6 $\mu W/m^2$ to 1.6 mW/m^2 in the 8–12 micron band in 40 seconds. The estimated power density of the background channel is 115 nW/m^2. The power density of the flare channel initially increases to 280 $\mu W/m^2$ in 3 s and then decreases exponentially to 5.0 $\mu W/m^2$ in the next 9 s. This is in accordance with the preferred values if the test system were to simulate realistic deployment conditions. Figure 16.8 shows the spatial profile of the target and flare signatures at two different time instants T1 and T2 respectively indicating time instants of launch of flare and some time delay after launch. As shown in the figure, the target signature grows in intensity and size to indicate missile approaching the target aircraft. The intensity of flare signature initially increases and then decreases. The flare increases in size to indicate an expanding flare and also moves away from the trajectory following a ballistic path.

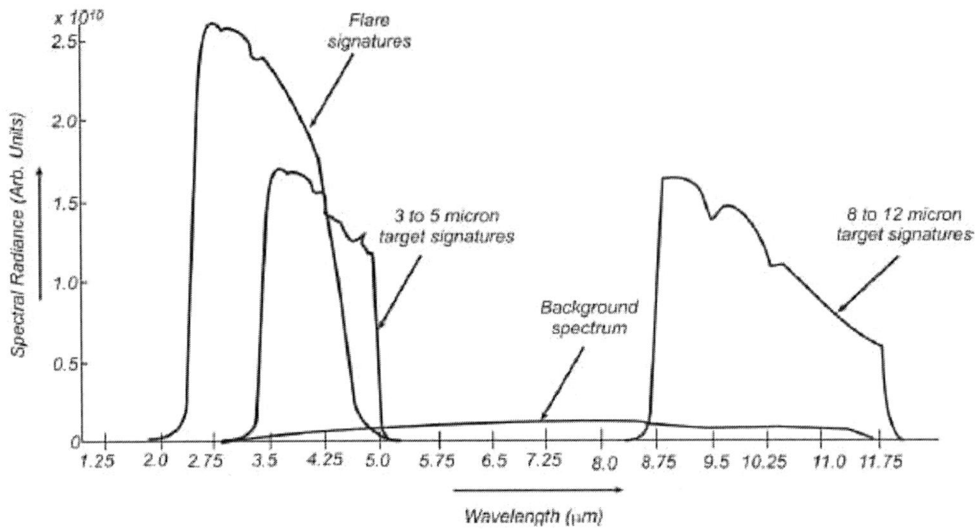

Figure 16.7 Estimated spectral profile of the output spot in one of the operational modes

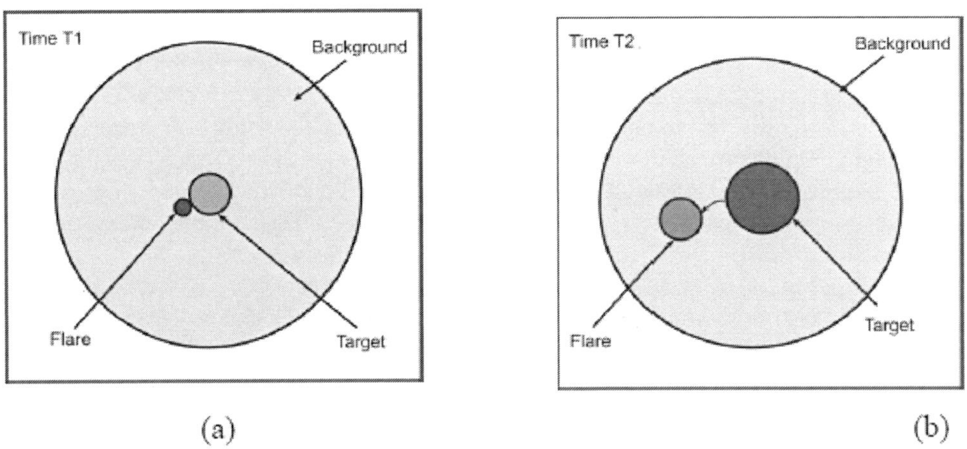

Figure 16.8 Spatial profile of the output spot (a) Time instant T1 (b) Time instant T2

16.1.5 Performance evaluation

The experimental hardware was evaluated for various performance parameters, which mainly includes spectral distribution and power density of 3–5 and 8–12 micron bands of the target channel, 3–12 micron band of the background channel and 3–5 micron band of the flare channel. In addition to these measurements, variations of temporal and spatial profiles of the flare channel are also measured.

In the first step, power density of the 3–5 micron and 8–12 micron bands of the target channel, 3–12 micron band of background channel and 3–5 micron band of the flare channel is measured by switching one channel at a time. The power density is measured with a power meter. The measured minimum and maximum values are 4 $\mu W/m^2$ and 7.3 mW/m^2 for a 3–5 micron band of target channel, 1.5 $\mu W/m^2$ and 1.55 mW/m^2 for 8–12 micron band of the target channel. The power density of the background channel is 110 nW/m^2. The flare channel has maximum and minimum values of power density as 275 $\mu W/m^2$ and 5.0 $\mu W/m^2$. The measured values are in close agreement with the estimated values. In the second step, a family of intensity versus wavelength characteristics of the output for different operational modes are plotted by activating the relevant channels. Figure 16.9 shows the family of curves. The wavelength spectrum was plotted with a spectroradiometer. The curves are self explanatory. It may be mentioned here that the dip near the 4.2 micron wavelength is due to CO_2 absorption and the dip near the 5.9 micron wavelength is due to detector changeover in the spectroradiometer.

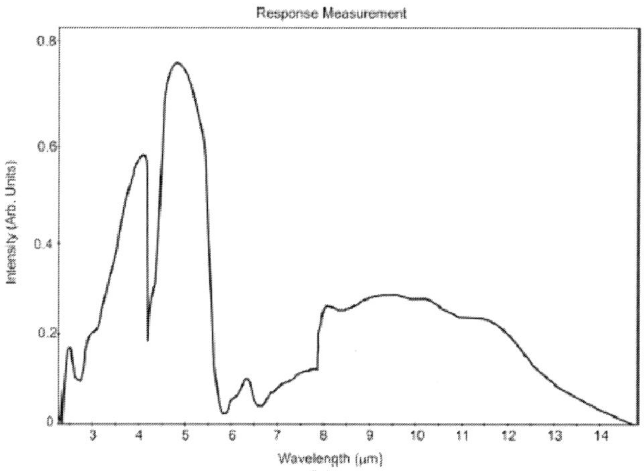

(a)

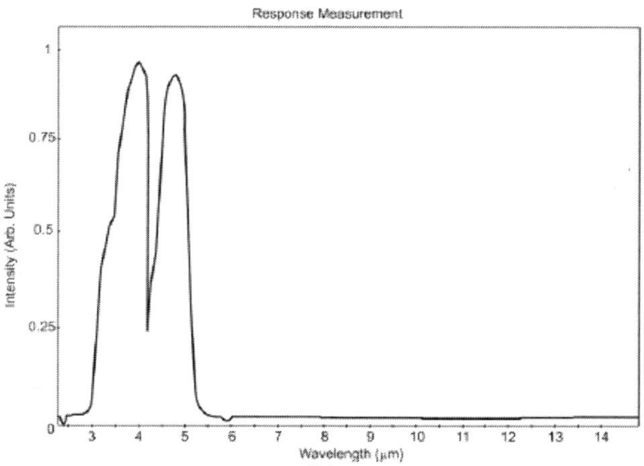

(b)

Figure 16.9 (a) Measured intensity versus wavelength characteristics of target channel (b) Measured intensity versus wavelength characteristics of the flare channel (c) Measured intensity versus wavelength characteristics of the background channel

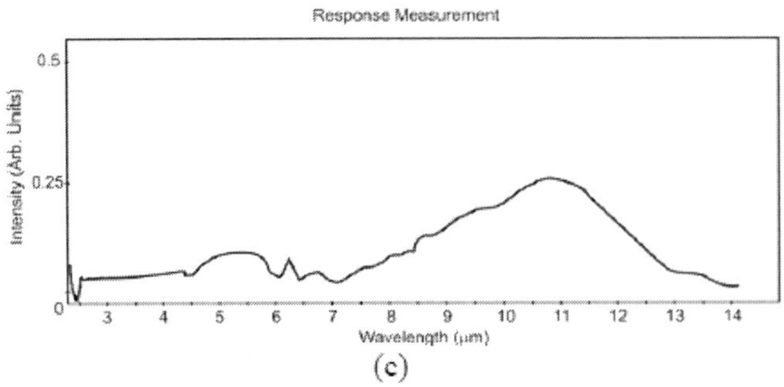

Figure 16.9 Contd...

16.2 Conclusion

If the IR guided missile seeker head were to be tested in a realistic deployment scenario, the target signatures need to be generated along with static background and dynamic flare. For comprehensive characterization of seeker heads, the test system should be capable of generating different target signatures to simulate different approach angles of the guided weapon with respect to the target. Not only this, the test system should also be capable of simulating variation in the target signature spot size with time. The design concept presented in this chapter addresses all these issues without making the test system too complex as is the case with most of the test systems available internationally for the purpose. The design makes use of simple devices such as low power consuming solid state black body emitters driven by embedded controller for generation of target, flare and background signatures with all the desired amplitude variations, motorized irises for generation of desired spatial and temporal profiles of target and flare signatures and band pass filters for spectral shaping and beam combiners for generating the composite radiation. The proposed beam combining optics topology is such that it leads to minimum transmission and reflection losses. The measurements carried out with the prototype hardware confirm the efficacy of the design and the measured power density figures and intensity versus wavelength profiles are in close agreement with the estimated values.

Bibliography

1. J. W. Chambers II. 2000. 'The oxford companion to American military history.' *Precision Guided Munitions.*
2. D.J. Higgins. April 2006. *Precision Guided Munitions in an Urban Environment.* Air Command and Staff College: Air University.
3. K. Chrzanowski. 2001. 'Testing of military optoelectronic systems .' *Optoelectronics Review* 9, 4: 377.
4. David H. Pollock. 'Countermeasures: Infrared and Electro-optical Handbook.' 7: 159–284.
5. G. A. Rao and S. P. Mahulikar. 2005. 'Effect of atmospheric transmission and radiance on aircraft infrared signatures.' *Journal of Aircraft* 42, 4: 1046–1054.

6. P. Jacobs. 1990. 'Review of the technology and research in the area of infrared signature of targets and backgrounds.' *The International Society for Optical Engineering: Characterization, propagation, and simulation of infrared scenes, SPIE.* 1311: 80–94.
7. S. P. Mahulikar, G. A. Rao, H. R. Sonawane and H. S. S. Prasad. August 18–21, 2009. 'Infrared signature studies of aircraft and helicopters.' PIERS Proceedings.
8. J. S. Accetta and G. H. Kelly. 1992. 'Targets, backgrounds and discrimination.' *International Society for Optical Engineering* 1725: 319.
9. S.P. Mahulikar, H. R. Sonawane and G.A. Rao. Oct - Nov 2007. *Infrared signature studies of aerospace vehicles Progress in Aerospace Sciences* 43, 7–8: 218–245.
10. G. A. Rao and S. P. Mahulikar. 2005. 'Effect of atmospheric transmission and radiance on aircraft infrared signatures.' *Journal of Aircraft* 42, 4: 1046–1054.
11. J. Knowles and C. Lawrence. 2006. 'New tricks for flares.' *Journal of Electronic Defence* 29, 10: 25.
12. R. J. Ritchie. 2003. 'Army advanced IRCM Flares.' *Aircraft Survivability* 2: 22–23.

17 Laser and Opto-electronics Technologies for Low Intensity Conflict (LIC) Applications

Anil Kumar Maini

17.1 Introduction

A low intensity conflict is the most common form of warfare today and is likely to be so in foreseeable future. Data suggests that more than 75% of the armed conflicts since World War II have been of low intensity variety. *Low intensity conflict operation* is a military term used to refer to the deployment and use of troops and/or assets in situations other than conventional war. As compared to a conventional war; in the case of low intensity conflict operations, armed forces engaged in the conflict operate at a greatly reduced tempo, perhaps with fewer soldiers, reduced range of tactical equipment and limited scope to operate in a military manner. Moreover, use of artillery is avoided in the case of conflicts in urban territories and use of air power is often restricted to surveillance and transportation of personnel and equipment. Low intensity conflicts pose an alarming threat to national security and is an area of concern for the whole of the international community today. Its scope extends from emergency preparedness and response to domestic intelligence activities to riot and mob control, from combating illegal drug trafficking to protection of critical infrastructure, from handling counter-insurgency and anti-terrorist operations to detection of nuclear and biological agents, from detection and identification of chemical, biological warfare and explosive agents to detection of concealed weapons. Laser and opto-electronics technologies play an important role in handling low intensity conflict situations. The key advantages of the use of laser technology in such applications are near zero collateral damage, speed of light delivery and potential for building non-lethal weapons. Some of the well established laser devices in low intensity conflict (LIC) applications include laser dazzlers for close combat operations, mob/riot control and protection of critical infrastructures from aerial threats; lidar sensors for detection of chemical, biological and explosive agents; Femto second lasers for imaging of concealed weapons; lasers for sniper and gun fire location identification and so on. Use of laser vibrometry and electron speckle interferometry techniques for detection of buried mines and high power lasers for disposal of unexploded ordnances are emerging applications of laser technology for homeland security. This chapter briefly describes both established and emerging applications of laser and opto-electronics technologies in homeland security in terms of salient features, potential usage and international developments.

17.2 Non-lethal Laser Weapons

Non-lethal weapons act as a force multiplier enabling friendly forces to discourage, delay or prevent hostile action. They are particularly effective in situations where use of lethal force is not preferred; examples being limiting escalation and temporarily disabling facilities and equipment. Laser based non-lethal weapons such as laser dazzlers can be used for counter-insurgency, anti-terrorism, crowd control and infrastructure protection applications. Laser dazzlers are emerging internationally as a new non-lethal alternative to lethal force for law enforcement, homeland security, border patrol, coastal protection, infrastructure protection and host of other low intensity conflict scenarios due to their proven efficacy and significantly minimized collateral damage.

Figure 17.1 Laser dazzler

17.2.1 Fundamentals

Laser dazzler emits a high intensity laser beam in the visible band, usually in the blue-green region, to temporarily impair the vision of the adversary without causing any permanent or lasting injury or adverse effect to the subject's eyes (Fig. 17.1). The blue-green region is the chosen wavelength band as the human eye's response is the highest in this band (Fig. 17.2). Laser dazzlers typically employ state-of-the-art diode pumped solid-state lasers. Non-linear optical devices are employed for shifting the laser wavelengths to the visible part of the spectrum. For achieving the temporary blindness effect any visible wavelength in principle can be used. However, green wavelength, falling at the peak response of the eye, is found to be more effective. Here a widely diverging laser beam is grossly pointed towards the object of interest. The diverging output also results in a larger spot at the intended target, making aiming the device over long distances, or at multiple subjects, much easier.

A laser dazzler system typically uses a laser power of a few hundreds of milli watts for short range handheld devices to few tens of watts for long range (up to 10 km) platform mounted devices. Table 17.1 presents the visual effects of laser on the human eye depending upon the laser power density.

The most commonly employed wavelength for the purpose is 532 nm usually generated by using either laser diodes or frequency double Nd-YAG laser modules. Laser dazzler is a non-lethal weapon specifically designed for applications where subject vision impairment is to be achieved at a distance ranging from a few tens of meters to several kilometres in bright ambient conditions.

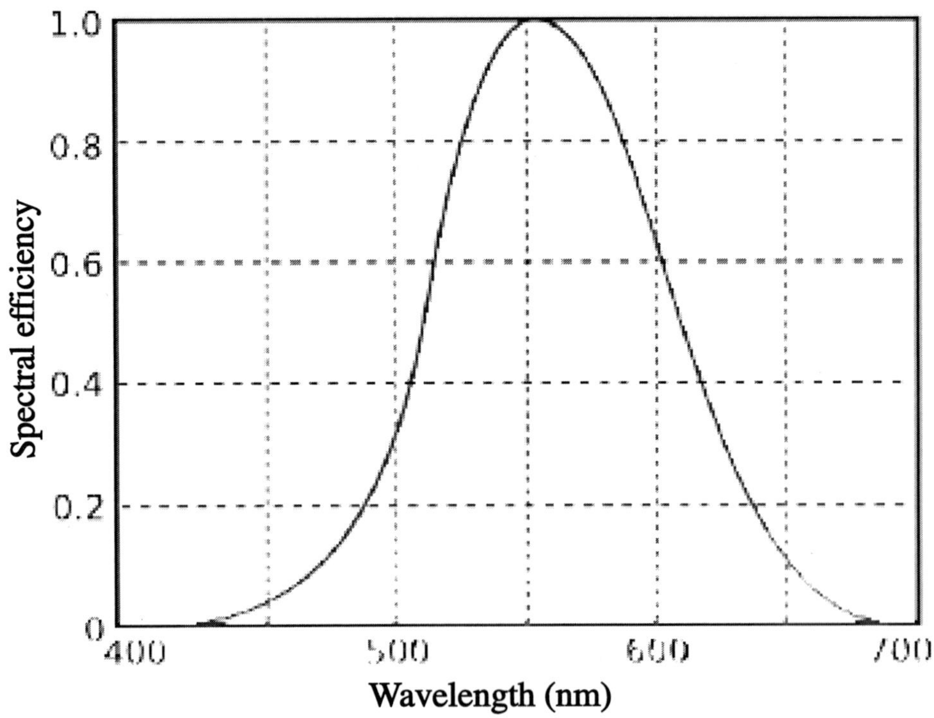

Figure 17.2 Human eye response

Table 17.1 *Visual effects of laser on human eye*

Laser power density	Phenomenon	Effect
0.5 μW/cm^2	Distraction	Distracting, but can see past the light
5 μW/cm^2	Glare	Interferes with vision
100 μW/cm^2	Temporary flash blindness	Blocks vision during and after exposure
> 2500 μW/cm^2	Haemorrhages/Lesions	Probability of irreversible eye damage

Choice of performance parameters such as laser power, spot size at the target, laser power density etc. are driven by nature of deployment. The beam shaping and directing optics is so designed to achieve desired value of nominal ocular hazard distance (NOHD) and a laser power density that does not exceed the maximum permissible exposure (MPE) figure dictated by ANSI (American National Standards Institute) standards for eye safety. MPE is dependent on wavelength and exposure time. At 532 nm wavelength, MPE value is 2.5 mW/cm^2 for an exposure time of 0.25 s and 1.0 mW/cm^2 for exposure time of 10 s. These devices usually produce a randomly pulsed output in the range of 10 to 20 Hz riding a DC level for better overall effect. DC level is usually kept at 30–50% percent of the peak intensity level. Night time maximum operational range is typically three to four times the maximum day time operational range.

17.2.2 International status

A large number of companies are offering short to medium range laser dazzlers with operational ranges from 10 s of meters to a few km for various application scenarios. These devices are particularly suited to counter insurgency and homeland security applications. Laser dazzlers are available in a variety of package and mounting configurations. Most devices are configured around frequency doubled Nd-YAG modules operating at 532 nm and are designed to produce power density in the range of 0.2–0.6 mW/cm^2 at the target plane. Advanced versions of these devices have an in-built laser spot diameter control that enables them to offer a very small NOHD. The laser spot diameter in this case is so varied as to maintain a nearly constant power density.

The first reported use of laser dazzlers was by the Royal Navy warships of England during Falkland's war of 1982. Russians have also been reported to use laser dazzlers against US aerial platforms in the straits of Juan de Fuca. One popular family of short range laser dazzlers including GLARE MOUT, GLARE GBD-IIIC and the GLARE LA-9/P have been developed by B.E. Meyers, USA to provide a non-lethal deterrent weapon which temporarily interferes with a suspect's vision without causing any ocular damage. The GLARE MOUT is a non-lethal visual disruption laser with an effective range of 150 m–2 km. The device is ideally suited for small arms integration as well as mobile crew-served applications. When a suspect individual approaches a restricted or controlled area, the GLARE MOUT is able to hail and warn the targeted person before shots have to be fired. Its highly intense green laser beam sends an irrefutable, multi-lingual, cross-cultural warning to the subject.

The GLARE GBD-IIIC (Fig. 17.3) is a medium to long range visual deterrent laser device for warning applications. Whether it is ship-to-ship signalling or airborne over-watch, the GLARE GBD-IIIC is the most effective green dazzler available. With twice the power and a more concentrated beam than the GLARE MOUT, the GLARE GBD-IIIC can effectively hail and warn as far away as 4000 meters. Like the GLARE MOUT, the GLARE GBD-IIIC produces an overtly bright green diffused laser spot to impair the visual ability of a suspect.

The GLARE LA-9/P is another visual deterrent laser device having an operational range of 300 m–4 km. It also incorporates eye safe technology that rapidly detects if a bystander is inside the NOHD (nominal ocular hazard distance), and shuts off the glaring laser output to prevent unintended eye injury.

Yet another short range laser dazzler is the compact high power (CHP) laser dazzler (Fig. 17.4) from LE systems, USA developed under the sponsorship of DARPA. The dazzler emits a 500 mW flashing green dazzling laser beam and is capable of engaging targets up to 200 meters.

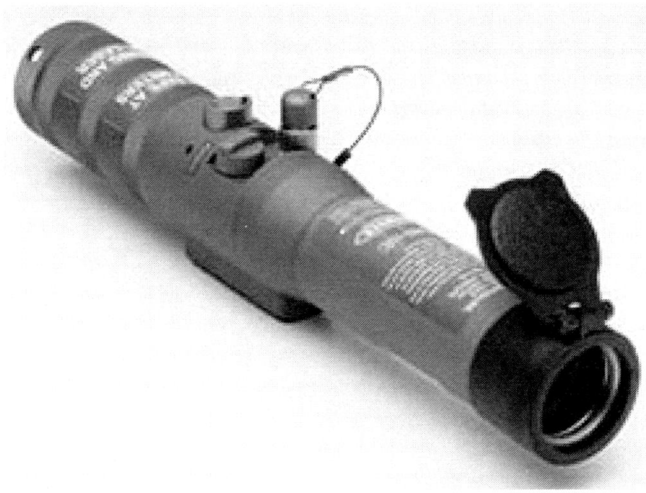

Figure 17.3 GLARE GBD-IIIC medium range laser dazzler

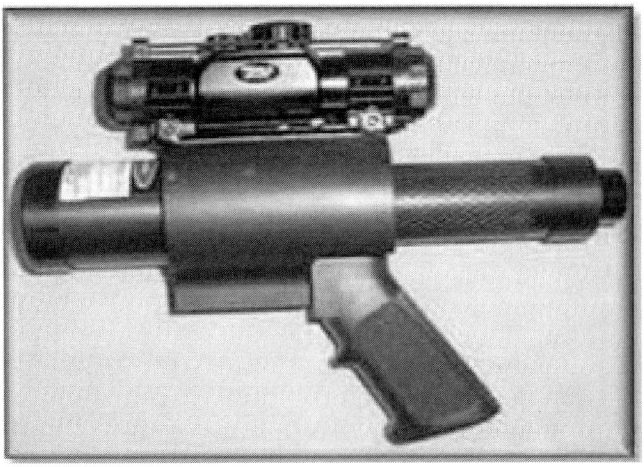

Figure 17.4 CHP laser dazzler

The *Photonic Disruptor*, classified as TALI (threat assessment laser illuminator) is a non-lethal high power green laser developed by Wicked Lasers, USA in cooperation with Xtreme Alternative Defense Systems. This tactical laser is equipped with a versatile focus-adjustable collimating lens to compensate for range and power intensity when used to either incapacitate an attacker in close range or safely identify threats from a distance.

Laser dazzler systems for crowd control and infrastructure protection applications are also under development. Crowd control laser dazzler systems with day/night time operational ranges of the order of 250–500 m are proposed to be vehicle mounted systems with the laser integrated with a gimbal platform and equipped with joystick control usually provided in front of the co-driver's seat for viewing, scanning and controlling laser beam position on the target scene. An electronically controlled laser beam projector assembly usually controls the laser beam spot size at the target. Fig. 17.5 shows the photograph of one such system.

Figure 17.5 KLEG laser dazzler on CROWS-II platform

Long range dazzler systems with maximum operational range approaching 10 km are also commercially available. The *ZM-87 Portable Laser Disturber* is a Chinese electro-optic countermeasure laser device. It can blind enemy troops at up to 2 to 3 km range and temporarily blind them at up to 10 km range. It has been reported that ZM-87 has been widely deployed including on naval vessels. Long range laser dazzlers intended for protection of critical infrastructure and strategic assets are usually integrated with a 360 gimbal mount and an electro-optic tracker. They are also being considered for applications such as critical infrastructure and strategic assets. Such a system would be capable of providing effective means to thwart suspect aircraft and other aerial vehicles from violating no fly zones and protected aerospace. The dazzler can also provide effective means for intention assessment of the suspect aircraft thereby facilitating quick decision for restoration of lethal forces.

17.2.3 Emerging trends

While short and medium range hand-held and weapon-integrated versions of laser dazzler are in widespread use for anti-terrorist and counter-insurgency applications; vehicle-mounted systems designed for operations against unruly and violent crowds are beginning to appear on the scene. Another interesting development in the recent past has been a keen interest in long range laser dazzler systems mounted on Gimbal platforms and equipped with electro-optic tracker and also integrated with a network of radars to provide 24×7 protection to strategic assets from airspace violation by aerial platforms. In the not too distant future, one would see deployment of laser dazzlers with global coverage. These systems are proposed to use a remotely controlled membrane reflector to receive the dazzling laser beam from the source station and guide it to the intended target location. Fig. 17.6 illustrates the concept.

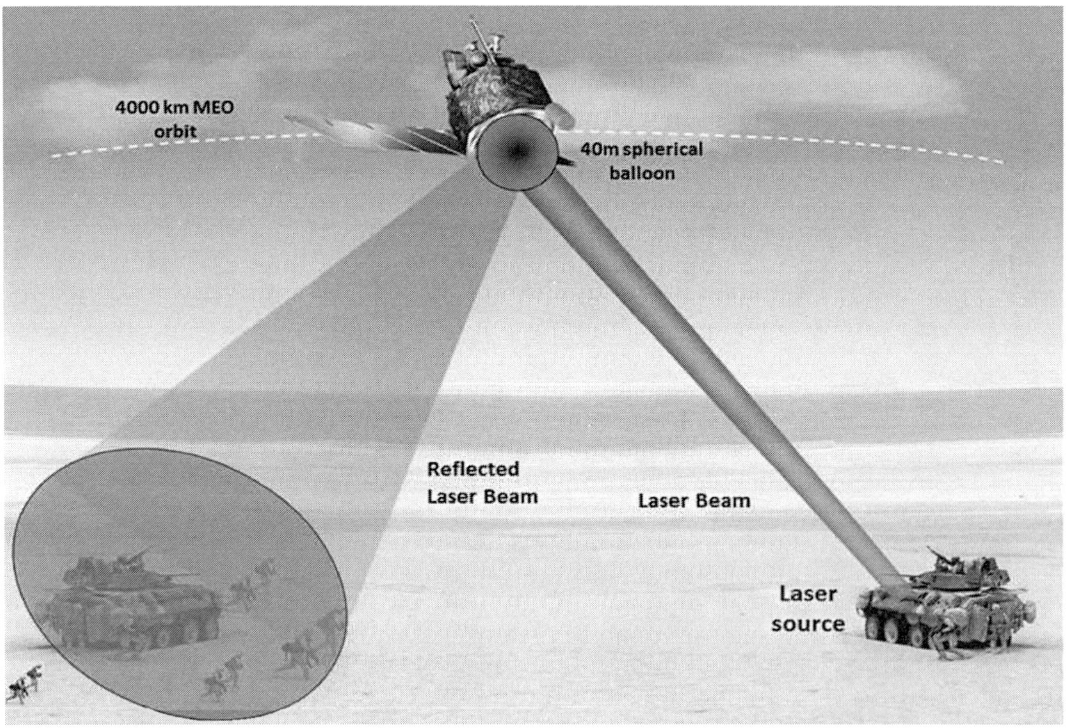

Figure 17.6 Laser dazzler for global coverage

17.3 Laser-based Ordnance Disposal

Disposal of unexploded ordnances including surface laid mines, improvised explosive devices (IED), grenade shells, artillery/mortar rounds, cluster bombs and so on from safe stand-off ranges using a high power laser beam is an emerging application of directed energy laser systems in homeland security related applications. Ordnance is disposed of by focusing high power laser beam on the ordnance casing thereby heating it until the temperature of the backplane of the casing exceeds the ignition temperature of the explosive filler. The explosive filler ignites and begins to burn and the process is independent of the type of fusing used by target explosives. This leads to a low level detonation or deflagration rather than full power detonation. The advantages of using of laser energy for ordnance disposal include large magazine, high precision, controllable effects with reduced collateral damage and assured and fast disposal from safe stand-off ranges.

One such international system is the ZEUS-HLONS (HUMMWV laser ordnance neutralization system) of USA (Fig. 17.7). The concept of neutralization of live ordnances using laser energy was first demonstrated in the field in 1994 with the development and field testing of Mobile Ordnance Disrupter System (MODS) that employed a 1.1 kW arc lamp driven solid-state laser mounted on an M113 A2 armoured personnel carrier. The ZEUS system initially employed a 500 watt laser. The latest version of the Zeus system is equipped with a 2 kW fiber laser mounted on to the roof of an adapted Humvee vehicle.

Figure 17.7 Zeus-HLONS laser ordnance neutralization system

Another laser system designed for disposal of unexploded ordnances is the *Laser Avenger* system from Boeing combat systems (Fig. 17.8). Boeing successfully tested a 1 kW solid state laser weapon mounted on a converted anti-aircraft vehicle in Redstone Arsenal in Huntsville Alabama by neutralizing multiple types of improvised explosive devices, including large calibre artillery munitions and mortar rounds. The system was operated at safe distances from the targets under a variety of conditions including different angles and ranges.

Figure 17.8 THOR laser IED neutralization system

Laser ordnance neutralization system is being considered for a range of application scenarios. One possible application relevant to homeland security is disposal of improvised explosive devices also called roadside bombs. Laser based disposal system offers a quick, safe, and reliable method of neutralizing these roadside bombs from a safe stand-off distances in excess of 150 m. Another possible application is in the disposal of friendly explosives that have outlived their shelf life. Since full-scale conventional wars are not there to be fought often there is a large inventory of ammunition stored in the ammunition depots which have exceeded their shelf lives and are waiting to be neutralized. Neutralization of such large quantities of ammunition is not only a cumbersome exercise; it also poses a safety hazard.

There can be various possible variants of the ordnance neutralization system. One configuration is the stand-alone high power laser system mounted on a vehicular platform. ZEUS-HLONS and Laser Avenger are examples of this category. In another possible variant, the system may be mounted on a remotely controlled vehicle with a robotic arm. While the laser system along with its power and thermal management systems could be mounted on the platform; the laser delivery system could be mounted on the robotic arm. In addition to these configurations, high power laser system may be integrated with other means such as ground penetration radar for detection of buried explosives. The integrated system allows detecting and neutralizing both surface-laid as well as buried explosives.

In some cases, the laser ordnance disposal system is mounted on a remotely controlled weapon station to get both defensive as well as offensive capability. RAFAEL has introduced a high-energy laser weapon system called THOR, developed in Israel to defeat improvised explosive charges (IEDs), roadside bombs, unexploded ordnance (UXO) and other potentially hazardous explosives. The system comprises a high-energy laser along with its beam director and a coaxial 12.7 mm machine gun. Since the laser is used to burn the IED rather than activate it, it can be used to avoid collateral damage which can result from other neutralization procedures. Figure 17.9 shows a photograph of the system.

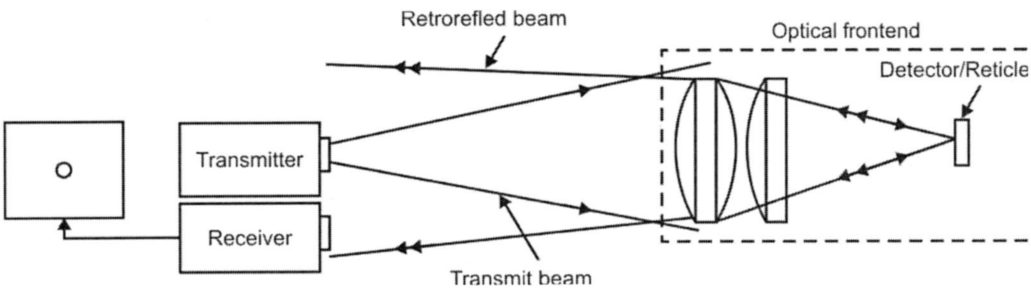

Figure 17.9 Cat eye effect

17.4 Optical Target Locator System : Counter-Sniper System

Yet another emerging application of laser technology is in detection and identification of battlefield optronic sighting systems which includes optical scopes, night vision devices, thermal imagers, laser range finders, target designators and so on. The device operates on the principle of cat eye effect (Fig. 17.9). The target optical device is illuminated by a laser beam, the optical system returns a fraction of it as back-scattered energy. Back-scattered energy is received by a sensitive receiver. With particular reference to homeland security, such a device could be very useful for detecting optical scopes employed by snipers. Another homeland security related application could be surveillance of sensitive areas particularly in urban territories. Figure 17.10 shows a typical deployment of such a device in an urban environment when used for area sanitization of the. The user illuminates the suspect area, a multi-storey building in this case, with a widely diverging laser beam emitted by the transmitter portion of the electro-optic target locator (Fig. 17.10). If there was an optical device looking towards the laser beam within its field of view; a retro-reflected laser beam would travel back towards the laser transmitter and reveal the location of the optical device.

Figure 17.10 Typical deployment scenario in an urban environment

One such international system is the Laser Sniper Detector (SLD-400, SLD-500) from CILAS, France. It can be fitted on a tripod or on a static vehicle. Its highly accurate threat detection and localization capability is compatible with the performance of the associated weapon or fire control system. SLD-500 is an improvement over SLD-400. Figure 17.11 shows the photograph of another similar system called the Optifinder-1200 from M/s Torrey Pines Inc.

Figure 17.11 Optifinder-1200 sniper detection system

Optifinder-1200 is an active system designed to detect the retro-reflected signal caused by the front end optics of the target system. It uses a combination of high sensitivity detector and optical filter to offer a maximum operational range of 1200 m. It also makes use of time-of-flight calculation of the transmitted signal and the retro-reflected signal to compute the target range. The chosen detector offers day/night operational capability. The system allows penetration through bad weather and vegetation by range gating. Three variants of optifinder system are available with operational ranges of 350 m, 1200 m and 2000 m.

17.5 Detection of Chemical and Biological Warfare Agents

Efficient and reliable technology for detection and identification of chemical, biological and explosive agents has become far more relevant and increasingly important in the present day scenario. Chemical and biological weapons are relatively inexpensive to produce and are capable of unleashing a bigger devastating effect as a terrorist weapon. The international community has already witnessed the devastating effects these weapons are capable of causing on more than one occasion. Millions have been affected by the destruction caused by these weapons of mass destruction in the past. The sarin attack by the Aum Shinrikyo cult in 1995 in a Tokyo subway, the hydrogen cyanide and mustard gas attack by Iraq in the Anfal campaign against the Kurds, most notably in the Halabja massacre in 1988 and the international dispersal of Anthrax spores against USA in 2001 are some examples of chemical and biological weapons attack. Use of improvised explosive devices by terrorists and antinational elements, killing soldiers and civilians alike has become a routine affair. The incidents have brought to the fore the need for stand-off detection and identification of chemical, biological and explosive agents. Laser technology has played an important role in providing an effective solution to this problem. In the paragraphs to follow are discussed some laser based methodologies/techniques that are either in use or being explored for the purpose of detecting and identifying chemical and bilogical weapons.

17.5.1 Detection of chemical warfare agents

Nerve agents such as organophosphonate compounds and blister agents such as mustard compounds because of their acute toxicity, rapidity of action, indetectability by human senses and economic viability are the two potentially very harmful classes of chemical warfare agents. Nerve agents are the organophosphoris easters which initially stimulate and then paralyze certain nerve transmissions by interfering with cholinesterase enzyme. Common chemical warfare agents belonging to this category include Tabun, Sarin, Soman, Vx, Dichlorvos and Malathion. Figure 17.12 shows a photograph of a victim of chemical warfare agent attack.

Lidar based detection is the most practically realizable technique of stand-off detection of chemical and biological warfare agents. Commonly exploited physical phenomena forming the basis of chemical and biological agent detection include elastic back-scattering, laser induced fluorescence and differential absorption. They are superior to point detection systems such as infrared spectroscopy and Raman spectroscopy because of their capability to range and discriminate the chemical and biological molecules in real time. Ultraviolet laser induced fluorescence could be used for detection of biological agents since most biological molecules fluoresce when excited by ultraviolet radiation.

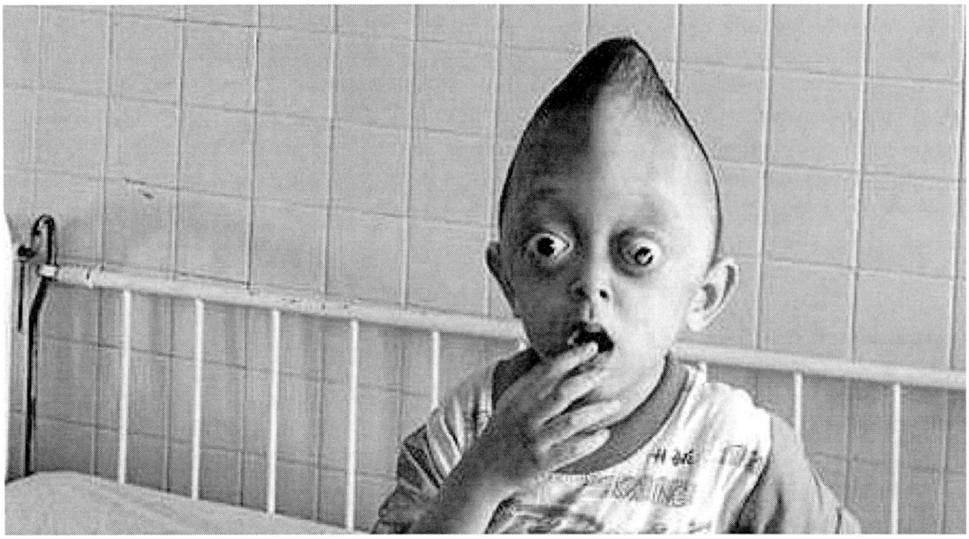

Figure 17.12 Effect of chemical weapon attack

Differential absorption technique known as DIAL uses two wavelengths—one corresponds to the wavelength of peak absorption of the targeted molecule and the other corresponds to the weak absorption of the targeted molecule. Ratio of the two received back-scattered signals measures the concentration of the targeted chemical warfare agent.

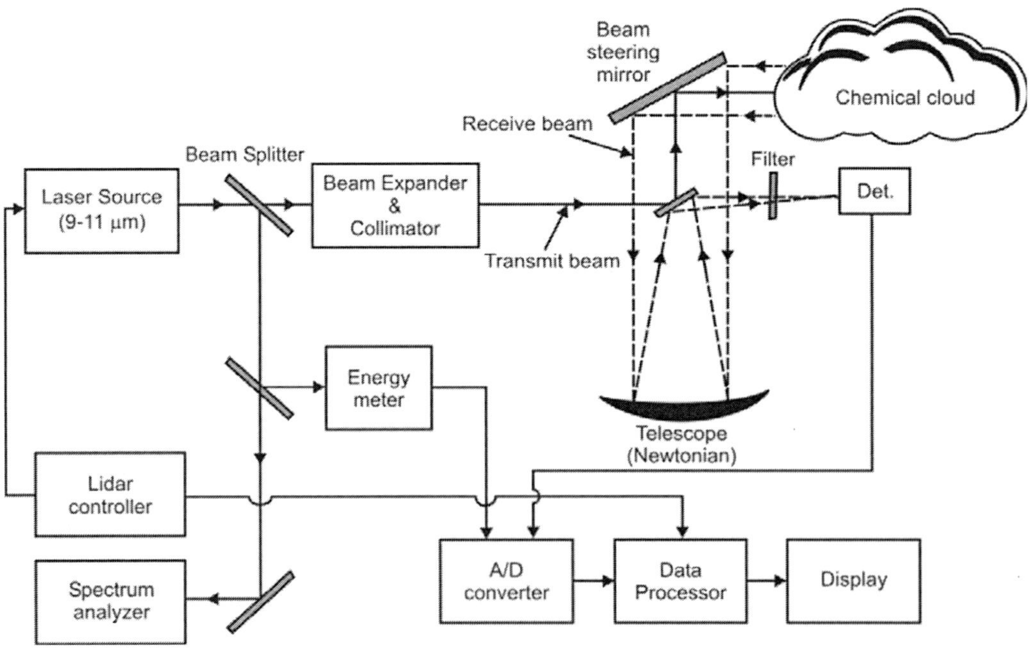

Figure 17.13 Block schematic of differential absorption lidar

Differential absorption lidar (DIAL) is the most frequently used technique used for detecting chemical warfare agents along with detection of toxic gases and pollutants. Figure 17.13 shows the block schematic arrangement of the DIAL system. Two laser pulses, called ON and OFF wavelengths, are transmitted towards the targeted area of interest in the atmosphere that is suspected to be contaminated with a potentially harmful chemical warfare agent. As outline earlier, the ON wavelength corresponds to the peak absorption of the suspected harmful species and the OFF wavelength is slightly detuned from the ON wavelength to encounter significantly lower absorption from the same species. It may be mentioned here that the choice of ON and OFF wavelengths is unique for a given chemical species and is determined by using a tunable laser that scans the area of interest by transmitting pairs of these ON and OFF wavelengths. Obviously, the On wavelength called λ_{ON} encounters maximum absorption with the result that the corresponding back-scattered signal is relatively weak as compared to the λ_{OFF} that encounters weak absorption and produces a relatively stronger back-scattered signal. This forms the basis of knowing the atmospheric constituents at that time. Ratio of the two back-scattered signals gives concentration of the molecule of interest.

Many of the chemical warfare agents have significant absorption bands in the 3–4 μm and 9–11 μm bands. CO_2 lasers emitting in the 9–11 μm band have been commonly used for the detection of a majority of the chemical agents. Some DIAL systems do make use of both these bands and therefore employ both tunable mid-IR as well as CO_2 laser sources. The schematic arrangement of the DIAL system shown in Fig. 17.13 however makes use of only the CO_2 laser. The transmitted laser beam after suitable collimation is directed towards the atmosphere in the desired direction with the help of a scanning gimbal mirror. The back-scattered signal is received by the receiving telescope and is further focused on to the detector subsystem. The interference filter is used to block the undesired radiation and allow only the radiation of wavelength of interest to pass through it. The detected signal is then digitized and fed to the data processor to extract the desired information on the type and concentration of the chemical species.

17.5.2 Detection of biological agents

Stand-off detection of biological warfare agents and their discrimination from background aerosols is extremely challenging as the distinction between innocuous ambient bacteria and virulent microbes amounts to subtle differences in their molecular make up. As these subtle differences involve a very small percentage of molecules; they produce a very small change in their optical signatures. This makes it extremely difficult to reliably detect and discriminate these potentially dangerous molecules. The processes of detection and discrimination are further complicated by the fact that there is a constant variation in the growth media and the contaminants associated with processing of biological warfare agents affect their optical signatures. Figure 17.14 shows the effects of a biological warfare agent attack.

The stand-off detection of biological warfare agents is based on the concept of laser induced fluorescence (LIF) effect. Laser induced fluorescence is the emission from atoms or molecules after they have been excited to go to higher energy levels by excitation by another laser. The emission takes place at a wavelength that is higher than the wavelength of laser light exciting it. The biological warfare agent molecules mainly constitute aromatic amino acids and coenzymes. Aromatic amino acids such as tryptophan, tyrosine and phenylanine absorb laser radiation at 280–290 nm and fluoresce in 300–400 nm bands. It is therefore possible to detect biological warfare agents by using UV laser excitation at suitable wavelengths. Discrimination of biological agents can also be achieved only from the LIF signal because the fluorescence cross-section of particles in the range of 1 to 10 μm size are sufficiently large to make single particle interrogation feasible.

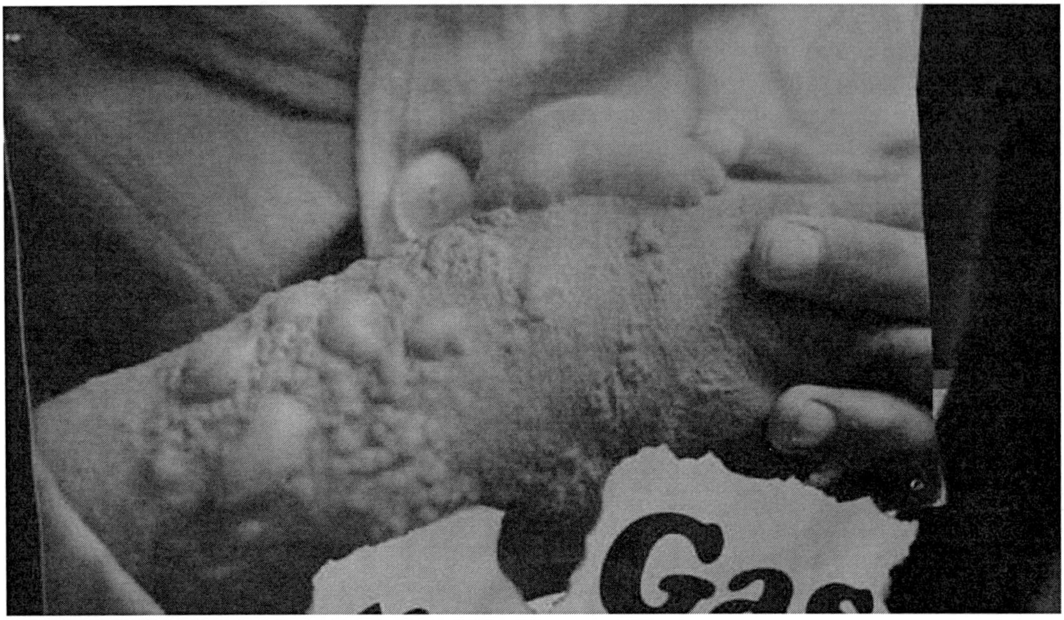

Figure 17.14 Effects of biological warfare attack

Figure 17.15 shows the block schematic arrangement of a typical monostatic UV LIF lidar system. The principle components of this lidar system include a frequency tripled or quadrupled ND-YAG laser respectively emitting at 355 nm (frequency tripled output) and 266 nm (frequency quadrupled output), a telescope used for both transmission and reception, a photomultiplier tube used for recording the back-scattered signal and the bio-fluorescence signal, a spectrograph with a gated ICCD array for recording the dispersed fluorescence spectra and the lidar controller. The fourth harmonic at 266 nm is transmitted towards biological cloud. The telescope receives the back-scattered and the bio-fluorescence signal, which is then fed to the PMT channel. Backscattered signal is received by the gated PMT channel and is present whenever there is a cloud along the beam path. This channel is used to measure the distance to the cloud. A solar blind PMT channel is used to receive the fluorescence signal and is activated only when there is a suspicious cloud. A spectrograph with a gated ICCD is used to identify the nature of the bio-molecule that is responsible for the fluorescence. The bandwidth of the receiver channel is usually kept smaller so as to disallow unwanted background radiation from entering the channel.

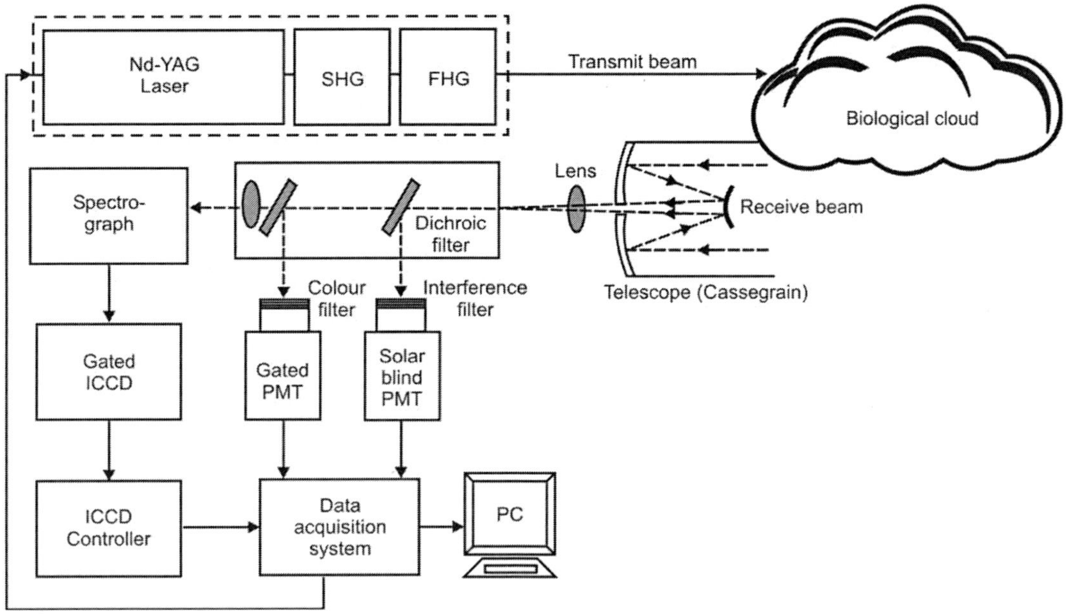

Figure 17.15 Block schematic arrangement of UV-LIF lidar system

17.6 Detection of Explosive Agents

Stand-off detection of explosives using lasers with trace detection method is one of the most widely researched technologies internationally. For the technology to mature to an extent where it can be transformed into a product usable in the kind of environment and field conditions usually encountered in homeland security related applications is a great technological challenge. One of the major problems comes from decrease in the intensity of back-scattered light signal due to wavelength dependent absorption and scattering losses and the intensity decreasing inversely with distance squared. The problem is compounded by the fact that the trace level associated with common explosive agents are extremely low, in the range of a fraction of ppb (parts per billion) to a few ppm (parts per million) in the case of common explosives. The second major problem pertains to unique identification of the targeted explosive agent in a background of interferents. Many chemical agents have atomic compositions including sulphur, phosphorus, fluorine and chlorine in addition to nitrogen, oxygen, hydrogen and carbon present in organic molecules. Thus, the detection methodology needs to be highly sensitive and selective. Laser based spectrometric methods have the potential of being fast, sensitive, selective, able to detect and identify a wide range of explosive agents and upgradeable to handle new threats.

Atmospheric transmission at the wavelengths concerned is an important factor while assessing suitability of a given stand-off detection methodology. Commonly used technologies are laser induced breakdown spectroscopy (LIBS), Raman spectroscopy and its variants; laser induced fluorescence (LIF) spectroscopy and IR spectroscopy. These are all trace detection methods; though there are bulk detection methods such as millimetre wave imaging and terahertz spectroscopy.

Laser induced breakdown spectroscopy focuses a high energy laser beam on the trace sample to break down a small part of the sample into a plasma of excited ions and atoms. The plasma emits light that is characteristic of emissions from ionic, atomic and small molecular species. These light emissions are detected by a spectrometer to identify the elemental composition (Fig. 17.16). One of the challenges in the use of LIBS is how to assess its efficacy in detecting and identifying explosive species in a real environment that is replete with many interfering substances. One way to get the desired selectivity is to use a double pulse LIBS. In a double pulse LIBS, the first pulse is used to create a so-called laser generated vacuum and the second pulse transmitted a few microseconds later generates the return signal. The double pulse LIBS is also observed to improve sensitivity in addition to enhancing selectivity. Selectivity can be further improved by adding temporal resolution to the LIBS emission analysis.

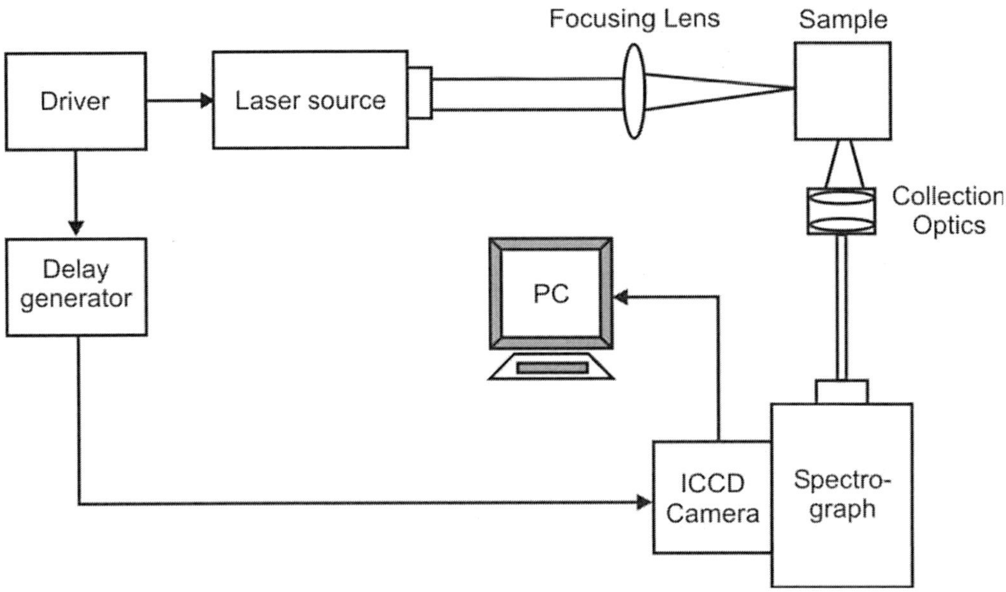

Figure 17.16 LIBS set-up

1064 nm has been widely employed for LIBS systems. Due to serious eye hazards posed by 1064 nm, scientists have also tried 266 nm. 266 nm also allows the designer to build Raman capability into the system. It may be mentioned here that 266 nm has 600 times higher maximum permissible exposure (MPE) limit as compared to 1064 nm.

Raman spectroscopy offers another method for stand-off detection of explosive agents. It has been extensively used for many years as a standard analytical tool for identification of chemical agents in the laboratory environment. The basis of detection in this case is the shift in the wavelength caused due to the inelastic Raman scattering by the target molecule. The inelastic scattering of impinging photons where some energy is lost to (or gained from) the target molecule returns scattered light with a higher (or lower) wavelength depending upon whether energy was lost to (or gained from) the target molecule. The difference is dictated by the energy of vibrational modes of the target molecule and therefore constitutes the fingerprint or the basis of identification. Complex mixtures are identified using algorithms for pattern recognition.

A major drawback of the Raman technique is its extremely poor sensitivity caused by the fact that Raman scattering occurs for one in about 10^7 photons impinging on the sample. Weak return signal intensity of Raman spectroscopy limits its use for trace detection as it also makes it sensitive to ambient light and fluorescence from the sample itself or other chemicals in the vicinity. The fluorescence masks the Raman signal. These problems are overcome by the use of resonant Raman spectroscopy. With a tunable laser, the wavelength can be chosen to match or nearly match a resonant absorption in the target molecule leading to intensity enhancement of the order of 10^6.

The problem of fluorescence masking Raman signal can be overcome by the use of either infrared or ultraviolet radiation. Infrared radiation does not have sufficient energy to cause fluorescence and ultraviolet radiation will cause fluorescence in the visible range, which is well separated from the Raman signal.

Laser induced fluorescence (LIF) is another important tool for similar applications. Though a very valuable tool in combustion diagnostics and for studying decomposition of explosives, it has not been found very useful for detection of explosive agents.

Almost all the laser based stand-off detection methods have detection limits too high to make them suitable for detection of explosives in field conditions, which requires the capability to detect traces in the vapour phase or in the form of particles. Another problem is insufficient selectivity to identify target explosives in the presence of interferents. In fact, none of the laser based spectroscopic methodologies available today is ready for full functionality prototype manufacturing. There is a lot of scope for further research to improve the performance of the more mature technologies in terms of both detection sensitivity and selectivity. The final solution would probably lie not in one type of sensor but integration of several technologies to derive the benefits of the best of each.

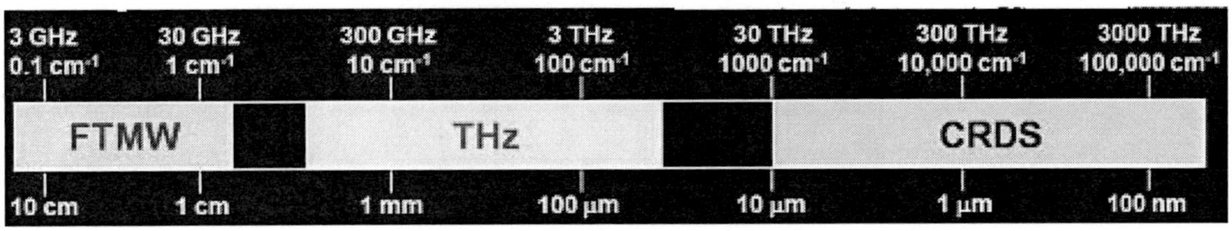

Figure 17.17 THz band

17.7 Detection of Concealed Weapons

A number of terror attacks around the globe have brought to public awareness the extent of security deficits at airports and other public spaces and the need for introducing body and baggage scanners as additional security measures. With suicide missions having become a common mode of terrorist attacks, equipping the security agencies with the capability to detect concealed weapons can be an important force multiplier. The need to enhance security infrastructure at these places and also at the sites of critical infrastructure and strategic assets has resulted in the strong interest of the science and technology community in sensor technologies capable of detecting concealed weapons and explosive devices. The objective of the sensor technology is to facilitate the law enforcement agencies in non-contact, reliable and rapid screening of individuals and groups without any significant disruption of their activities.

Guns, knives and explosives are the common potential threats usually carried by terrorists and anti-social elements concealed by them underneath their clothing or hidden in the baggage. The solution lies in the use of advanced sensor technologies that would allow individual and baggage screening as well as area screening.

The increased threats from terrorists and anti-social elements in recent years have led to the development of many technologies intended for detection of concealed weapons, contraband, explosives etc. These include X-ray scanners, trace detection systems including both point and stand-off detection systems capable of detecting explosive vapours or particles left behind while handling, neutron activation and other systems based on energetic radiation. Millimeter wave (MMW) imaging and terahertz (THz) spectroscopy are two very potent techniques relevant to the detection of concealed weapons and explosives. The MMW region of electromagnetic spectrum (1–10 mm, 30–300 GHz) penetrates clothes and many other materials allowing imaging of objects under clothes. In addition, MMW radiation is also safe at moderate intensities to use on people. However at these wavelengths, the spectra do not have any features that can be used to identify materials.

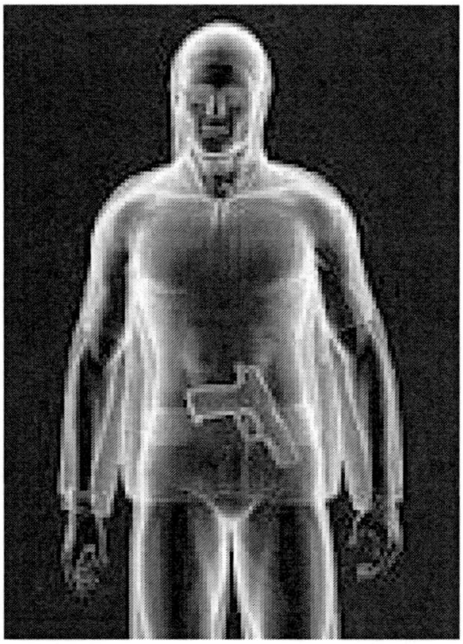

Figure 17.18 THz imaging

THz band lies between infrared and microwave bands typically extending from 0.1 mm to 1.0 mm (Fig. 17.17). At shorter wavelengths, (0.1–1.0 mm, 300 GHz–3.0 THz), one can get spectroscopic information. THz radiation being non-ionizing in nature is also very safe at moderate intensities for use on people. Frequencies above 3.0 THz yield still better spectroscopic information but these are not suitable for stand-off detection due to heavy absorption in moist air. In fact, moist air is nearly opaque to electromagnetic radiation above 3.0 THz. Another significant challenge to the use of THz imaging technology for detection of concealed weapons is the required level of imaging resolution that would enable discrimination between hand guns, knives or explosive belts from commonly carried objects such as cell phones, wallets, pens etc. It may be mentioned here that the technology of THz imaging has advanced to a level that allows imaging of concealed objects with acceptable resolution. Figure 17.18 shows a typical image obtained with THz imaging.

Figure 17.19 Terahertz security camera, type T5000 from ThruVision

There are two categories of techniques for generation of THz radiation. One of them uses resonant optical rectification. THz radiation in this case is generated from the polarization change that follows transport of excited carriers in an applied electric field. The other technique is that of non-resonant optical rectification based on non-linear dielectric crystals and is an efficient method of generating THz radiation. Without going into the micro details of generation of THz radiation, it would suffice to say here that THz can penetrate many non-metallic materials such as paper, cloth and leather, which allows it to be used effectively in building imaging systems suitable for screening individuals and baggage for any weapons, contraband, explosives etc at airports, railway stations and other similar public places. Figure 17.19 shows the photograph of one such THz imaging security camera, type T5000 from M/s ThruVision. The camera can detect weapons, drugs or explosives hidden under people's clothes from up to 25 meters and is effective even when people are moving. It does not reveal physical body details and the screening is safe for use on human beings. The technology has a range of military and civilian applications. Figures 17.20 and 17.2 depict such application scenarios.

17.8 Conclusion

Laser and optoelectronics technology is destined to play a pivotal role in the future when it comes to strengthening the security and law enforcement agencies engaged in low intensity conflict operations including counter-insurgency, anti-terrorist, infrastructure protection and law enforcement operations. Laser based non-lethal weapons including different variants of laser dazzler to cover various application requirements; stand-off explosive detectors and laser based neutralization of ordnances and THz imaging technology for detection of concealed weapons and explosives are going to be particularly attractive. Stand-off explosive detection technology would need to mature faster to achieve the desired levels of sensitivity, selectivity and immunity to false alarm. The capability to detect explosives would need to be taken to the next level to be able to detect explosives such as mines buried underneath the earth's surface or under the ocean bed. Once mature, the technology would need to be integrated with the ordnance disposal system as detection must precede disposal, though disposal alone also has tremendous application potential. Portable anti-sensor laser devices to incapacitate observation aids such as NVDs, thermal imagers used by the adversary would soon become a common device for LIC applications. There could be many more laser devices based on directed energy property of the lasers finding application in homeland security related applications.

Figure 17.20 THz security camera deployment at airport

Bibliography

1. Yun-Shik Lee. 'Principles of terahertz science and technology.' *Springer.*
2. D. Sayeedkia. 'Handbook of terahertz technology for imaging, sensing and communications.' *Woodland Publishing Limited.*
3. Dwight. L. Woolard, James. O. Jens, R. Jennifer Hwu and Michael. S. Shur. 'Terahertz science and technology for military and security applications.' *World Scientific.*
4. Jeff Hecht. 'Diode pumped solid state lasers: Laser dazzlers are deployed.' *Laser Focus World*
5. Wikipedia. 'Dazzler (weapon).'
6. Wikipedia. 'Zeus-Hlons (laser ordnance neutralization system).'
7. 'Directed energy: A solution looking for a problem or an answer to the improvised explosives devices threat.' *Rusi Defence Systems*, Spring 2011.
8. Bartle Bull. October 2005. 'The IED tussle.' *Prospect Magazine, Issue* 115.
9. '*Improvised Explosive Devices (IEDs) In Iraq And Afghanistan: Effects and Countermeasures.*' Congressional Research Service Report for Congress, February 2008.
10. E. Colin, Webb, Julian and D. C. Jones. *Handbook of Laser Technology and Applications: Volume-III.* Institute of Physics Publishing.

18 Development of Laser Cavity and Resonator for High Power Chemical Oxygen Iodine Laser (COIL)

A. K. Varshney, Ashwani Kumar, Gaurav Singhal, Mainuddin and R. K. Tyagi

18.1 Introduction

COIL is a chemical laser in which the required pumping energy for population inversion is released via a chemical reaction. This property makes the COIL attractive for defence application because it eliminates the need for electrical power supply at remote locations. Among other chemical lasers, COIL has the advantages of power scalability, short wavelength (1.315 μm) compatible with fiber (Grunewald et al.) for remote operation and also better laser material interaction.

In COIL, a gas–liquid phase reaction between basic hydrogen peroxide and chlorine gas at sub-atmospheric pressure (Azyazov et al.) produces the pumping source, singlet oxygen. This is diluted with sufficient nitrogen buffer gas to reduce the various loss mechanisms. Part of the pump energy contained in the singlet oxygen is used in the dissociation of iodine molecules into iodine atoms and the rest is used to excite these iodine atoms by near resonant energy transfer reaction. The interaction of singlet oxygen with atomic iodine at appropriate flow conditions results in the generation of laser gain medium inside the laser cavity from where the laser output power is extracted using an optical resonator.

To develop a high power COIL, it is important to study the gain characteristics, i.e., the small signal gain and the saturation intensity of the active medium under different flow conditions and to evaluate the optimum cavity coupling for achieving maximum output power. In this chapter, different COIL input parameters required for optimal gain medium formation in the laser cavity have been analyzed and gain characteristics using simplified saturation model (SSM) (Hager et al.) for the development of high power COIL have been estimated. The resonator parameters and output mirror coupling are evaluated keeping in view the resonator stability, diffraction loss, utilization of mode volume and laser beam divergence. On the basis of these parameters, the laser cavity and optical resonator for high power COIL have been developed and tested.

18.1.1 COIL gain medium formation

COIL operates on the $I^*(^2P_{1/2}) - I(^2P_{3/2})$ transition of atomic iodine at a wavelength of 1.315 μm as Eq. (18.1)

$$I^*\left(^2P_{1/2}\right) + h\nu \rightarrow I\left(^2P_{3/2}\right) + 2h\nu(1.315\mu\text{m}). \tag{18.1}$$

The population inversion between spin orbit states of atomic iodine is achieved through near resonant energy transfer from singlet oxygen $O_2(^1\Delta_g)$ which is an excited energy state of oxygen molecule as shown in Fig. 18.1.

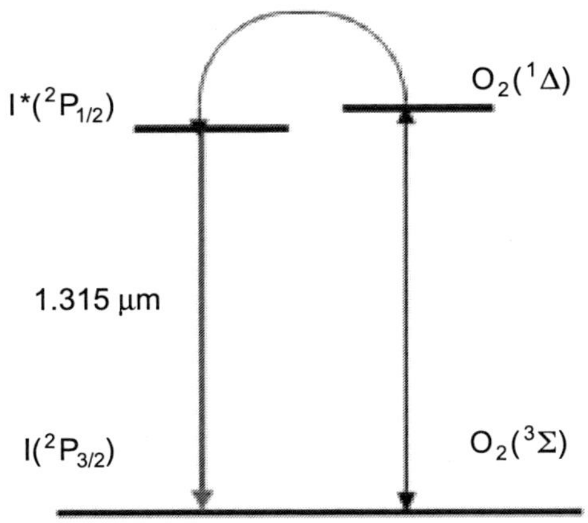

Figure 18.1 Energy level diagram

Among the various methods for generating the singlet oxygen, the chemical method is the best till date for the large-scale production of singlet oxygen with maximum yield. The chemical method is based on the reaction between chlorine gas Cl_2 and basic hydrogen peroxide (BHP) according to the Eq. (18.2).

$$Cl_2 + 2KOH + 2H_2O_2 \rightarrow 2KCl + 2H_2O + H_2O_2 + O_2\left(^1\Delta_g\right) \tag{18.2}$$

The schematic of COIL based on chemical method is shown in Fig. 18.2. The singlet oxygen is produced in a singlet oxygen generator (SOG).

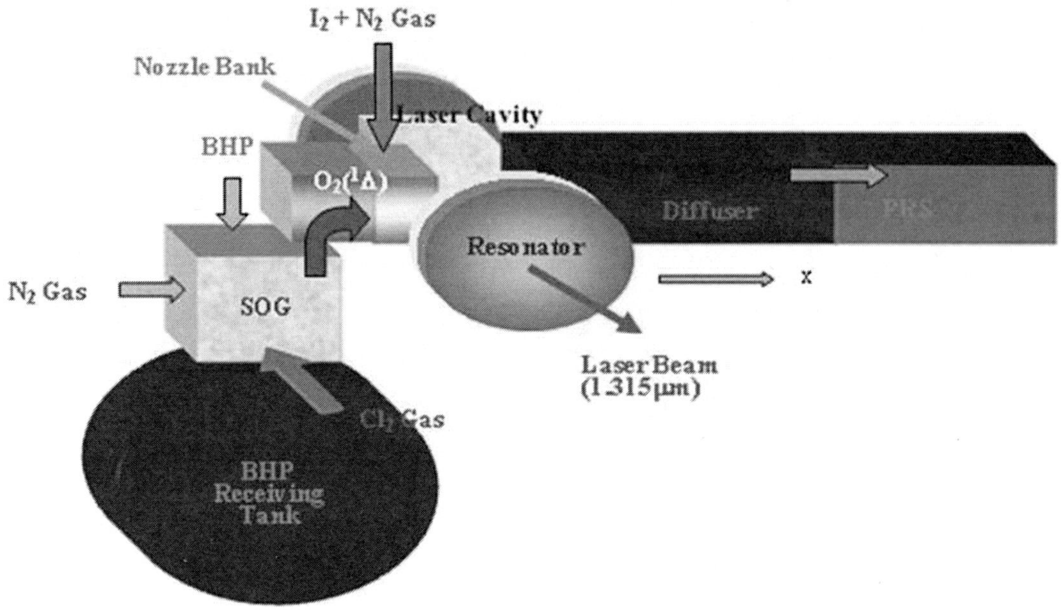

Figure 18.2 Schematic of COIL

The molecular iodine is injected into the primary flow containing the excited state $O_2(^1\Delta_g), O_2(^1\Sigma)$ and the ground state $O_2(^3\Sigma)$ molecular oxygen. The dissociation of molecular iodine is a complex process involving energy pooling but can be described globally according to Eq. (18.3).

$$nO_2(^1\Delta_g) + I_2 \rightarrow nO_2(^3\Sigma_g) + 2I + \Delta H(n) \tag{18.3}$$

where n represents the number of $O_2(^1\Delta g)$ molecules required to dissociate one I_2 molecule and $\Delta H(n)$ is the heat released as a function of n.

This is followed by rapid energy transfer and equilibration with the establishment of the population inversion in atomic iodine as Eq. (18.4).

$$O_2(^1\Delta_g) + I(^2P_{3/2}) \Leftrightarrow I^*(^2P_{1/2}) + O_2(^3\Sigma_g) \tag{18.4}$$

The ratio of excited and ground state atomic iodine concentration is determined by the equilibrium constant for the reaction k_{eq} from Eq. (18.5).

$$\frac{\left[I^*(^2P_{1/2})\right]}{[I(^2P_{3/2})]} = \frac{\left[O_2(^1\Delta_g)\right]}{\left[O_2(^3\Sigma_g)\right]} k_{eq} \tag{18.5}$$

where $k_{eq} = k_f/k_r = 0.75\ \exp(402/T)$ and forward reaction rate constant $k_f = 2.3 \times 10^{-8}/T\ cm^3/mol.s$, where T is the cavity temperature.

Since the concentration of $O_2(^1\Sigma)$ is very small, it is assumed negligible so that the total oxygen concentration

$$[O_2] = \left[O_2(^1\Delta_g)\right] + \left[O_2(^3\Sigma_g)\right]. \quad (18.6)$$

The ratio of singlet oxygen to total oxygen concentration is called singlet oxygen yield and at the upstream of the cavity before I_2 dissociation, it can be given as Eq. (18.7).

$$Y_P = \frac{\left[O_2(^1\Delta_g)\right]}{[O_2]} \quad (18.7)$$

Singlet oxygen yield (Y_0) upstream of the cavity after dissociation of molecular iodine may be expressed as Eq. (18.8).

$$Y_0 = Y_P - Y_{\text{diss}}, \quad (18.8)$$

where the singlet oxygen yield Y_{diss} is used in dissociation of molecular iodine

$$Y_{\text{diss}} = \frac{n[I_2]F}{[O_2]} \quad (18.9)$$

where $n = 2-5$ and iodine molecule dissociation fraction is 90–95% in case of grid type nozzle and total oxygen concentration $[O_2] = U^*[Cl_2]$ in which chlorine utilization (U) is greater than 90% in case of counter-flow type jet SOG and $[Cl_2]$ is the chlorine concentration.

After iodine dissociation, the ratio of iodine atoms to total oxygen is very small, typically 0.02–0.04. Therefore, each iodine atom is repumped and cycled many times along the flow during the lasing process. The rate of change of the $[O_2(^1\Delta_g)]$ and $[I^*]$ concentrations with respect to the flow direction x in the laser cavity can be expressed by the time dependent rate Eqs (18.10) and (18.11).

$$u\frac{d\left[O_2(^1\Delta_g)\right]}{dx} = -k_f\left[O_2(^1\Delta_g)\right][I] + k_r\left[O_2(^3\Sigma_g)\right][I^*] \quad (18.10)$$

$$u\frac{d[I^*]}{dx} = k_f\left[O_2(^1\Delta_g)\right][I] - k_r\left[O_2(^3\Sigma_g)\right][I^*] - \frac{g\tilde{I}}{h\upsilon} \quad (18.11)$$

where u is the gas flow velocity which is supersonic in case of high power COIL. Therefore, the cavity temperature, gas density and gas flow velocity are assumed constant during lasing. $\tilde{I}$ is two way intra-cavity circulating intensity and $h\nu$ is the laser photon energy.

The laser gain (g) in the laser cavity in the presence of radiation field is generally given by Eq. (18.12).

$$g = \sigma\left[N_2 - \frac{g_2}{g_1}N_1\right], \quad (18.12)$$

where σ is the stimulated emission cross-section. N_2 and N_1 are the concentration and g_2 and g_1 are the degeneracy for upper and lower laser levels respectively.

In case of COIL, the atomic iodine $5^2P_{1/2} - 5^2P_{3/2}$ lasing transition composed of six hyperfine components is shown in Fig. 18.3. The nuclear spin is $I = 5/2$. The $5^2P_{1/2}(F' = 3) - 5^2P_{3/2}(F'' = 4)$ component provides the highest gain and lasing occurs on this individual transition.

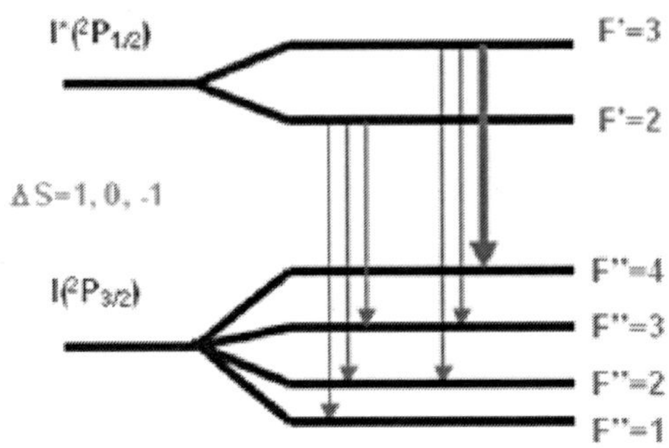

Figure 18.3 Hyperfine atomic structure of iodine

The degeneracy of the hyperfine components are $g = 2F + 1$ so that the degeneracy of the upper and lower laser levels are $g_2 = 7$ and $g_1 = 9$ respectively. Now if all the hyperfine levels are populated statistically,

$$I^*\left({}^2P_{1/2}, F' = 3\right) = \frac{7}{12} I^*\left({}^2P_{1/2}\right). \tag{18.13}$$

$$I\left({}^2P_{3/2}, F'' = 4\right) = \frac{9}{24} I\left({}^2P_{3/2}\right). \tag{18.14}$$

The stimulated emission cross-section (σ) of I^* under low cavity pressure condition (typically 3 torr) may be expressed using Doppler broadened line width

$$\sigma = \frac{A\lambda^2}{8\pi} g_D(\lambda), \tag{18.15}$$

where A is the rate of spontaneous emission and line shape function $g_D(\lambda)$ is

$$g_D(\lambda) = \frac{2}{\Delta\upsilon_D}\left(\frac{\ln 2}{\pi}\right)^{1/2} \text{ at center wavelength,} \tag{18.16}$$

where Doppler line width

$$\Delta\upsilon_D = \left(\frac{8kT\ln 2}{mc^2}\right)^{1/2} \cdot \frac{c}{\lambda}, \tag{18.17}$$

where k is the Boltzmann's constants, T is gas temperature in Kelvin in the laser cavity and m is the mass of iodine atom. In the absence of a radiation field in the laser cavity, pumping reaction sets the small signal gain.

$$g_0 = \sigma_0 [I]_0 \frac{[2Y_0 k_{eq} - 1 + Y_0]}{2[1 + Y_0 k_{eq} - Y_0]} \quad (18.18)$$

where $\sigma_0 = (7/12)\sigma$ and $[I]_0 = [I^*] + [I] = 2[I_2] * F$.

When the equilibrated oxygen–iodine gas mixture with y_0 and g_0 enters into the laser cavity, the circulating intensity rapidly builds up and seeks whatever value is required to satisfy the threshold condition. Because of its increase in the circulating intensity, gain (g_0) is either reduced or saturated. As a result, forward and reverse rates are no longer equal. Net energy transfer occurs with a rapid decrease in the yield in the flow direction.

An approximation to the loaded gain distribution may be obtained using the geometrical optics model. In this case, the round trip loop gain equilibrium condition gives

$$g^* = \alpha + g_T, \quad (18.19)$$

where α is the distributed loss and the threshold gain g_T is

$$g_T = -\ln r_1 r_2 / 2l_g, \quad (18.20)$$

where r_1 and r_2 are the mirror reflectivities and l_g is the gain medium length.

Since the loaded gain g^* is constant in flow direction, I^* is also constant and its derivative vanishes in Eq. (18.11). Integrating Eq. (18.10) gives the singlet delta oxygen yield and the intensity as a function of flow direction (x).

$$Y = (Y_0 - Y_{\text{sat}}) e^{-\frac{x}{u\tau}} + Y_{\text{sat}}. \quad (18.21)$$

$$\tilde{I} = \frac{h\nu [O_2] (Y_0 - Y_{\text{sat}}) e^{-\frac{x}{u\tau}}}{g * \tau}. \quad (18.22)$$

The saturation yield Y_{sat} and time constant τ are defined by

$$Y_{\text{sat}} = \frac{2g * + \sigma_0 [I]_0}{\sigma_0 \left(1 + 2k_{\text{eq}}\right) [I]_0 + 2g * \left(1 - k_{\text{eq}}\right)} \quad (18.23)$$

$$\tau = \frac{3\sigma_0 k_{\text{eq}}}{k_f \left[\sigma_0 \left(1 + 2k_{\text{eq}}\right) [I]_0 + 2g * \left(1 - k_{\text{eq}}\right)\right]} \quad (18.24)$$

where time constant (τ) is the transit time (x/u) required for e-fold reduction in the extractable yield.

The total output power is obtained by integrating the output intensity over the mirror length (l)

$$P = \frac{A}{l_g} \int_0^l \tilde{I}_{\text{out}} dx \quad (18.25)$$

where A is the cross-sectional flow area.

The output intensity is related to the average two way circulating intensity

$$\tilde{I}_{\text{out}} = \frac{1}{2}\Omega\tilde{I}, \tag{18.26}$$

where $$\Omega = \frac{(1 - r_1 - s_1)\sqrt{r_2}\ln r_1 r_2}{\left(\sqrt{r_1} + \sqrt{r_2}\right)\left(\sqrt{r_1 r_2} - 1\right)}, \tag{18.27}$$

s_1 is the mirror scattering loss. The aperture diffraction loss is assumed negligible.

Therefore, the output power is obtained as

$$P = \frac{\varepsilon\dot{x}(O_2)\Omega(Y_0 - Y_{\text{sat}})(1 - e^{-\gamma})}{2l_g g*}, \tag{18.28}$$

where $\varepsilon = Nh\nu$, N is the Avogadro number and $\dot{x}(O_2)$ is the oxygen molar flow rate.

The residence time parameter (γ) is defined as

$$\gamma = \frac{l}{u\tau} \tag{18.29}$$

The saturation intensity is defined such that $g^* = g_0/2$

$$I_{\text{sat}} = \frac{h\nu u[O_2](Y_0 - Y_{TH})\left[1 - e^{-\gamma_0\left(1 - \frac{1}{2}\phi\right)}\right]}{g_0 l\left(1 - \frac{1}{2}\phi\right)}, \tag{18.30}$$

where

$$Y_{TH} = \frac{1}{1 + 2k_{\text{eq}}}. \tag{18.31}$$

τ_0 is the value of the τ for $g^* = 0$ and ϕ is called COIL chemistry parameter defined as

$$\phi = \frac{\left(k_{eq} - 1\right)(Y_0 - Y_{TH})}{1 + Y_0 k_{eq} - Y_0} \tag{18.32}$$

18.1.2 Laser gain estimation

On the basis of these formulations, laser gain has been estimated depending upon the cavity temperature, singlet oxygen yield at the upstream of the laser cavity and the number of singlet oxygen molecules used in the dissociation of each iodine molecule as shown in Figs 18.4 and 18.5. Figure 18.4 shows the small signal gain variation depending upon the cavity temperature for the fixed value of yield (60%). The small signal gain increases with the decrease in the cavity temperature and has less effect on the number of singlet oxygen molecules used in dissociation.

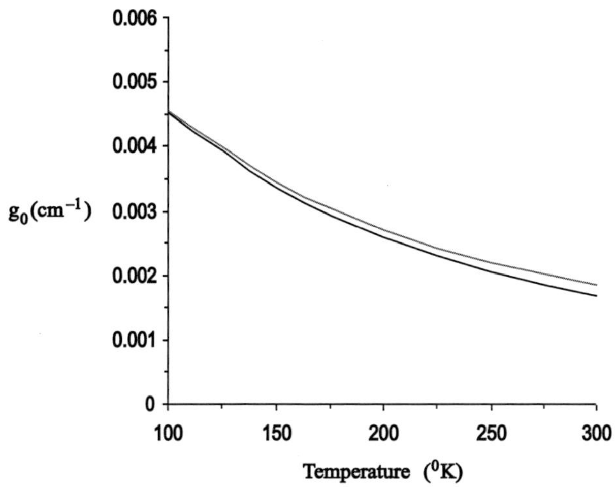

Figure 18.4 Gain vs cavity temperature for $n = 2$ (Grey curve) and 5 (Black curve)

Figure 18.5 shows the small signal gain variation for different values of singlet oxygen yield upstream of the cavity before the dissociation of iodine molecules for $n = 5$. It is obvious from the curves that the small signal gain ranges between 0.002 – 0.004 cm^{-1} for a yield greater than 60% and cavity temperature below 250°K.

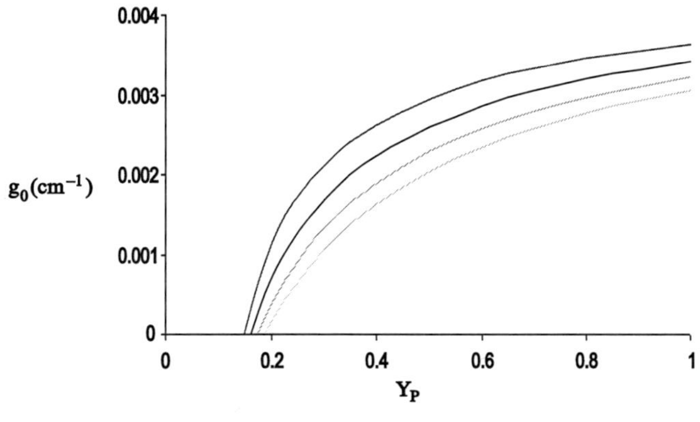

Figure 18.5 Gain vs yield for temperature = 150°K, 200°K, 250°K, 300°K (Top to Bottom)

18.1.3 COIL cavity and resonator design aspects

Laser beam inside the gain medium begins when an excited iodine atom emits radiation spontaneously, triggering the process of stimulated emission. The optical resonator provides sufficient feedback for the accomplishment of this process in order to have enough radiations for useful laser output beam. Therefore, the development of an efficient high power COIL development depends on the proper design of laser cavity containing the gain medium and optical resonator which contributes to the power and quality of the laser output beam.

Advanced COIL operation occurs at sub-atmospheric pressure (3–5 torr) with supersonic flow of laser gas species in the laser cavity. Therefore, it is essential to consider mechanical aspects like boundary layer problem and vibrations affecting the resonator alignment. In high power COIL, the gain medium cross-section is sufficiently large which supports the higher order laser modes in the cavity. This poses a compromise between laser output power and beam quality. The resonator design includes mirror size, shape, curvature and spacing between them and output mirror coupling which must be optimized keeping in view the resonator stability criteria, diffraction properties of the light rays, utilization of gain mode volume, beam divergence and misalignment sensitivity.

An optical resonator consists of two mirrors positioned to reflect and amplify the light between them before the light leaves the laser as output beam. The most commonly used optical resonators are composed of two flat or curved mirrors facing each other. The laser beam properties such as beam stability, beam size and shape, beam divergence and number of modes oscillating in the laser cavity strongly depend on resonator parameters like curvatures and sizes of the mirrors with respect to the spacing between them (Koechner).

A resonator may be regarded as a periodic focusing system in which the light ray bouncing back and forth between mirrors will have its trajectory periodically redirected by the two mirrors. If the transverse displacement of a ray's trajectory oscillates periodically, but remains bounded within the dimensions of the mirrors after many passes, the resonator is called a stable resonator. On the other hand, if it diverges out after multiple passes, the resonator is known as an unstable resonator. If the distance between mirrors is denoted by L and the mirror radii of curvature by R_1 and R_2, the effect on the ray may be analyzed using matrix optics and the condition for stable resonator may be expressed in terms of g parameters,

$$0 \leq g_1 g_2 \leq 1$$

where

$$g_1 = 1 - L/R_1$$

and

$$g_2 = 1 - L/R_2.$$

To show graphically the stability criterion, the stability diagram may also be used where g_1 and g_2 parameters are drawn as the coordinate axes as shown in Fig. 18.6. The shaded region shows the stability of the resonator.

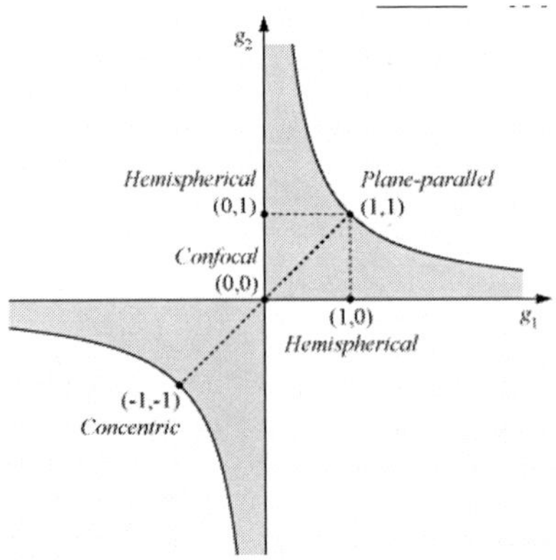

Figure 18.6 Stability diagram

In order to analyze the effect of resonator parameters on the laser beam characteristics, we must consider not only the stability criterion, but also the diffraction properties of the light rays, utilization of the gain volume and the divergence of the laser beam. If the resonator lies well inside the shaded stable region of the stability diagram, the lowest order modes in such resonators are narrowly confined along the axis of the resonator and have Gaussian transverse amplitude dependences. The diffraction losses for such lower order modes are generally small. However, for higher order modes, these losses increase with the increase in the mirror spacing. The resonators provide high divergence to the laser beam. Moreover, such resonators are also less effective from the gain volume utilization point of view. However, the resonators located near the boundaries of the stable region provide low beam divergence and utilize more gain volume.

18.1.4 Development of laser cavity and resonator

A high power COIL has been developed at LASTEC which employs a twin jet type SOG configuration with the extraction of singlet oxygen combined. This is carried to the nozzle by nitrogen buffer gas. Iodine molecules carried by the secondary nitrogen buffer gas are injected into the pumping medium in the subsonic regime followed by supersonic expansion into the laser cavity via a supersonic grid nozzle array. The laser output power is extracted by employing the optical resonator across the laser gas flow direction.

The laser cavity has been developed (Fig. 18.7) for this laser. It has an active medium length of 1000 mm with cavity width of 82 mm. It is a combine one for both the laser gas flow columns from the individual SOG modules. At one side, it is connected with the nozzle bank (1) which ensures the entry of the laser gases with cross-section of 1149 mm × 74 mm. At the other end, it is connected to the supersonic diffuser (2) for exhaust of the gases with cross-section of 1149 mm × 86 mm. The laser cavity is connected with the resonator using rubber bellow that helps in the alignment of the optic axis of resonator mirrors with the center of the laser gain medium. These rubber bellows absorb the mechanical vibrations and keep the resonator insensitive to these vibrations. The cavity limb (3) has provision to provide an air curtain for the protection of the resonator mirrors from the laser gases. The viewing window arrays (4) with 7 view ports on each side of the laser cavity are used to observe the uniform mixing of the gases in the laser gain medium.

Figure 18.7 Developed laser cavity

The developed resonator assembly and mirror mounts are shown in Figs 18.8 and 18.9 respectively.

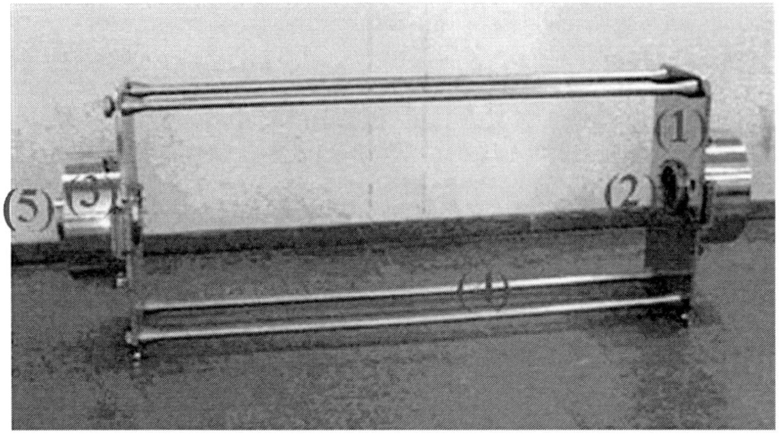

Figure 18.8 Resonator assembly

The resonator assembly consists of two mirror mounting plates (1) with inter-mirror spacing of 2000 mm. The mounting plates have passage (2) at the center for laser beam extraction and bellow coupler to connect the laser cavity with the help of the rubber bellow. The resonator lids (3) protect the mirrors from the surrounding environment and shocks occurring during the start of laser operation. The resonator rods (4) provide support to the mounting plate and keep the separation between the mirrors fixed. The optical window (5) at the center of the resonator lid provides the passage for laser beam extraction. The whole resonator assembly is placed on the vibration isolated metallic platform using jacks which help in adjusting the mounting plates in order to align the resonator axis with the laser cavity.

Figure 18.9 Developed resonator mirror mounts

In order to extract the power from the COIL gain medium, plano-concave stable resonator configuration as shown in Fig. 18.10 is considered. The resonator includes the concave rear mirror and the flat semi-transparent output coupler. Both mirrors are made of fused silica with multilayer dielectric coating. The separation between the two mirrors is 2000 mm. Both mirrors of the stable resonator are circular with diameter 5 inches and thickness 15mm. The radius of curvature (ROC) of the concave rear mirror is 10 m and reflection is > 99.9%. For a COIL system having gain coefficient 0.002–0.004 cm^{-1}, the optimum output mirror coupling lies between 10–15%.

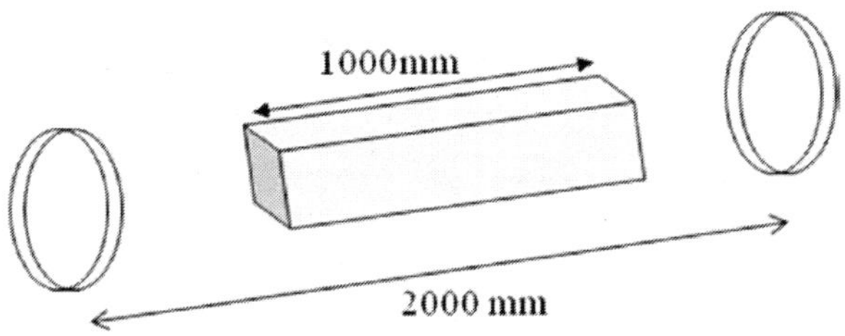

Figure 18.10 Resonator configuration

18.2 Results and Discussion

The developed laser cavity has been integrated with other COIL subsystems like nozzle bank and supersonic diffuser. The COIL has been operated with the help of a dedicated date acquisition and control system (DACS) (Mainuddin et al.) and the required laser cavity conditions are achieved. The resonator including the resonator assembly and mirror mounts has been mounted and connected with the laser cavity. Mechanical and optical alignments of the resonator have been carried out in order to extract the laser output power. Test runs for various values of COIL input parameters are conducted and COIL power is demonstrated. The laser cavity pressure is 3 – 5 torr for optimized COIL parameters. The laser gain ranges between 0.002 – 0.004 cm^{-1}. The output mirror reflectivity is 10–15% depending upon the COIL input parameters. The beam size is around 70×70 mm^2 and the experimentally determined value of the beam divergence in stable resonator configuration is around 25 m radian.

18.3 Conclusion

In this chapter, the various physical and chemical aspects related to gain formation in the laser cavity have been discussed. The COIL input parameters are analyzed and gain characteristics for the development of high power COIL are estimated. On the basis of estimated parameters, the laser cavity and optical resonator have been developed and integrated with the other developed COIL subsystems. Test runs have been given and COIL power has been demonstrated.

Bibliography

1. V. N. Azyazov, M. V. Zagidullin, V. D. Nikolaev, M. I. Svistun and N. A. Khvatov. 1994. 'Jet $O_2(^1\Delta)$ generator with oxygen pressures upto13.3 kPa.' *Quantum Electronics* 24, 2: 120.
2. A. S. Boreysho. 2005. 'High power mobile chemical lasers.' *Quantum Electronics* 35, 5: 393.
3. D. L. Carroll, D. M. King, L. Focker, D. Stromberg, T. J. Madden, W. C. Solomon and L. N. Sentman. 1998. 'COIL for industrial applications.' *AIAA*. 98:2992.
4. E. A. Duff and K. A. Truesdell. 2004. 'Chemical oxygen iodine laser (COIL) technology and development.' *Proceedings of SPIE* 5414: 52.
5. K. Grunewald, W. L. Bdin, F. Duschek, J. Hankle and W. O. Schall. 2004. 'High power COIL fiber transmission for D & D application.' *Proceeding of SPIE* 5479: 35.
6. G. D. Hager, C. A. Helms, K. A. Truesdell, D. Plummer, J. Erkkila and P. Crowell. 1996. 'A simplified analytical model for gain saturation and power extraction in the flowing chemical oxygen iodine laser.' *IEEE J. of QE* 32, 9: 1525.
7. W. Koechner. 1999. *Solid-State Laser Engineering Fifth Edition*.
8. Mainuddin, Gaurav Singhal, A. K. Varshney, R. K. Tyagi and A. K Maini. 2011. 'Implementation of safety measures for chemical oxygen iodine laser, proceedings of the DRDO sponsored eight control instrumentation system conference.' *CISCON*2011: 212.
9. W. E. McDermott, N. R. Pchelkin, D. J. Benards and R. R. Bousek. 1978 . 'An electronic transition chemical laser.' *Applied Physics Letter* 32: 469.

19 Study of Laser Induced Dusty Plasmas with Micro-Raman Scattering Spectroscopy

Mukesh Kumar and K. B. Singh

19.1 Introduction

Laser induced dusty plasma have wide applications in plasma spectroscopy because of their fast response times and ability to generate plasmons were at a relatively long distances. Most of the previous fundamental studies of laser induced dusty plasma at or near the atmospheric domain of the ionosphere. The aim of this study is to evaluate the effects of micro-Raman scattering spectroscopy to generate the transparent dusty plasma in the field of a relativistically laser induced dusty plasma medium. The complete analysis will be performed by the basic mathematical equations for plasma under micro-Raman scattering spectroscopy. In this chapter, theoretical considerations about the problems encountered will be investigated.

19.2 Basic Equations

Consider a dusty plasma medium with extremely massive highly negatively charged dust grains electrons and non-thermal ions charge conservations which, implies the following equilibrium equations

$$n_{\text{ions}} = Z_{eo} n_{dp} + n_e \tag{19.1}$$

where n_{ions}, n_e and n_{dp} are the unperturbed ions, electrons and dust particles number and Z_{eo} is the unperturbed number of charges of the dusty plasma with micro-Raman scattering spectroscopy. The following equations are the normalized continuity momentum and Poisson's equations that govern the dynamics of the dusty plasma.

$$\frac{\partial n_{dp}}{\partial t} + \frac{\partial (n_e.v_d)}{\partial x} = 0, \tag{19.2}$$

$$\frac{\partial u_d}{\partial t} + v_d \frac{\partial v_d}{\partial x} = Z_{do} \frac{\partial \phi}{\partial x}, \tag{19.3}$$

$$\frac{\partial^2 \phi}{\partial x^2} = Z_{eo} n_{dp} + n_e - n_{di}, \tag{19.4}$$

where n_d,v_d,ϕ and Z_{eo} are the dust number density, dust particles velocity, electrostatic potential and the variable charge number of dust particles respectively. We know that electrons and ions obey the Boltzmann distribution and non-thermal distribution respectively, as follows,

$$n_e = \frac{n_{eo}}{Z_{dp} n_{dp}} \exp(B_1 \phi), \tag{19.5}$$

also

$$n_i = \frac{n_{io}}{Z_{dp} n_{dp}} (1 + B\phi + B\phi^2) \exp(-\phi). \tag{19.6}$$

Parameters B_1 and β are defined as, follows,

$$B_i = \frac{T_i}{T_e}, \quad \beta = \frac{4\alpha}{(1+3\alpha)}. \tag{19.7}$$

It must be noted that the charging currents originates from the electrons and ions reaching the induced dusty plasma under Raman spectroscopy. The dusty plasma charging time is of the order of 10^{-9} second, while the dust motion time under induced plasma medium quickly reach the local equilibrium at which the currents flow from the electron and ions to the dusty plasma. Then the balanced equation implied under this conditions for micro-Raman scattering spectroscopy,

$$I_e + I_i \approx 0. \tag{19.8}$$

Suppose that the streaming velocities of electrons and ions are much smaller than the thermal velocities. The electron and ion currents are

$$I_e = -e\pi r^2 \left(\frac{8T_e}{\pi m_e}\right)^{1/2} n_e \exp\left(\frac{e\phi}{T_e}\right). \tag{19.9}$$

$$I_i = e\pi r^2 \left(\frac{8T_i}{\pi m_i}\right)^{1/2} n_i \exp\left(1 - \frac{e\phi}{T_i}\right). \tag{19.10}$$

19.3 Emission Spectra-Based Measurements

The spectra was taken at different dusty plasmons level in the same set of spectral lines. The general appearance of the spectra and their time dependence varied significally with ionization level of concentration in terms of particle density. At atmospheric conditions, the development of the emission spectra was reported and calculated. It has been noted that when atmospheric moisture is reduced, the line widths also becomes reduced to each respective time intervals. Qualitative values of moisturing temperature and electron number density are extracted from the spectra. The optical depth, the uncertainty in these particular parameters, fit. Therefore, it is quite possible that the small discrepancy in temperature at very early times is an artifact of the optical depth correction in the spectral fit. For all of the above comparisons, the incident laser power was kept constant while the temperature was varied.

19.4 Discussion

On the basis of above facts, perhaps the least surprising, is the reduction in absorbed scattered power energy fraction with temperature. Since the electromagnetic field at the laser induced plasma focus is of first order, independent of the number density of the gas, we can envision the laser focus as a constant volume region of space. A secondary effect which may be the reason for the deviation from linearity is the transition in the mode of induced plasma generating from a cascade process to milt photon ionization. Of possible importance is the fraction of light that is absorbed versus that which is scattered in these power measurements. The power meter only detects light along the axis of the incident laser beam; though the collection solid angle of the power meter covers a region larger than the expected volume. The dependence of the electron number density on temperature is very strong, while the dependence on absorbed laser power is much weaker. However, the weak but significant dependence of neon absorbed laser power is considered, than a slightly weaker dependence on absorbed laser power is absorbed. Assuming that the temperature at each time interval for each temperature zone is same, then at a given time, energy per unit's mass of the induced plasma will be the same for every atmospheric layer of temperature zone at that time interval. If this is true, then the total energy in the plasma is proportional to the gas density time *s* the plasma volume for times of a few hundred nanoseconds. The nature of variation has been plotted in the form of graph.

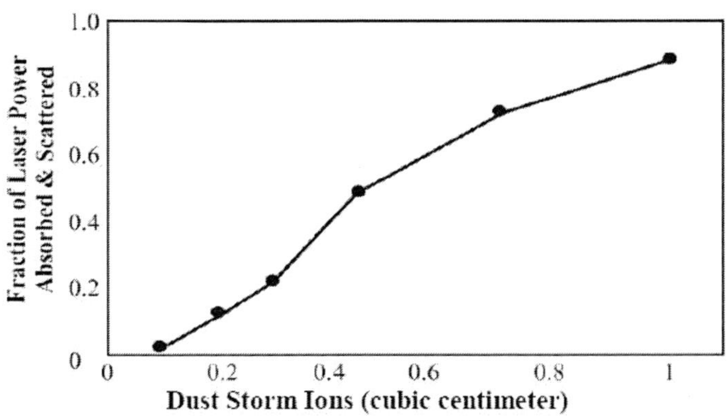

Figure 19.1 Absorbed and scattered laser-induced plasma dust particles for dust storm ions in the atmosphere

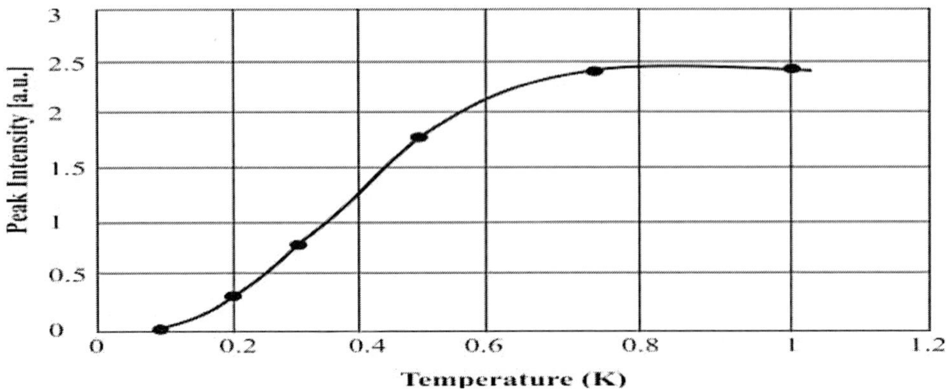

Figure 19.2 Peak emission of the emission from the laser-induced plasma over a range of dust storm ions with temperature

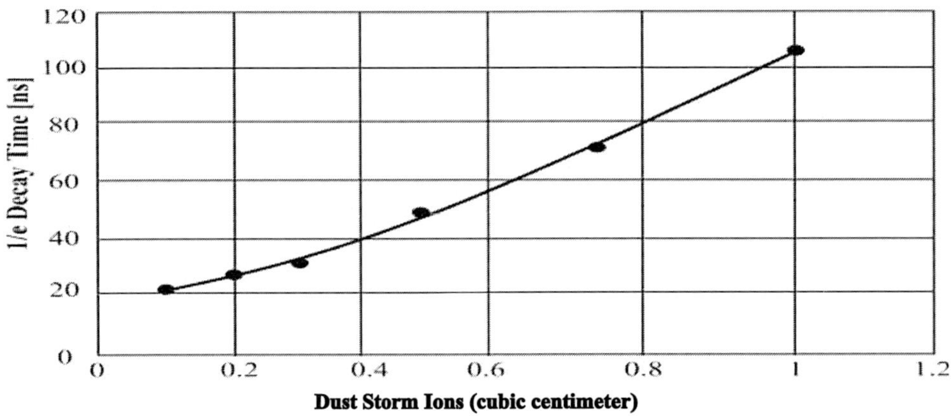

Figure 19.3 Time for emission from the laser-induced plasma in a dust storm due to decay to a level of $1/e$ of the peak intensity for a range of dust storm ions

19.5 Conclusions

A detailed investigation of this effect of micro-Raman scattering spectroscopy has been performed. The major conclusion of this study is as follows.

(i) At reduced temperature, the decay of intensity in the laser induced dusty plasma ionic absorption is markedly faster than at normal atmospheric condition, most likely due to reduced optical depth in the plasma.

(ii) Electron number density in the induced dusty plasma and also in the plasma decreases roughly linearly with ambient ions for fixed input laser energy and slightly less than linearly for fixed absorbed laser energy.

Bibliography

1. P. K Shukla and A. A. Maum. 2002. 'Introduction to dusty plasma physics.' *Bristal institute of physics*.
2. N. N Rao and P. K. Shukla. 1990. *Planet Space Science* 38: 543.
3. Roy, R. Choudhry and S. Mukherjee. 1997. *Phys. Plasmas* 3: 2610.
4. A. A. Mamum. 1997. *Phys. Rev. E.* 55: 1852.
5. S. K. El-Labany and W. F El-talbany. 2003. *Phys. Plasmas* 10: 989.
6. P. H Sananka and P. K. Shukla. 2002. *Phys. Scripta* 784: 181.
7. TX Phuoc. 2005. 'An experimental and numerical study of laser induced spark in air, opt. lasers.' *Eng*. 43: 113119.
8. J. Hermann, C. Boulmer-Leborgne and D. Hong. 1998. 'Diagnostics of the early phase of an ultraviolet laser induced plasma by spectral line analysis consider self absorption.' *J. Appl. Phys*. 83: 69196.

20 Enhanced Collisionless Absorption in a Cluster with Extreme Few-cycle Laser Light

M. Kundu and P. K. Kaw

20.1 Introduction

The construction of laser based particles (electrons, ions) and light sources (x-rays) require efficient coupling of laser energy with a target material. It is experimentally established that nano-meter size atomic clusters (hydrogenic and rage gas) can absorb laser light very efficiently. Compared to the size of a cluster, the wavelength λ of light is usually very large (typically $\lambda \sim 248{-}1064$ nm). For this reason, on one hand, light can completely illuminate a cluster having solid-like atomic density, and on the other hand, the reflection of laser energy becomes negligible leading to more than 80% laser absorption[1]. Laser heated clusters were shown to produce (i) near MeV ions[1–8], (ii) keV electrons[9–13], (iii) harmonic radiation[14–18], and (iv) x-rays[19–22].

The laser energy absorption in clusters is commonly studied with a pump-probe experimental setup[24–26]. After the ionization by the pump, the charge density $\rho(t)$ and the corresponding Mie-plasma frequency $\omega_M(t) = \omega_p/\sqrt{3} = \sqrt{4\pi\rho(t)/3}$ increases above the laser frequency ω (ω_p is the plasma frequency, atomic units $e = m = 4\pi\varepsilon_0 = 1$ are used unless mentioned explicitly). Due to the higher mobility of electrons than the massive ions, electrons first absorb laser energy. Their displacement from the quasi-static ion background leads to the formation of a space charge field which may exceed the applied laser field. The space charge field when added to the laser field may cause further ionization of ions, known as ionization ignition[27–32] which is not possible by the laser field alone. At a given laser intensity, this ionization ignition and ionization of atoms/ions (inner ionization) depend upon the number of electrons who have left the cluster completely. The complete removal of an electron from the cluster is called outer ionization. The outer ionization leaves behind a non-neutral ionic background which expands due to Coulomb repulsion between the ions. When the Mie-frequency of the expanding plasma meets the condition $\omega_M(t) = \omega$ at a later time, typically > 50 fs[24] for 800 nm laser, enhanced energy absorption takes place from the probe. This linear resonance (LR) was explained with a nano-plasma model[33] where the expanding cluster was assumed to be homogeneously charged. However, in reality, ions near the cluster boundary leave the cluster with higher kinetic energy much earlier than the relatively slower ions at the cluster core. Therefore, cluster plasma can be assumed to be homogeneous only in the core region where $\omega_M(t)$ may be uniquely defined for the LR to happen. A large fraction of ions outside the homogeneous ion core are not influenced by this LR.

However, during the early time of a long pulse or when the pulse duration is short < 50 fs[34], the expansion of over-dense ionic background is not sufficient to meet the above conventional LR condition $\omega_M(t) = \omega$. However, absorption and outer ionization still persist otherwise cluster can—not expand. In this case, it was shown that non-linear resonance (NLR) can occur to lead an efficient laser absorption[34–38] when the time-dependent frequency of a driven electron in the self-consistent, anharmonic, cluster potential meets the laser frequency ω.

The conventional LR (with two pump-probe pulses, mentioned above) in the expanding stage was shown less efficient[39] than the LR which can occur in the early cycles of a single short laser pulse of shorter wavelength when cluster, ions are compact and the nano-plasma density is relatively higher. For xenon clusters such an early LR requires an intense short pulse of wavelength ~ 125 nm[39]. However, experiments close to this ultraviolet wavelength with intensity > 10^{15} Wcm^{-2} are rare except for a few experiments with free electron lasers[40–42] and x-ray lasers[43] with intensities < 10^{15} Wcm^{-2}. Recently, sub-cycle (SC)[44, 45], single cycle[46–49], and extreme few-cycle laser pulses with near relativistic peak intensity[50, 51] were reported which have already stimulated research activities for direct acceleration of electrons[52]. Interestingly, these ultra-broadband pulses show another encouraging route for efficient absorption in a cluster without significant expansion. On one hand, wider range of ω_M (due to charge inhomogeneity) may overlap on some part of the broad frequency spectrum of the near-single cycle pulse to yield higher absorption via the early LR. On the other hand, the decrease of the relative density $\rho/\rho_c(\omega) > 1$ [$\rho_c(\omega) = \omega^2/4\pi$ is the critical density at ω] w.r.t. higher ω component may drive NLR with a greater efficiency[35, 36] together with LR.

In this chapter, laser–cluster interaction with a pulse duration τ below one optical period $T = \lambda_0/c$ at the central wavelength $\lambda_0 = 800$ nm is studied. For such a broadband pulse, the ponderomotive energy $U_p = E_0^2/4\omega^2$ (it is the average kinetic energy of a free electron in a monochromatic laser field of frequency ω and field strength E_0) can not be uniquely defined, and the energy absorption by a charge particle with extremely short pulses and relatively longer pulses are presumed to be quite different.

The condition $R_0 \ll \lambda$ allows the use of dipole approximation[37] for the sub-cycle pulses. The effect of $\boldsymbol{v}\times\boldsymbol{B}$ Lorentz force is also negligible for the intensities < 10^{18} Wcm^{-2} considered in this work. However, previously used conventional $\sin^2$ pulses[35–37, 53] or the Gaussian pulses[25, 26, 34, 54–56] for the laser electric field $E(t)$ can not be used in the near single-cycle and sub-cycle regime since they yield non-zero value of the integral $\int_{t_i}^{t_f} \mathrm{E(t)}\, \mathrm{d}t$ over the pulse duration $t_i \to t_f$ and results in an unphysical energy gain by a free charge[57]. In Section 20.2, two valid pulses namely, the $\sin^2$-pulse and the Gaussian pulse are considered. Then laser energy absorption is studied by a rigid sphere model (RSM)[37] and more realistic three-dimensional PIC simulations[35, 36] in Section 20.3 for a Deuterium cluster. Results are summarized in Section 20.4.

20.2 Model of a Sub-cycle Laser Pulse

An accurate model of the laser pulse is required for laser–cluster interaction in the SC regime. First, a widely used conventional model of an n-cycle laser pulse $E(t) = E_0 \sin^2(\omega t/2n)\cos(\omega t)$, is discussed for $0 < t < nT$ (the $\sin^2$-pulse)[35–37, 53] which is not universally applicable, and in some cases leads to $\int_0^{nT} E(t)\, \mathrm{d}t \neq 0$. For example, $n = 1$ yields $\int_0^{nT} E(t)\, \mathrm{d}t = -T/4$. To overcome this limitation, vector potential $A(t) = (E_0/\omega)\sin^2(\omega t/2n)\cos(\omega t)$ for $0 < t < nT$ is considered; giving $E(t) = -\mathrm{d}A/\mathrm{d}t$ as

$$E(t) = (E_0/\omega)\begin{cases} \sum\limits_{k=1}^{3} d_k\omega_k \sin(\omega_k t) & 0 < t < nT \\ 0 & t > nT \end{cases}, \tag{20.1}$$

where $d_1 = 1/2$, $d_2 = d_3 = -1/4$, $\omega_1 = \omega$, $\omega_2 = (1 + 1/n)\omega$, $\omega_3 = (1 - 1/n)\omega$. Since the integral $\int_0^{nT} E(t)\, \mathrm{d}t = -A(t)$ vanishes irrespective of the pulse duration $\tau = nT$, Eq. (20.1) is useful in the sub-cycle regime.

Another example of a widely used laser pulse is the Gaussian pulse of the conventional form $E(t) = \mathrm{Re}[E_0 \exp(-t^2/2\delta^2)\exp(\mathrm{i}\omega t + \mathrm{i}\phi_0)]$. Following[57], the corresponding $E(t)$ which is also valid in the sub-cycle regime, can be found as

$$E(t') = \mathrm{Re}\left[-\mathrm{i}E_0 \frac{(1 + \mathrm{i}t'/\omega\delta^2)^2 + 1/(\omega\delta)^2}{1 + 1/(\omega\delta)^2}\frac{p(t')}{p(0)}\right]. \tag{20.2}$$

where $p(t) = p_0 \exp(-t^2/2\delta^2)\exp(\mathrm{i}\omega t + \mathrm{i}\phi_0)$, $t' = t - t_0$, t_0 is the time when $E(t') = E_0$, and δ is related to FWHM $\tau = 2\delta\sqrt{2\ln 2}$. It can be seen that[57] the integrals of the conventional Gaussian pulse and the corrected pulse vanish when $\tau > 2T$. For this reason, conventional Gaussian pulse was safely applied[25, 26, 34, 54–56] for laser–cluster interaction with long pulses, where $\tau \gg 2T$. However, when $\tau = T$, the integral of the conventional Gaussian pulse deviates from zero[57]. This deviation increases further for sub-cycle pulses, e.g., $\tau = T/2$. However, Eq. (20.2) always leads to $\int_{t_i}^{t_f} E(t)\, \mathrm{d}t = 0$.

One can apply either of the pulses, Eq. (20.1) or Eq. (20.2), for laser–cluster interaction. However, for simplicity, only Eq. (20.1) is used in the studies.

20.3 Deuterium Cluster at Various Pulse Duration

Laser energy absorption in a cluster depends on many facts, e.g., (i) cluster size[18, 24, 55, 58], (ii) laser wavelength[39, 54, 55, 58], (iii) peak intensity of the laser[24, 34–36, 55, 58], (iv) also pulse duration $\tau \gg 2T$[24, 53, 55, 56, 58, 59], and (v) the atomic constituents in a cluster. With so many different parameters it is difficult to know which combination of parameters will lead to an enhanced absorption in a cluster. Two-dimensional PIC simulations[56] and experiments[24] with rare gas clusters for a fixed pulse intensity and cluster size showed that energy absorption increases initially with increasing pulse duration τ (with τ between $\sim 100T - 150T$ at 800 nm), reaches a peak value, and decreases for τ in the pico-second regime. These results were explained on the basis of the conventional dual pulse LR. However, τ is restricted below 50 fs where this conventional LR does not work. The results for $\tau > 2T$ though agreeing qualitatively with previous works[24, 56], are *entirely* different for $\tau < 2T$.

A Deuterium cluster of radius $R_0 = 2.2$ nm (with the number of atoms $N = 2176$) and charge density $\rho = 27.86\rho_c(\omega)$ is considered. For this charge density, one obtains $\omega_{\rm M} \approx 3\omega$. The laser pulse duration is varied from multi-cycle to the sub-cycle regime keeping peak intensity at a given value.

20.3.1 Energy absorption with a rigid sphere model

In a simple approximation, cluster can be approximated as a homogeneously charged sphere of pre-ionized plasma. In the absence of an external laser field, the center of the negatively charged electron sphere initially coincides with the center of the positively charged massive ion sphere which may be considered immobile during the short duration of the pulse. First energy absorption is studied with this rigid sphere model (RSM) of cluster[35–37]. It is important to mention that RSM supported many results of realistic PIC simulations[35, 36] in the past, despite of its simplicity.

The equation of motion of the electron sphere, in a linearly polarized (along x) laser field $E(t)$ reads

$$\ddot{\boldsymbol{x}} + g(r)\,\boldsymbol{x}/r = -\,\hat{\boldsymbol{x}}E(t), \tag{20.3}$$

where $r = |x|$ is the displacement of the electron sphere and $g(r)$ is the electrostatic restoring force given by

$$g(r) = \omega_{\rm M}^2 \begin{cases} r + \alpha r^2 + \beta r^4 & 0 \le r \le 2R_0 \\ \gamma/r^2 & r \ge 2R_0, \end{cases} \tag{20.4}$$

Here $\alpha = -9/16R_0$, $\beta = 1/32R_0^3$, and $\gamma = 1$. As the electron sphere moves away from the center (at $r = 0$) of the ion sphere, $g(r)$ changes from harmonic to anharmonic and finally Coulombic for $r \ge 2R_0$[35–37]. The absorbed energy $E_{\rm a}$ by the cluster (of mass m_s and total charge q_s of the electron sphere) at the end of the pulse is calculated as $E_{\rm a} = E_{\rm a}(\tau) = m_s\dot{x}^2/2 + q_s \int g(r)\,{\rm d}r$.

Before absorption by the above anharmonic oscillator is studied, it is worthwhile to study the corresponding harmonic oscillator (HO) with sub-cycle pulses by setting $\alpha = \beta = \gamma = 0$. Using Eq. (20.1) for $E(t)$,

$$x = \sum a_k[\sin(\omega_k t) - \omega_k \sin(\omega_{\rm M} t)/\omega_{\rm M}]/(\omega_{\rm M}^2 - \omega_k^2), \tag{20.5}$$

where $x(0) = \dot{x}(0) = 0$ and $a_k = -E_0 d_k \omega_k/\omega$. The absorbed energy is simply $E_{\rm a} = m_s v^2/2 + q_s\omega_{\rm M}^2 x^2/2$. Normalized $E_{\rm a}/NU_{\rm p}$ vs τ/T is plotted in Fig. 20.1 for the HO (dashed line). It shows that $E_{\rm a}/NU_{\rm p}$ increases with decreasing τ/T from $\tau/T = 2$ (with a faster rise of $E_{\rm a}/NU_{\rm p}$ for $\tau/T \lesssim 1$), reaches a peak between $\tau/T \approx 0.3 - 0.4$, and then drops with further decrease of τ/T. Energy absorption by HO in Fig. 20.1 can be explained from the dynamics of the electron sphere in the (x, v_x) phase space (x and v_x are normalized by R_0 and quiver velocity $v_0 = E_0/\omega$ respectively) in Fig. 20.2 for different values of pulse duration $n = \tau/T = 4, 2, 1, 0.5$. For large $n \to \infty$, the electron sphere is driven with a single frequency ω far below the linear resonance $\omega_M \approx 3\omega$, it returns to its initial point after the pulse is over, and no net energy absorption takes place by the HO. Situation becomes different as τ/T shortens. Already at $\tau/T = 4$ (in a), one sees that phase is not returned to its initial value! The distance of the initial point to the final point also increases with gradual decrease of $\tau/T = 4, 2, 1, 0.5$ leading to the increased energy absorption in Fig. 20.1 for HO. As τ/T is decreased from $1 \to 0.5$, the frequency component ω_2 changes from $2\omega \to 3\omega$, and the harmonic oscillator (electron sphere) passes from near resonance to resonance where both excursion x and velocity v_x is substantially increased (c, d). For the lowest value $\tau/T = 1/8$ considered in this work, both frequency components ω_2, ω_3 exceed ω_M (plasma becomes under-dense w.r.t. ω_2, ω_3) and absorption drops as seen in Fig. 20.1. Contrary to the known long pulse scenario (where HO can not gain energy), an HO driven by ultra-broadband pulses may *gain* energy by retaining phase different from its initial value.

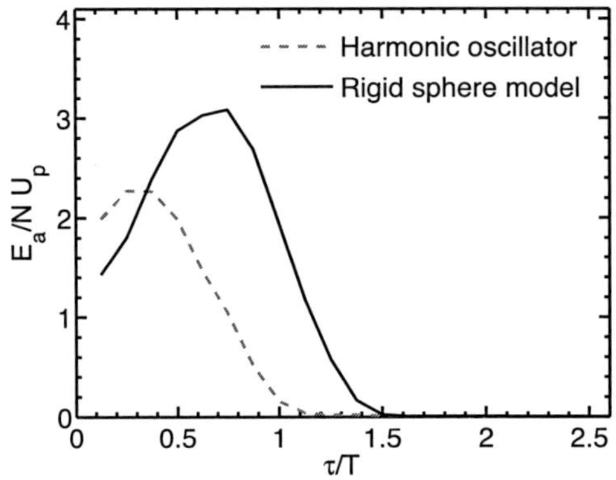

Figure 20.1 Absorbed energy E_a/NU_p vs pulse duration τ/T in case of RSM (solid line) and harmonic oscillator (dashed line, with $\alpha = \beta = \gamma = 0$ in Eq. (20.4)) using $E(t)$ of Eq. (20.1). The deuterium cluster of radius $R_0 = 2.2$ nm ($N = 2176$), charge density $\rho = 27.86\,\rho_c$ is irradiated by laser pulses of central wavelength $\lambda_0 = 800$ nm and peak intensity 5×10^{15} Wcm^{-2}.

For the above HO approximation, the whole electron cloud will continue to oscillate with a fixed ω_M in the infinitely deep parabolic potential after the laser pulse. While in reality, electrons may leave the cluster (outer ionization), and at the time of leaving, they must sense the Coulomb tail of the potential due to the quasi-static ion background. Equation (20.4) shows this property and allows LR and/or NLR depending upon the frequencies in the driver. When the electron sphere passes the resonance (either LR or NLR), the final (x, v_x) becomes different than its initial value. Solving Eqs (20.3) and (20.4) numerically, we compare E_a/NU_p vs τ/T of the full RSM (solid line) with the harmonic oscillator (dashed line) in Fig. 20.1. Both RSM and HO show that E_a/NU_p increases as τ/T decreases, reaches a maximum below $\tau/T = 1$ and then drops for very low τ/T. For the RSM, phase space (x, v_x) trajectory starts becoming more chaotic due to the non-linearity in $g(r)$, leading to the faster rise in E_a/NU_p already at a longer $\tau/T \sim 1.5$ and shifting of peak position towards $\tau/T = 1$ compared to the HO. The E_a/NU_p vs τ/T curve, and the peak position for HO is independent of the peak intensity of the laser pulse $I_0 = E_0^2$, but not for the RSM due to the intensity dependent eigen-period. The anharmonicity in the RSM leads to higher absorption than the HO.

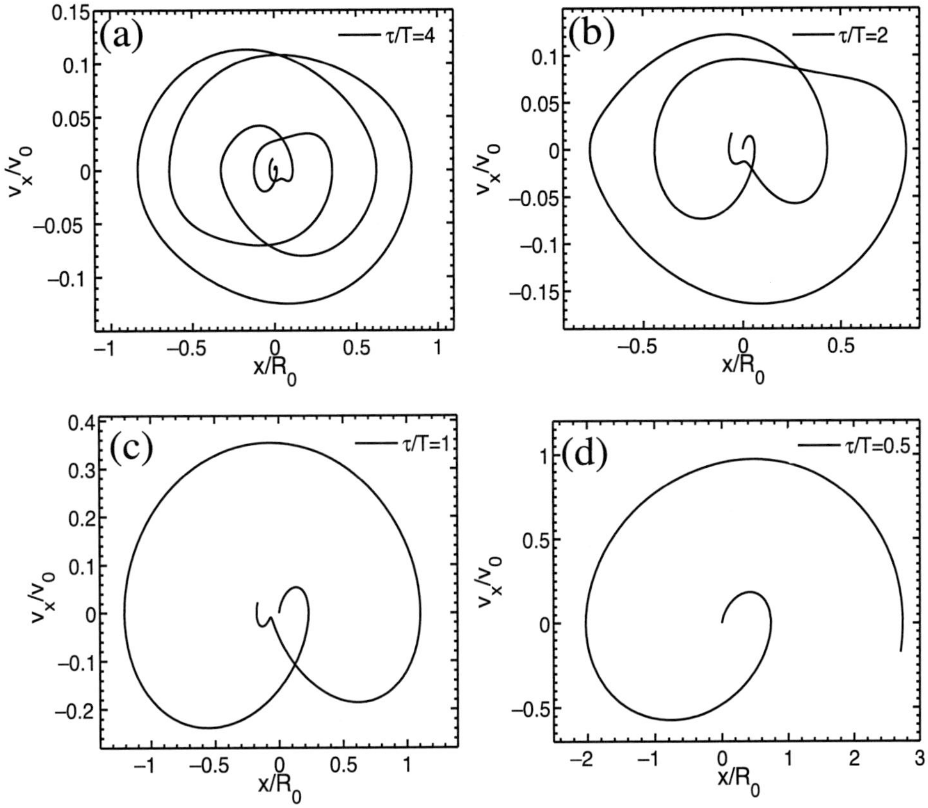

Figure 20.2 Phase space (x, v_x) diagram for the harmonic oscillator at various pulse duration $\tau/T = 4,2,1,0.5$ corresponding to the parameters of Fig. 20.1.

20.3.2 Energy absorption with PIC simulation

Energy absorption with realistic three-dimensional particle-in-cell (PIC) simulations will be discussed in this section. The details of PIC simulation were already described in[18, 19]. In the PIC simulation, initially, laser field $E(t)$ ionizes all neutral atoms A to A^+ (within the cluster) via over-the-barrier ionization[60] when the laser field reaches a critical value $E(t) \geq I_p(Z)^2/4Z$ (where $I_p(Z)$ is the ionization potential for charge state Z). The displacement of electrons by the laser field, in the subsequent interval, creates a space charge field $\boldsymbol{E}_{sc}$. The total field $\hat{\boldsymbol{x}}E(t)+ \boldsymbol{E}_{sc}(t,R_i)$ at a radial location R_i of the ith ion decides further ionization if the condition $|\hat{\boldsymbol{x}}E(t)+ \boldsymbol{E}_{sc}(t,R_i)| \geq I_p^2(Z)/4Z$ is satisfied. The equation of motion of the ith PIC electron (having the same charge-to-mass ratio of a real electron) is $\ddot{\boldsymbol{r}}_i = -$E(t) -$\boldsymbol{E}_{sc}(t,r_i)$, whereas the jth ion of mass M_j and charge Z_j obeys $M_j\,\ddot{\boldsymbol{R}}_j = Z_j\,[E(t)+ \boldsymbol{E}_{sc}(t,R_j)]$. The total absorbed energy $E_a(t)$ is obtained by summing over the kinetic energy of all particles (PIC electrons and PIC ions) plus the electrostatic field energy for a given $\tau = nT$ and a peak intensity I_0.

The deuterium cluster considered in Fig. 20.1 is now irradiated with different peak intensities, and pulse durations. Figure 20.3 shows $E_a(\tau)/NU_p$ vs τ/T for intensities 5×10^{15} Wcm^{-2} and 10^{16} Wcm^{-2}. Figure 20.3(b) is the enlarged view of Fig. 20.3(a) for $\tau/T < 2$. The vertical dashed line at $\tau/T = 1$ distinguishes the sub-cycle regime from the multi-cycle regime. A distinct behavior of $E_a(\tau)/NU_p$ is seen for $\tau/T < 2$ and $\tau/T > 2$. At the lower intensity, 5×10^{15} Wcm^{-2} the absorbed energy $E_a(\tau)/NU_p$ initially increases from $1.5 \rightarrow 2.3$ as τ/T varies from $2 \rightarrow 6$. Between $\tau/T = 6-12$, it is seen that $E_a(\tau)/NU_p$ increases at a slower rate and finally saturates near ≈ 3 for $\tau/T > 12$. Similar variation of $E_a(\tau)/NU_p$ vs τ/T is also seen at the higher intensity, 10^{16} Wcm^{-2}. The energy absorption for pulse durations beyond $\tau/T > 2$ matches qualitatively with the results in [56] for deuterium cluster. Contrary to the above conventional long pulse scenario, $E_a(\tau)/NU_p$ (more clear in Fig. 20.3(b) increases to 2.8 and 2.3 (for respective intensities 5×10^{15} Wcm^{-2} and 10^{16} Wcm^{-2}) as τ/T decreases to ≈ 0.8 from $\tau/T = 2$. When τ/T is decreased further below 0.8, $E_a(\tau)/NU_p$ drops similar to the RSM results in Fig. 20.1. The peak value $E_a/NU_p \simeq 2.8$ is comparable to E_a/NU_p at $\tau/T = 10$ for the intensity 5×10^{15} Wcm^{-2}, and the maximum value $E_a/NU_p \approx 1.6$ reached between $\tau/T = 8-10$ for 10^{16} Wcm^{-2} is even lesser than the corresponding peak value $E_a/NU_p \approx 2.3$. The lesser energy content in a sub-cycle (or a near single-cycle pulse) pulse but relatively higher absorption indicates higher absorption efficiency with decreasing τ/T.

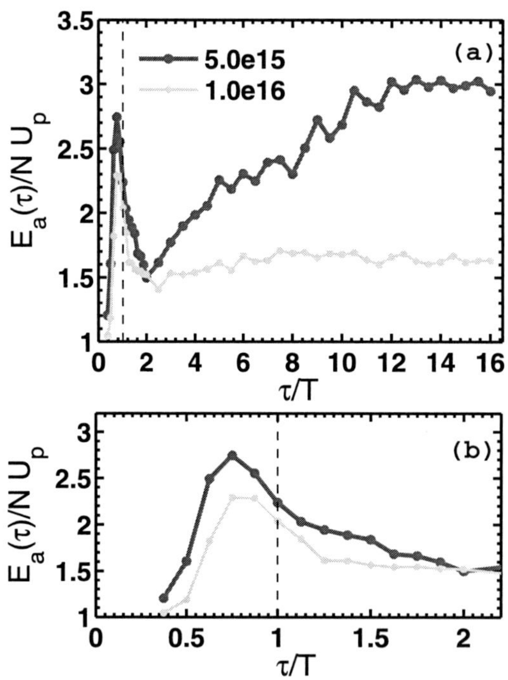

Figure 20.3 PIC results for energy absorbed E_a/NU_p per cluster ion N in units of U_p vs pulse width τ/T for peak intensities 5.0×10^{15} Wcm^{-2} (dark line) and 10^{16} Wcm^{-2} (light line). (b) is the expanded view of (a) for $\tau/T < 2$. Other laser and cluster parameters are the same as in Fig. 20.1. Above $\tau/T = 2$, absorbed energy increases with increasing τ/T, may reach a saturation for higher $\tau/T > 12$ due to saturation of inner and outer ionization. From $\tau/T = 2$, E_a/NU_p turns, increases with decreasing τ/T, and reaches a peak about $\tau/T \approx 0.8$

Enhanced absorption is possible when there is an efficient outer ionization of electrons. Figure 20.4 shows outer ionization n_{out}/N (n_{out} is the number of electrons that have left the cluster) vs τ/T corresponding to Fig. 20.3. Outer ionization also shows a distinct nature for $\tau/T < 2$ and $\tau/T > 2$ as the energy absorption in Fig. 20.3. For $\tau/T > 2$, outer ionization gradually increases and saturates close to $n_{\text{out}}/N \approx 1$ (meaning complete removal of all electrons from the cluster) beyond $\tau/T \approx 6$ and 12 for intensities 10^{16} Wcm^{-2} and 5×10^{15} Wcm^{-2} respectively. The more the peak intensity, the shorter is the required pulse duration for 100% outer ionization. At a pulse duration $\tau/T = 12$, almost all electrons are freed from the cluster for both the intensities. For a higher $\tau/T > 12$, although pulse energy increases (for a fixed intensity), further absorption is not possible by those free electrons. It leads to a saturation in $E_a(\tau)/NU_p$ when $\tau/T > 12$. In the short pulse case when $\tau/T < 2$, outer ionization increases with decreasing τ/T (more clear in Fig. 20.4(b), reaches a peak $n_{\text{out}}/N \approx 1$ at $\tau/T = 0.8$ exactly where $E_a(\tau)/NU_p$ is maximum and then drops for $\tau/T < 0.8$ similar to $E_a(\tau)/NU_p$ of Fig. 20.3(b). Similar values of outer ionization (e.g., $n_{\text{out}}/N \approx 1$) requires a pulse duration $\tau/T > 10$ in the long pulse regime. Thus, efficient energy absorption is the result of efficient outer ionization.

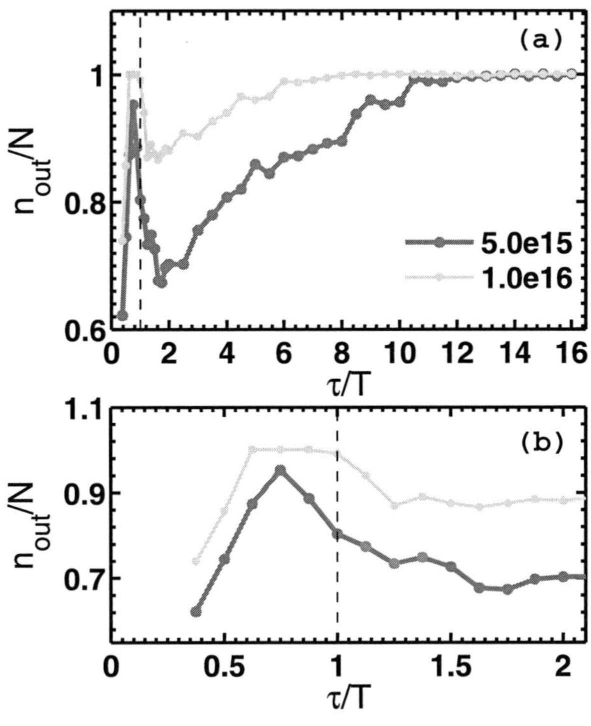

Figure 20.4 PIC results for outer ionization n_{out}/N vs pulse duration τ/T for the same peak intensities 5.0×10^{15} Wcm^{-2} (dark line) and 10^{16} Wcm^{-2} (light line) corresponding to Fig. 20.3. (b) is the expanded view of (a) for $\tau/T < 2$. At a higher intensity, 10^{16} Wcm^{-2}, n_{out}/N reaches unity at a relatively shorter $\tau/T \approx 6$ compared to the longer $\tau/T \approx 12$ required for the same saturation (i.e., $n_{\text{out}}/N \approx 1$) at a lower intensity, 5×10^{15} Wcm^{-2}. The saturation in E_a/NU_p in Fig.20.3 occurs after those $\tau/T > 2$ where n_{out}/N begins to saturate. Below $\tau/T = 2$, n_{out}/N increases, and reaches a peak value at $\tau/T = 0.8$ exactly where E_a/NU_p is peaked

To explain the mechanism of energy absorption and the peak position at $\tau/T \approx 0.8$ for $E_a(\tau)/NU_p$ in Fig. 20.3 (also n_{out}/N in Fig. 20.4), the different frequencies $\omega_i = \omega_1, \omega_2, \omega_3$ in Eq. (20.1) may be compared with the Mie frequency $\omega_M \approx 3\omega$. For pulse durations $\tau/T = 2-16$, plasma remains overdense, i.e., $\omega_M > \omega_i$, and the absorption process is solely due to the NLR for all ω_i. When $\tau/T = 1$, one finds $\omega_i = \omega, 2\omega, 0$; plasma is marginally overdense w.r.t. $\omega_2 = 2\omega$, and NLR becomes more efficient due to the decrease of the relative density $\rho/\rho_c(\omega_2)$. Thus, in passing from

$\tau/T = 2$ to $\tau/T = 1$, increased absorption and outer ionization occur because of the gradual decrease of the ratio $\rho/\rho_c(\omega_2)$ w.r.t. ω_2 that drives NLR more efficiently (note that NLR at frequency ω always occurs). For $\tau/T = 0.5$, the frequencies are $\omega_i = \omega, 3\omega, \omega$ and the laser pulse becomes $E(t) = E_0[\sin(\omega t)/4 - 3\sin(3\omega t)/4]$. In this case, LR (for 3ω component) and NLR (for ω component) works together, but the opposite phase (sign) of the two parts of the field (one with $\sin(\omega t)/4$ and the other with $-3\sin(3\omega t)/4$) seems to counteract, resulting in absorption and outer ionization below the respective peak values. Also, for such a short pulse, the temporal duration of this LR is *too* small to have a great effect. One should recall that just meeting the resonance condition is not enough. The time elapsed near the resonance and a higher driver strength are also important for a greater absorption. With a much shorter $\tau/T = 0.25$ (where $\omega_i = \omega, 5\omega, 3\omega$), plasma becomes under-dense w.r.t. $\omega_2 = 5\omega$ for which neither NLR nor LR works, and the LR at $\omega_3 = 3\omega$ (and the NLR at $\omega_1 = \omega$) is less efficient than the LR at $\omega_2 = 3\omega$ for a higher $\tau/T = 0.5$. From the above discussion, a peak in the absorbed energy and outer ionization for some value of τ/T between $1 - 0.25$ (in the present case $\tau/T \approx 0.8$), where plasma goes from marginally over-dense to critically dense to under-dense w.r.t. some frequency components of the driver, is justified. It is also surprising that the peak position at $\tau/T \approx 0.8$ and peak value of E_a/NU_p closely agree with the RSM result in Fig. 20.1 for the intensity 5×10^{15} Wcm^{-2}.

20.4 Summary

Laser–cluster interactions are studied by varying the laser pulse duration τ below the optical period T (at 800 nm) with a simple rigid sphere model (RSM) and self-consistent three-dimensional PIC simulations. In this sub-cycle regime *not all* models of the laser pulse are valid. With a valid model of the laser pulse, it is shown that qualitative variation of energy absorption vs τ for a given peak intensity of the laser pulse although obeying previously reported works (i.e., increase of absorbed energy with increasing τ) for $\tau > 2T$, is completely opposite for $\tau \lesssim 2T$ where both absorbed energy and outer ionization increases with decreasing τ (i.e., inspite of decreasing pulse energy). For a deuterium cluster, the absorbed energy and outer ionization is shown to reach a peak value near $\tau \approx 0.8T$ due to the quicker saturation of inner ionization. This new feature of energy absorption with decreasing τ below $2T$ is justified with a harmonic oscillator (HO) and an anharmonic oscillator (i.e., RSM) model. In this sub-cycle regime, linear and non-linear resonances may work together leading to higher absorption.

Bibliography

1. T. Ditmire, R. A. Smith, J. W. G. Tisch and M. H. R. Hutchinson. 1997. *Phys. Rev. Lett.* 78: 3121.
2. T. Ditmire, J. W. G. Tisch, E. Springate, M. B. Mason, N. Hay, J. Marangos and M.H.R. Hutchinson. 1997. *Nature* (London). 386: 54.
3. T. Ditmire, J. W. G. Tisch, E. Springate, M. B. Mason, N. Hay, J. P. Marangos and M. H. R. Hutchinson. 1997. *Phys. Rev. Lett.* 78: 2732.
4. V. Kumarappan, M. Krishnamurthy and D. Mathur. 2001. *Phys. Rev. Lett.* 87: 85005
5. M. Krishnamurthy, D. Mathur and V. Kumarappan. 2004. *Phys. Rev. A.* 69: 033202.
6. M. Lezius, S. Dobosz, D. Normand and M. Schmidt. 1998. *Phys. Rev. Lett.* 80: 261.

7. V. Kumarappan, M. Krishnamurthy and D. Mathur. 2002. *Phys. Rev. A*. 66: 033203.
8. Y. Fukuda, K. Yamakawa, Y. Akahane, M. Aoyama, N. Inoue, H. Ueda and Y. Kishimoto. 2003. *Phys. Rev. A*. 67: 061201.
9. E. Springate, N. Hay, J. W. G. Tisch, M.B. Mason, T. Ditmire, M. H. R. Hutchinson and J. P. Marangos. 2000. *Phys. Rev. A*. 61: 063201.
10. E. Springate, S. A. Aseyev, S. Zamith and M. J. J. Vrakking. 2003. *Phys. Rev. A*. 68: 053201.
11. Y. L. Shao, T. Ditmire, J. W. G. Tisch, E. Springate, J. P. Marangos and M. H. R. Hutchinson. 1996. *Phys. Rev. Lett*. 77: 3343.
12. V. Kumarappan, M. Krishnamurthy and D. Mathur. 2003. *Phys. Rev. A*. 67: 043204.
13. L. M. Chen, J. J. Park, K. H. Hong, I. W. Choi, J. L. Kim, J. Zhang and C. H. Nam. 2002. *Phys. Plasmas* 9: 3595.
14. J. W. G. Tisch, T. Ditmire, D. J. Fraser, N. Hay, M. B. Mason, E. Springate, J. P. Marangos and M. H. R. Hutchinson. 1997. *J. Phys. B: At. Mol. Opt. Phys*. 30: 709.
15. T.D. Donnelly, T. Ditmire, K. Neuman, M. D. I Perry and R. W. Falcone. 1996. *Phys. Rev. Lett*. 76: 2472.
16. C. Vozzi, M. Nisoli, J-P. Caumes, G. Sansone, S. Stagira, S. De Silvestri, M. Vecchiocattivi, D. Bassi, M. Pascolini, L. Poletto, P. Villoresi and G. Tondello. 2005. *Appl. Phys. Lett*. 86: 111–121.
17. B. Shim, G. Hays, R. Zgadzaj, T. Ditmire and M. C. Downer. 2007. *Phys. Rev. Lett*. 98: 123902.
18. M. Kundu, S. V. Popruzhenko and D. Bauer, S. V. Popruzhenko, M. Kundu, D. F. Zaretsky and D. Bauer. 2007, 2008. *Phys. Rev. A*. 76: 033201. *Phys. Rev. A*. 77: 063201.
19. J. Jha, D. Mathur and M. Krishnamurthy. 2006. *J. Phys. B: At. Mol. Opt. Phys*. 38: L291–L299.
20. J. Jha, D. Mathur and M. Krishnamurthy. 2006. *Appl. Phys. Lett*. 88: 041107.
21. V. Kumarappan, M. Krishnamurthy, D. Mathur and L. C. Tribedi. 2001. *Phys. Rev. A*. 63: 023203.
22. L. M. Chen, F. Liu, W. M. Wang, M. Kando, J. Y. Mao, L. Zhang, J. L. Ma, Y. T. Li, S. V. Bulanov, T. Tajima, Y. Kato, Z. M. Sheng, Z. Y. Wei and J. Zhang. 2010. *Phys. Rev. Lett*. 104: 215004.
23. J. Zweiback, R. A. Smith, T. E. Cowan, G. Hays, K. B. Wharton, V. P. Yanovsky and T. Ditmire, 2000. *Phys. Rev. Lett*. 84: 2634.
24. J. Zweiback, T. Ditmire and M. D. Perry. 1999. *Phys. Rev. A*. 59: 3166.
25. U. Saalmann and J.M. Rost. 2006. *J. Phys. B: At. Mol. Opt. Phys*. 39: R39–R77.
26. V. P. Krainov and M. B. Smirnov. 2002. *Phys. Rep*. 370: 237.
27. C. Rose-Petruck, K. J. Schafer, K. R. Wilson and C. P. J. Barty. 1997. *Phys. Rev. A*. 55: 1182.
28. D. Bauer and A. Macchi. 2003. *Phys. Rev. A*. 68: 033201.
29. K. Ishikawa and T. Blenski. 2000. *Phys. Rev. A*. 62: 063204.
30. F. Megi, M. Belkacem, M. A. Bouchene, E. Suraud and G. Zwicknagel. 2003. *J. Phys. B: At. Mol. Opt. Phys*. 36: 273.

31. C. Jungreuthmayer, M. Geissler, J. Zanghellini and T. Brabec. 2004. *Phys. Rev. Lett.* 92: 133401.
32. D. Bauer. 2004. *J. Phys. B: At. Mol. Opt. Phys.* 37: 3085.
33. T. Ditmire, T. Donnelly, A. M. Rubenchik, R. W. Falcone and M. D. Perry. 1996. *Phys. Rev. A.* 53: 3379.
34. T. Taguchi, T. M. Antonsen, Jr., and H. M. Milchberg, Thomas M. Antonsen, Jr. Toshihiro Taguchi, Ayush Gupta, John Palastro and Howard M. Milchberg. 2004, 2005. *Phys. Rev. Lett.* 92: 205003. *Phys. Plasmas* 12: 056703.
35. M. Kundu and D. Bauer. 2006. *Phys. Rev. Lett.* 96: 123401.
36. M. Kundu and D. Bauer. 2006. *Phys. Rev. A.* 74: 063202.
37. P. Mulser and M. Kanapathipillai, P. Mulser, M. Kanapathipillai and D. H. H. Hoffmann. 2005, 2005. *Phys. Rev. A.* 71: 063201. *Phys. Rev. Lett.* 95: 103401.
38. I. Kostyukov and J. M. Rax. 2003. *Phys. Rev. E.* 67: 066405.
39. M. Kundu and D. Bauer. 2008. *Phys. Plasmas* 15: 033303.
40. H. Wabnitz, L. Bittner, A. R. B. de Castro, R. Döhrmann, P. Gürtler, T. Laarmann, W. Laasch, J. Schulz, A. Swiderski, K. von Haeften, T. Möller, B. Faatz, A. Fateev, J. Feldhaus, C. Gerth, U. Hahn, E. Saldin, E. Schneidmiller, K. Sytchev, K. Tiedtke, R. Treusch, and M. Yurkov. 2002. *Nature* (London). 420: 482.
41. T. Laarmann, M. Rusek, H. Wabnitz, J. Schulz, A. R. B. de Castro, P. Gürtler, W. Laasch and T. Möller. 2005. *Phys. Rev. Lett.* 95: 063402.
42. C. Bostedt, M. Adolph, E. Eremina, M. Hoener, D. Rupp, S. Schorb, H. Thomas, A. R. B. de Castro and T Möller. 2010. *J. Phys. B: At. Mol. Opt. Phys.* 43: 194011.
43. S. Namba, N. Hasegawa, M. Nishikino, T. Kawachi, M. Kishimoto, K. Sukegawa, M. Tanaka, Y. Ochi, K. Takiyama and K. Nagashima. 2007. *Phys. Rev. Lett.* 99: 043004.
44. A. E. Kaplan, S. F. Straub and P. L. Shkolnikov. 1997. *Opt. Lett.* 22: 405.
45. Wei-Jan Chen, Zhi-Ming Hsieh, Shu Wei Huang, Hao-Yu Su, Chien-Jen Lai, Tsung-Ta Tang, Chuan-Hsien Lin, Chao-Kuei Lee, Ru-Pin Pan, Ci-Ling Pan and A. H. Kung. 2008. *Phys. Rev. Lett.* 100: 163906.
46. P. B. Corkum and F. Krausz. 2007. *Nature Phys.* 3: 381.
47. T. Brabec and F. Krausz. 2000. *Rev. Mod. Phys.* 72: 545.
48. M. Y. Shverdin, D. R. Walker, D. D. Yavuz, G. Y. Yin and S. E. Harris. 2005. *Phys. Rev. Lett.* 94: 033904.
49. T. Brabec and F. Krausz. 1997. *Phys. Rev. Lett.* 78: 3282.
50. A. J. Verhoef, J. Seres, K. Schmid, Y. Nomura, G. Tempea, L. Veisz and F. Krausz. 2006. *Appl. Phys. B.* 82: 513.
51. T. Fuji, N. Ishii, C. Y. Teisset, X. Gu, Th. Metzger, A. Baltuska, N. Forget, D. Kaplan, A. Galvanauskas and F. Krausz. 2006. *Opt. Lett.* 31: 1103.
52. B. S. Xie, M.P. Liu, N. Y. Wang and M. Y. Yu. 2007. *Appl. Phys. Lett.* 91: 011118.
53. C. Prigent, C. Deiss, E. Lamour, J. P. Rozet, D. Vernhet and J. Burgdörfer. 2008. *Phys. Rev. A.* 78: 053201.

54. G. M. Petrov and J. Davis. 2006. *Phys. Plasmas* 13: 033106.
55. U. Saalmann and J. M. Rost. 2003. *Phys. Rev. Lett.* 91: 223401.
56. A. Gupta, T. M. Antonsen, T. Taguchi and J. Palastro. 2006. *Phys. Rev. E.* 74: 046408.
57. Q. Lin, J. Zheng and W. Becker. 2006. *Phys. Rev. Lett.* 97: 253902.
58. T. Martchenko, Ch. Siedschlag, S. Zamith, H. G. Muller and M. J. J. Vrakking. 2005 *Phys. Rev. A.* 72: 053202.
59. K. W. Madison, P. K. Patel, M. Allen, D. Price, R. Fitzpatrick and T. Ditmire. 2004. *Phys. Rev. A.* 70: 053201.
60. H. A. Bethe and E. E. Salpeter. 1977. *Quantum Mechanics of One- and Two-electron Atoms*. New York: Plenum Publishing Corporation.

21 LIDAR: A Precision Long Range Standoff Sensor of Atmospheric Pollutants, Chemical, Biological and Explosive Agents

A. K. Razdan and S. Veerabuthiran

21.1 Introduction

Remote sensing of the atmosphere can be broadly classified into two categories, namely passive remote sensing and active remote sensing methods. In passive remote sensing, the source is beyond the control of the observer, e.g., radiometer, photometer, spectrometer etc. In active remote sensing, the source can be controlled by the observer, e.g., LIDAR, RADAR, SODAR, SONAR etc. In the field of remote sensing, lasers play an important role due to their inherent capability of generating well-collimated beams with outstanding characteristics of high coherence, monochromaticity, and directionality. Laser remote sensing techniques provide powerful tools for scientific studies of atmosphere, environmental monitoring, measurement of air quality parameters, remote sensing of oceans and rivers, remote assessment of vegetation etc. Developments in laser technology such as second harmonic generation, high power compact diode pumped solid-state lasers, and tunable solid-state lasers etc. have opened new possibilities of laser remote sensing to explore the earth's atmosphere. Laser remote sensing of the atmosphere is generally referred to as LIDAR, the acronym for Light Detection And Ranging. Similar to radar, in lidar, a laser pulse is sent into the atmosphere and is used as a spectroscopic probe of its physical state and chemical composition. The emitted laser beam interacts with the atmospheric constituents causing alterations in the intensity, polarization and wavelength of the backscattered light. From the measurements of these parameters of the received backscattered light, one can deduce the properties of the atmosphere and its constituents. Lidar methods allow range resolved measurement to obtain a vertical profile of the atmospheric parameters. The distance of the scattering medium can be deduced with high accuracy from the time delay of the return signal. Lidar systems can be operated in the wavelength range extending from the ultraviolet to the infrared (UV to IR) by using different types of lasers. Using the techniques of scattering, absorption, resonance and fluorescence, lidar can measure solid particulate matter such as aerosols, species of very low concentrations such as ozone, water vapour and metal atoms such as Na, K etc. Lidar measurements of the Rayleigh scattering from neutral atmosphere can be used to determine neutral density, pressure, temperature, and wind speed. Experimental measurements made by many groups all over the globe in the past decade have demonstrated the reliability of various lidar techniques[1–3].

During the last few years, terrorism has emerged as a prominent threat to society. There is an increased threat scenario because of the possible use of chemical, biological and explosive (CBE) materials against public and military[4]. Remote detection of such chemical, biological, and explosive materials will be of great use to both military and civilian agencies[5–7]. Chemical warfare agents (nerve and blister agents), biological agents (anthrax) and explosive materials (triacetone triperoxide,with high vapor pressure) have strong optical absorption spectra in 3–4 micron and 9–11 micron spectral bands. Out of all currently available techniques, no single technique provides efficient detection against such threats at significant standoff distances. Light detection and ranging (LIDAR) technologies, based on the transmission of laser pulses and analysis of the return signals, have demonstrated impressive capabilities in remote detection of such toxic materials. LIDAR is a highly sensitive tool which can detect extremely low concentrations of various toxic materials present in the form of vapors at distances of a few hundred meters to kilometers. Differential absorption lidar (DIAL) system operates in any of the above spectral bands with suitable system configuration and can detect chemical materials up to a distance of few hundred meters. This system uses two infrared laser wavelengths to detect chemical materials: one corresponds to peak absorption of the molecule and the other corresponds to weak absorption. The ratio of these two signals gives concentration. Knowledge of which wavelength has been absorbed gives information about the specific absorbing molecule.

21.2 Principle of Lidar Operation

A fraction of the light in the transmitted pulse is scattered back to the receiver by the molecules and particles (aerosols) in the atmosphere, which act as distributed reflectors. The received radiation is collected by an optical system (telescope) and is focused on to an optical detector. The electrical signal from the detector is amplified, digitized, stored, and analyzed using a computer. Since the transmitted pulse has a finite duration τ, it illuminates a finite geometrical length $c\tau$ (c is the velocity of light) of the atmosphere at any instant. However, since the received energy must travel a two-way path, the atmospheric length (or range increment) from which signals received at any time is $c\tau/2$. This method allows a range resolved measurement. All the backscattered photons arriving in the time range $t \pm \Delta t/2$ are stored in one range bin. A typical range resolution of 75 m requires a 500 ns range bin. A single profile is recorded for each laser shot. The single shot profiles are added together at each altitude to build up the sufficient signal at the corresponding altitude. For example, if a laser fires 1000 shots at the rate of 10 pulses per second, then time required to build up a single statistically significant profile is 100 s.

21.3 Description of Typical Lidar Systems

The lidar system essentially comprises a transmitter, receiver, and lidar controller. The transmitter is a laser source, whereas the receiver is an optical telescope. The lidar controller provides accurate timing and control pulses for the operation of different channels, while simultaneously providing synchronization with the laser transmitter pulse.

21.3.1 Laser transmitter

A typical Nd: YAG laser is the main transmitting source in the lidar system. Its fundamental wavelength is 1064 nm in the near infrared. To operate the system in visible and UV range, the second, third and fourth harmonics of the fundamental wavelengths are used. The operating wavelengths at these harmonies are 532 nm, 355 nm, and 266 nm. The pulse width of the laser is 10 ns and its pulse repetition frequency (PRF) is 10 Hz with a typical energy of 500 mJ. The laser beam is transmitted into the atmosphere through 45° mirrors in co-axial configuration. The laser beam with a nominal divergence of 0.1 mrad is used.

21.3.2 Receiver

In general, the receiver system consists of a large aperture (typically 500 or 300 mm) Cassegrain telescope depending upon the altitude coverage. Using a post-optics system, the received signal is delivered to the optical detector (eg., photomultiplier tubes/APD). The detector channel receives all the signals covering from near ground to upper atmospheric region. The interference filter pass bandwidth of 0.5 nm centered on the laser line wavelength reduces the system background noise almost to a negligible value.

21.3.3 Detector

The purpose of a detector is to convert electromagnetic energy into an electric signal. Detectors are classified into thermal types and quantum types. Thermal detectors use infrared energy as heat and their photosensitivity is independent of wavelength. They do not require cooling, but have the disadvantage of slow response and low detection capability. In contrast, quantum detectors offer higher detection performance and a faster response speed, although their photosensitivity is dependent on wavelength. In general, quantum detectors must be cooled for accurate measurement, except for detectors used in the near infrared region. A number of different semiconductor materials are also in common use as optical detectors in lidar. These include silicon APDs in the visible, near ultraviolet and near infrared, germanium and indium gallium arsenide in the near infrared, and indium antimonide, indium arsenide, mercury cadmium telluride, and germanium doped with copper or gold in the mid-infrared.

21.3.4 Preamplifier for detection channel

The return signal produces a fast decaying current pulse at the output of the detectors. The gain provided by the detectors is not sufficient to amplify the lidar return echo from the lower altitude. It has to be further amplified and converted into a voltage pulse for further processing using the preamplifier. The current pulse is automatically converted into a fast transient voltage pulse by the input impedance of the preamplifier in the DC coupled mode.

21.3.5 Data acquisition system

The preamplifier output from the detection channel is recorded using appropriate digitizer boards with suitable sampling rate. The digitized samples are initially stored in the onboard memory of the digitizer and transferred to the hard disk for further processing.

21.4 Backscattering LIDAR

Backscattering Mie Lidar system has been designed and developed at Laser Science and Technology Centre (LASTEC), Delhi[8]. This system uses a pulsed Nd: YAG laser (Surelite I-10 Laser from Continuum, USA) with a fundamental emission wavelength of 1064 nm, as the main transmitting source. Laser pulse energy is variable from 25 mJ to 400 mJ. The pulse width of the laser is 7 ns and its pulse repetition frequency is 10 Hz. The laser beam diameter is 6 mm at the output aperture with a divergence of 0.6 mrad. The system is configured in a coaxial mode by folding the laser beam using two reflective mirrors so that the laser beam comes within a telescope field of view (FOV). Figure 21.1 describes the block diagram of the experimental setup. This setup is installed on a movable tripod stand, which is manually controlled. This tripod stand is designed and fabricated in-house. It can scan the atmosphere through 360° in azimuth and −10° to 45° in elevation. The scattered radiation from the atmosphere is collected by the 200 mm diameter Cassegrain telescope and its field of view is less than 3 mrad. A passband interference filter with bandwidth (FWHM) of 3 nm centered on the laser wavelength is used to reduce the atmospheric background noise. The IR optical signal from the telescope is focused onto a high quantum efficiency Si: avalanche photo diode (APD) detector module (Licel, Germany). The Si: APD module consists of an integrated TE cooler and temperature controller, preamplifier, focusing lens and HV power supply. NI's PCI bus based DAQ card is used as data acquisition hardware in this system. The return signal from the detector is connected to an analogue input channel of the DAQ and the trigger signal from Laser source is connected to the digital trigger input of the card. The DAQ is configured to a sampling rate of 10 MHz to have a range resolution of 15 m, which corresponds to 0.1 μs. The number of samples is set to 500 so as to collect data up to a range of 7.5 km. Figure 21.2 shows the LIDAR controller software developed in-house in LabView platform, which is used to control the laser source and data acquisition hardware. The laser system is switched on by invoking controls in the user interface and it starts firing pulses at a rate of 10 Hz. The data acquisition process is started when the start DAQ button is pressed in the graphical user interface and data for every 500 samples at 15 m range bins is stored to the computer hard disk for every laser pulse transmitted. A typical backscattered signal versus range is presented on the left side of Fig. 21.2. The backscattered signals from the atmosphere were collected from the near field to maximum range of 7–8 km. The Si: APD signal reached a maximum value of about 1100 mV, which is the saturation level of the detector. Thereafter, the signal started falling steadily with respect to range. It showed clearly that the experimentally measured multiple cloud signals (cloud signal strength is higher than the background signal) at a distance of 3.5 km and 5 km. The image shown in the right side of Fig. 21.2 represents the temporal variation of the received signals. Signals received from the nearby region are very strong and accordingly a colour coding is assigned (red:strong signals and blue:weak signals). From the LIDAR signal information pertaining to various parameters such as extinction coefficient, visibility has also been obtained by using suitable LIDAR inversion methods in real time. The complete LIDAR system is connected to a UPS system, which can keep the LIDAR operational with captive power for nearly one hour. This system has been tested and validated by measuring return signals from atmospheric aerosols and clouds in real time under different atmospheric conditions.

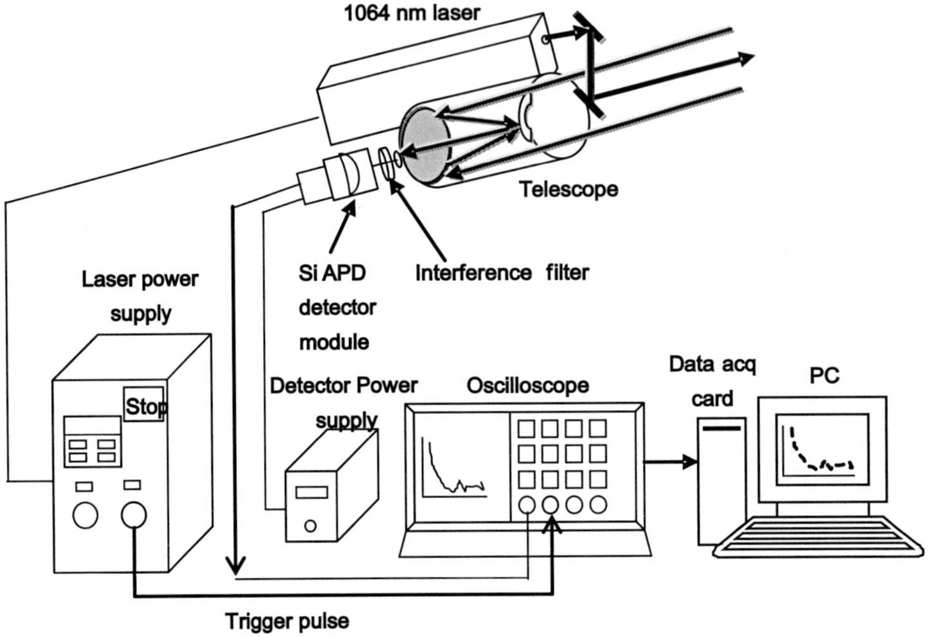

Figure 21.1 Block diagram of Mie LIDAR system developed at LASTEC, Delhi

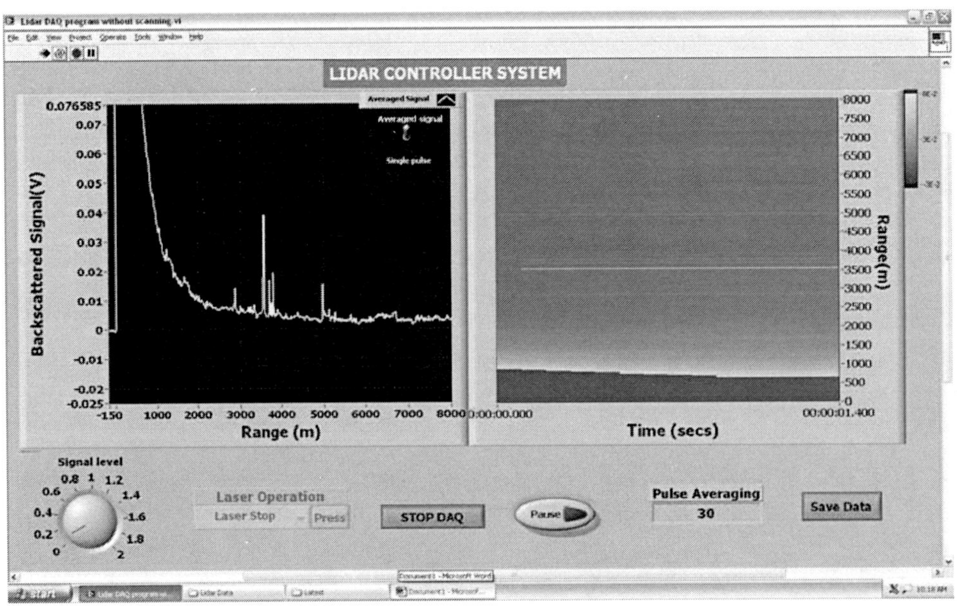

Figure 21.2 LIDAR controller and data acquisition display panel

21.4.1 Backscattering lidar data processing

From the LIDAR signal, information pertaining to various parameters has been obtained by using suitable LIDAR inversion methods. The single scattering equation for a monostatic pulsed LIDAR may be written as[9–10]:

$$P(R) = P_t \frac{c\tau}{2}(\beta_a(R) + \beta_g(R))\eta \frac{A}{R^2}.\exp\left[-2\int_0^R (\alpha_a(R) + \alpha_g(R))dR\right], \tag{21.1}$$

where $P(R)$ is the received signal at range R, P_t is the transmitted power, η summarizes all system constants such as optical efficiency, detection efficiency etc. $\beta(R)$ is the volume backscattering coefficient of aerosols, $(\beta_a(R))$ and air molecules $(\beta_g(R))$, $\alpha(R)$ is the volume extinction coefficient of aerosols $(\alpha_a(R))$ and air molecules $(\alpha_g(R))$, c is the speed of light and τ is the pulse duration. This equation indicates that in a homogeneous atmosphere, $P(R) \times R^2$ decreases with range as an exponential function. In the presence of a cloud layer, the signal level undergoes a rapid increase because of enhanced backscattering of cloud droplets, and then reaches its maximum value before decreasing to the level of backscatter from ambient air or disappearing into the background noise level. The lidar inversion method was developed using the analytical solution of the single scattering Lidar equation shown in Eq. (21.1). To invert Eq. (21.1), another relationship between the two unknowns α and β is needed. These two coefficients are dependent on the nature of the scattering particles. The relation between these two parameters is:

$$\beta = C\alpha^k. \tag{21.2}$$

Here constants C and k depend on the LIDAR wavelength and various properties of the aerosol particles. Constants, $k \approx 1$ and $C \approx 0.05\ sr^{-1}$ appear to be the values most often reported. They are generally taken to be constant over the optical path. The Lidar Eq. (21.1) is solved with the use of Eq. (21.2) to obtain the final backward integration equation to derive the atmospheric extinction coefficient, which is:

$$\alpha(R) = \frac{[P(R)R^2]^{1/k}}{\frac{[P(R_m)R_m^2]^{1/k}}{\alpha_m} + \frac{2}{k}\int_R^{R_m}[P(R)R^2]^{1/k}dR}. \tag{21.3}$$

The boundary value α_m is chosen at the far end near the maximum range R_m at which a usable signal is available before it reaches the constant background noise level ($\alpha_m = \alpha(R_m)$). The advantage of this method is the fact that the system constant no longer has to be determined. However, a drawback of this technique is to have a boundary value at the far end range R which is in general more difficult to obtain than that at the near range R. The influence of α_m on the resulting $\alpha(r)$ profile decreases strongly with the optical depth between R and R_m. A typical value in the order of 10^{-5} m^{-1} as a reference value to invert the LIDAR signal downward from a range of 5 km was assumed. The influence of a wrong starting value is mainly limited to the far end (last third) of the measurement path. The first two-thirds of the inverted measurement path can be seen as very good estimates. The reason for this is that the second term in the denominator of Eq. (21.3) rapidly becomes large compared with the first term, especially for large boundary values of α_m. Hence, evaluating the LIDAR profile for a cloudy and foggy atmosphere is straight forward. It is also noted that the backscattering-to-extinction ratio (C) is not uniform over different regions of the troposphere. Detailed analysis is required to study the influence of C as a function of relative humidity of the atmosphere and aerosol composition over Delhi. In the analysis described in this chapter, a uniform C is assumed over the entire range and multiple scattering factors have not been incorporated.

21.4.2 Determination of aerosol extinction coefficient

Experiments were conducted using this system under different atmospheric conditions ranging from clear sky conditions to a cloudy and overcast sky. A laser beam with pulse energy of 50 mJ was transmitted into the atmosphere at an elevation angle of 12° along the slant path. Since most of the experiments were conducted during daytime, regular monitoring of the sky condition was carried out visually. The backscattered signals from the atmosphere were collected from the near field to a maximum detectable range of up to 7–8 km using this system. Figure 21.3 shows the typical example of the Lidar signal obtained in the experiment conducted at LASTEC, Delhi on a clear day (winter month December 3, 2008, 15:30 hrs local time). Due to strong sky background noise, fluctuations will be seen in the return signals across the entire range. The return signals do not show any significant variation with respect to distance beyond 2 km, in other words, the signal merges with background noise levels. The fluctuations in the Lidar signal can be overcome easily by averaging a larger number of pulses. However, for the present studies, a total number of 128 pulses were averaged to obtain a single noise free smooth signal profile. A spatial moving averaging of 10 to 20 samples was applied to remove random noise depending upon the signal behavior (Fig. 21.3). The aerosol extinction coefficient profile was derived using the above method. The exponent k in Eq. (21.3) was set to unity since the resulting profile depends only weakly on k. Figure 21.4 describes the aerosol extinction coefficient obtained for single pulse along slant path. Extinction coefficient values mostly varied between $6.52 \times 10^{-5}\ m^{-1}$ and $1.21 \times 10^{-4}\ m^{-1}$. Results are in good agreement with the reported literature values. It can be seen that extinction values do not show much variations along the slant path. This is due to the fact that the laser beam travels in the homogeneous region of the atmosphere.

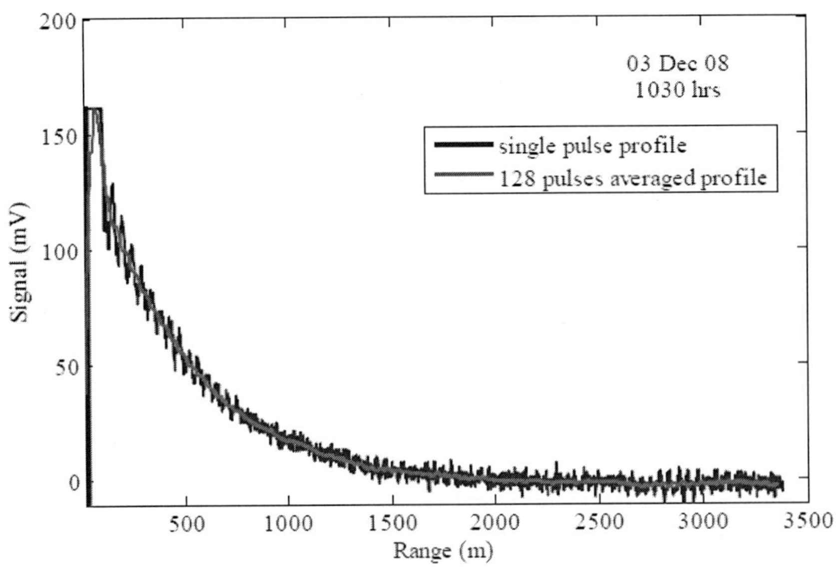

Figure 21.3 The backscattered signal versus range for single and average of 128 pulses

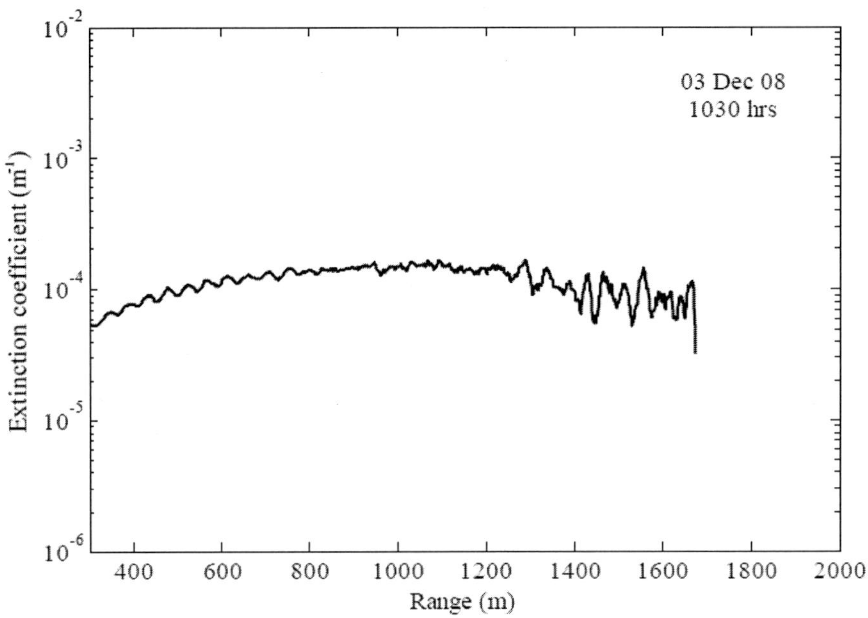

Figure 21.4 Aerosol extinction coefficient profile obtained at 1064 nm along the slant path

21.4.3 Detection of atmospheric clouds

One of the potential applications of Lidar is in the studies of characterization of clouds. Usually, strong backscattering arises from clouds due to the relatively large scattering cross-sections of cloud particles and the huge number density of scatterers. The same experimental setup (discussed in the beginning of the section) was used for detecting clouds in the atmosphere. In general, the sky condition was inspected visually before firing the laser beam into the atmosphere. The first information on the presence of a cloud is obtained by simply looking at the signal profile with altitude in the visual display on the oscilloscope and the computer. The Lidar data shows a sudden upsurge in signal strength where the clouds are present from its normal fall with range in the presence of background aerosols. The visual display gives clearly a qualitative nature of the cloud layer indicating its altitude and information whether the layers are single or multiple. The presence of cloud is determined by the following quantitative criteria: (1) a sudden upsurge, (2) a rapid decrease just after a maximum, and (3) the actual value of the extinction coefficient at the peak. Figure 21.5 shows the behavior of the Lidar return signals obtained on August 11, 2008, 16:21 hrs from the clouds. It clearly revealed the strong return signals from the clouds in the range between 7.15 km and 7.47 km. As a case study, an experiment was conducted on a rainy day. The backscattered signal obtained on August 13, 2008, 15:50 hrs from low-level clouds is presented in Fig 21.6. The total signal is basically characterized by the nature of cloud and other species in the atmosphere, i.e., humidity, drizzle and aerosols contribute to increases in the backscattered signal. The cloud signal-to-noise ratio is large and the retrieval of cloud heights (i.e., base and top) from a single profile is not difficult. The cloud base and heights have been derived from the measured waveforms at different times. The cloud thickness is then derived from the difference between the apparent cloud top and base. The cloud was detected between the 1.87 km and 2.62 km region.

The geometrical thickness of the cloud is about 550 m. Visual inspection revealed the presence of low-level rain bearing clouds on this day; the experimental site also received intermittent rainfall.

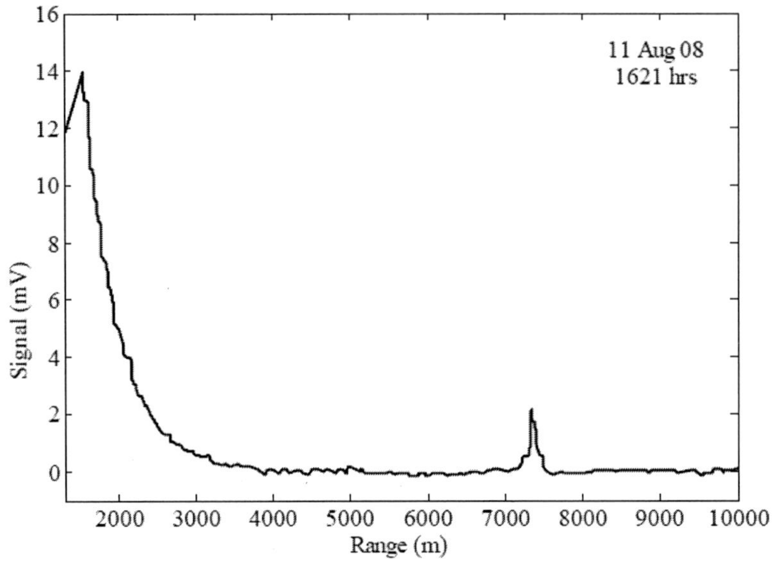

Figure 21.5 Multiple cloud returns

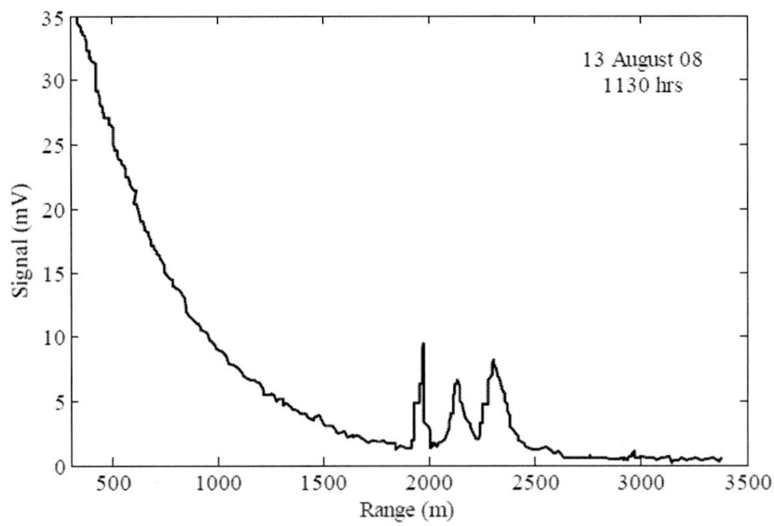

Figure 21.6 The backscattered signal received from multiple clouds on August 13, 08, 1130 hrs

21.4.4 Determination of visibility

Lidar offers advantages for determining the atmospheric visibility over extended and remote paths. The measurement of the volume extinction coefficient (α) is the preferred approach for determining visibility. When the atmospheric particle size distribution is unknown, empirical expressions may be used to predict the wavelength dependence of particulate extinction. A frequently quoted empirical relation for visually clear air is given below

$$V = \frac{3.91}{\alpha[km]}\left(\frac{0.55}{\lambda[\mu m]}\right)^q \tag{21.4}$$

here $q = 1.3$ for 'average seeing condition', α is the mean extinction coefficient value derived from a Lidar signal over the range interval from the near field to 2–5 km. The mean extinction coefficient value is derived from each profile over range intervals from near field to the maximum detectable range. This value is used in the above relation to calculate the atmospheric visibility at 1064 nm. Experiments were conducted for determining the visibility using the current system. Laser radiation was transmitted along the horizontal path for this purpose. The backscattered radiation was collected and processed to obtain a mean extinction coefficient using the above method.

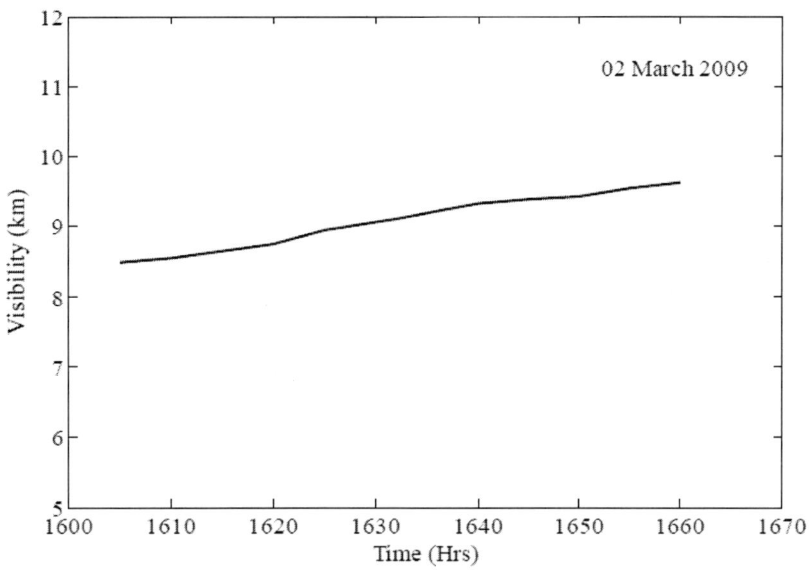

Figure 21.7 Temporal variation of visibility during 1 hour observation period on March 2, 2009

Figure 21.7 illustrates the temporal variation of visibility. Data collected on March 2, 2009 during the period 16:19 hrs to 16:40 hrs was used in this calculation. It shows the increasing trend in the visibility with respect to time due to local changes in the atmospheric processes. The average visibility obtained on this day was 9.1 km.

21.4.5 Measurement of depolarization signal

Lidar measurements of depolarization ratio provide highly reliable information to discriminate the nature of biological aerosols in the atmosphere[11]. The basic polarization Lidar application involves the transmission of a linearly polarized laser pulse and detection via a beam splitter of the orthogonal and parallel planes of polarization of the backscattered light. The ratio of two signals is known as the depolarization ratio (δ). The depolarization method is used to differentiate the solid, liquid, or mixed phase of biological aerosol or clouds. Spherical particles like water droplets and air molecules cause no depolarization, so that $\delta \sim 0$, whereas non-spherical particles, such as ice clouds, mineral dusts, biological aerosols produce depolarization, so that $\delta > 0$. Fluorescence Lidar coupled with depolarization measurement is suggested as one of the approaches for effective discrimination between biological aerosols and non-hazardous aerosol clouds. It has been used to measure the depolarization ratio of biosimulants (*Bacillus anthracis*), interferent aerosol clouds released in the field using 1.54 μm and 1.064 μm elastic lidar. In view of the importance of depolarization measurements, an experimental setup has been established at LASTEC to measure the parallel and perpendicular polarized signals from the atmosphere. In the present setup, a new mechanical adaptor has been designed, fabricated and fitted at the end of the receiving telescope. This can accommodate two individual detector boxes that contain an interference filter, Si: APD detector modules and a polarizing beam splitter. The backscattered radiation from the atmosphere is allowed to pass through polarizing cube beam splitter which then separates the parallel and perpendicular polarization components before it reaches the detectors. The detected signals are passed through A/D converters and data processors etc. for further processing. Preliminary experiments have been conducted recently to measure the depolarized signals from known near range targets. The linearly polarized laser radiation at 1064 nm with pulse energy of 50 mJ was transmitted into the atmosphere, which interacted with tree leaves. The multiple return signals from the target were measured simultaneously using two detectors. Detectors were operated at the same operating voltage in order to have the same gain at both channels.

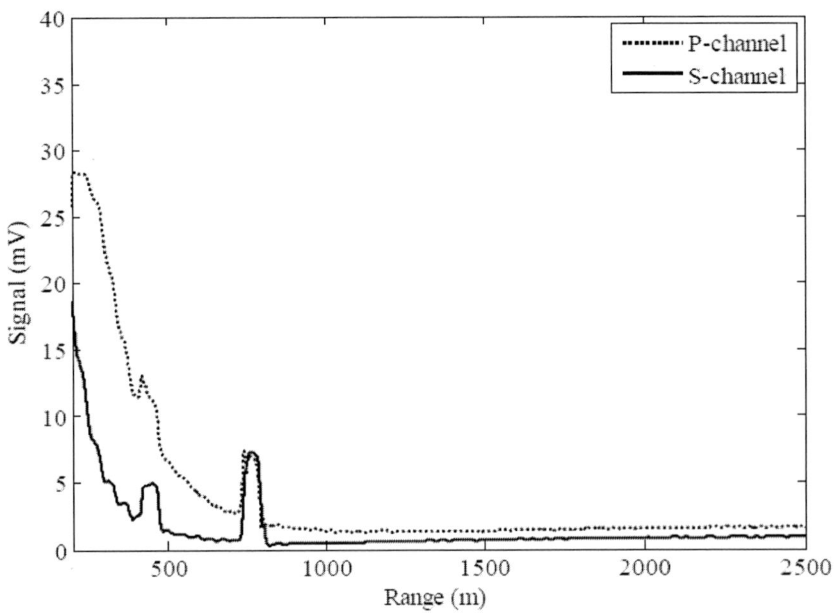

Figure 21.8 Depolarization signal measured from the two different targets (*S* channel: Perpendicular component, *P* channel: Parallel component)

Figure 21.8 shows a typical example of the measured return signals at the parallel and perpendicular channels. Return signals at both detectors reduce with range after reaching the saturation level.

There is an enhancement in signal strength observed at about 420 m and 720 m, this is due to tree leaves. The signal pattern and amplitudes are different for both detectors. This is due to the fact that the non-sphericity of the target changes the state of polarization. Further experiments are planned to measure the depolarization signals from various targets such as biosimulants, diesel, pollen etc.

21.5 LIF Lidar System

Laser induced fluorescence (LIF) is the emission from atoms or molecules that have been excited to higher energy levels by absorption of laser radiation. When excited with a laser, the excited atoms or molecules will after some time, usually in the order of a few nanoseconds to microseconds, de-excite and emit light at a wavelength larger than the excitation wavelength. The standoff detection of biological warfare (BW) agents is based on this concept. In general, most of the BW molecules fluoresce when they are excited by a suitable wavelength. These BW molecules are mostly constituted of aromatic amino acids and coenzymes. Aromatic amino acids, such as tryptophan, tyrosine and phenylalanine absorb light at 280–290 nm and they fluoresce in the spectral band[11] between 300 nm and 400 nm. Biogenic chemicals associated with cell metabolism, such as reduced nicotinamide adenine dinucleotide (NADH) and riboflavin have their maximum absorption cross-section at around 340 nm and the resulting fluorescence peaks between 450 nm and 560 nm. Hence, it is possible to detect the BW agents by using suitable UV excitation wavelength. Also, the discrimination of biological agents can be achieved only from the LIF signal because the fluorescence cross-sections for particles 1–10 micron size range are sufficiently large enough to make single particle interrogation feasible [12–14]. Currently, most prototype LIF LIDAR use either 266 or 355 nm UV light; both these wavelengths being easily derived from an Nd:YAG laser, which has a small footprint, relatively low maintenance and is readily available as a commercial source. 266 nm UV excites fluorescence primarily from the tryptophan within the bacterial cell wall and tyrosine (also NADH and flavins found in abundance in growth medium) and 355 nm UV excites fluorescence primarily from NADH (and also flavins) but not tryptophan. Therefore, it could be argued that 266 nm light detects proteins present in the bacterial cell wall whereas 355 nm light detects compounds found in large quantities in growth medium and which are less abundant in bacteria. However, the attenuation of 266 nm light by atmospheric ozone is approximately 10 times greater than that of 355 nm and so 355 nm LIDAR systems may have a longer detection range.

The block diagram of a typical UV laser induced fluorescence lidar (UVLIF) setup is shown in the Fig. 21.9[15]. This system uses Nd: YAG quadrupled wavelength at 266 nm as a transmitter. A collimated laser pulse is transmitted into the atmosphere towards the bioaerosol cloud. The Divergence of the laser beam after collimation is approximately 0.2 mrad. The backscattered signal is collected using a Cassegrain telescope, which focuses the return signals onto the detector box that contains beam splitter, interference filter, lens and two photomultiplier tubes (PMT). A beam splitter divides the incoming backscattered signal and allows the radiation to reach two PMTs. PMT1 is used to collect the elastic backscattered signals from the atmosphere at 266 nm whereas PMT2 collects induced florescence signals from the bioaerosol cloud. Two interference filters are used in this setup to block unwanted background light. PMT2 (fluorescence channel) is a time gated detector and receives the return signal only if any bioaerosol cloud is present. Gating time (equal to distance of cloud) is determined from the PMT1 (Mie channel) and the same information is fed to PMT2 to receive the signal from only the cloud. If the cloud is biological in nature, then it will fluoresce at 350 nm due to the presence of tryptophan. Further, the system will discriminate the biological nature of the cloud by processing the signals received at both detectors. An interference filter with FWHM of 1 nm centered at 266 nm allows scattered signals to reach PMT1 and other filter with FWHM of 20 nm centered at 350 nm permits induced fluorescence signal to reach PMT2. The detected signals are passed through an A/D converter data processors etc., for signal analysis. Lidar master controller controls the laser transmitter, data acquisition and signal processing sequence in photon counting mode.

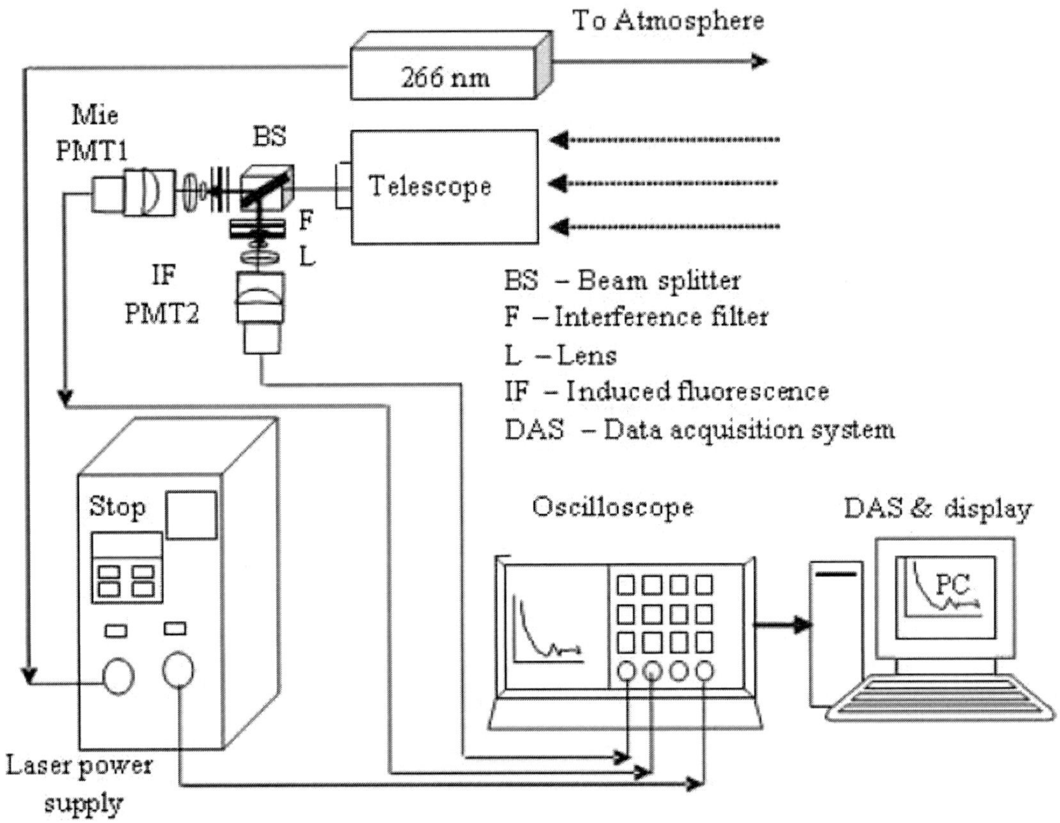

Figure 21.9 Experimental scheme of UV LIF lidar

Tryptophan is believed to be primarily responsible for the fluorescence of bacterial agents. It emits fluorescence signal, which peaks around 350 nm when it is excited by 266 nm UV radiation. The intensity of fluorescence signals from the tryptophan is stronger compared to other fluorophors of the biological aerosols. One can exploit this fluorescing property of tryptophan at excitation wavelength 266 nm for standoff detection of bioaerosol clouds. Critical analysis has been carried out for the Lidar setup, which comprises the system parameters such as E_t = 100 mJ, τ = 10 ns, PRF = 20 Hz, telescope diameter = 300 mm and atmospheric parameters such as $\alpha_{atm} = 5.0 \times 10^{-4}$ m^{-1}, $\beta_{atm} = 1.09 \times 10{-}4\ m^{-1}$ at 266 nm. Solar radiance equal to 1 $W/m^2/nm$ for daytime operation and 0.01 $W/m^2/nm$ for night time operation were used in calculation. Return signals have been simulated for transmission of a single laser pulse during day and nighttime operating conditions.

Figure 21.10 illustrates the range dependent return signal at an excitation wavelength of 266 nm for a clear atmosphere, which decreases up to a maximum range of 3 km (line curve in Fig. 21.10). The return signals as a function of range are also simulated by introducing the bioaerosol cloud (dotted curve in Fig. 21.10). The BG cloud of 200 m thickness is introduced between the ranges 1000 and 1200 m with varying concentrations. Analysis showed that this system can detect the BW aerosol cloud in daytime if its concentration varies from 9×10^6 ppl to 5×10^8 ppl. Fluorescence signals received at PMT2 were also simulated for the presence of BG cloud located at distance of 1000 m, which receives signals in the spectral band 350 nm ± 10 nm only if there is any induced fluorescence emitted from the BG cloud. A discrimination method is also developed to evaluate the detection performance of the system. This method is based on the ratio of the fluorescence signal to the UV elastic signal for biological aerosols. Figure 21.11 shows the range profile of the ratio of fluorescence to elastic scattering signals. Strong enhancement in this ratio is observed in the bioaerosol cloud region, which confirms the fluorescence nature of the cloud. However, it is mentioned that the discrimination algorithms are based on the spectrally resolved fluorescence signature of biological aerosols and interferents (rather than the broadband fluorescence signal) and may result in improved discrimination performance.

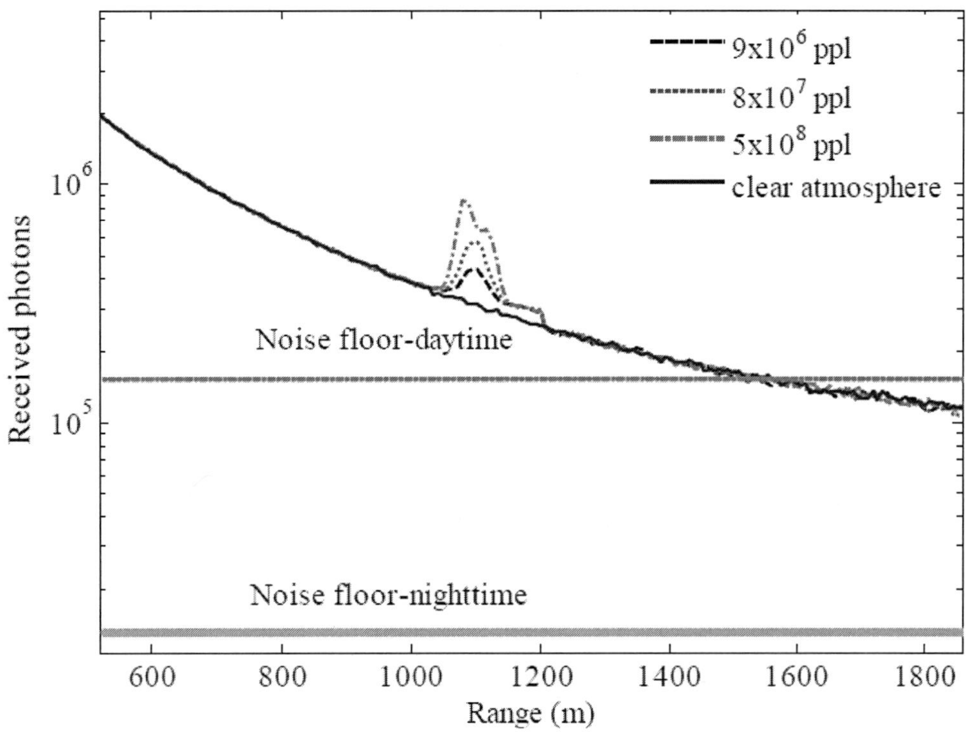

Figure 21.10 Simulated backscattered photon counts versus range for varying concentration of BG cloud located at 1000 m

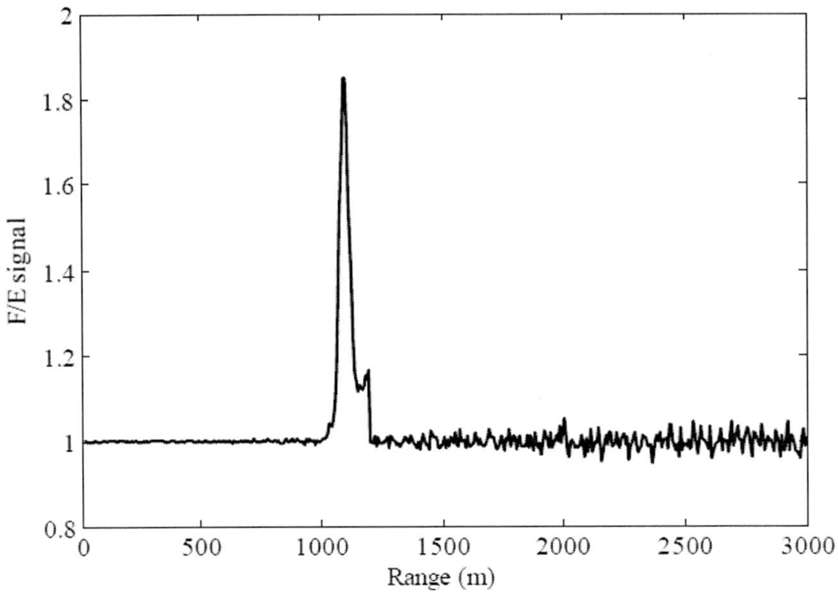

Figure 21.11 Ratio of fluorescence to elastic signal versus range

21.6 Differential Absorption LIDAR

The ability of differential absorption lidar (DIAL) measurements to determine and map the concentrations of the selected molecular species in ambient air is a powerful and useful technique. With the DIAL technique, one can investigate most important manmade pollutants and their concentrations as low as few hundred parts per billion both in free atmosphere and polluted areas such as cities or near industrial plants. Range resolved DIAL systems are sensitive enough to measure the ambient air concentrations and distribution of most of the important pollutant gases, including SO_2, NO_2, NH_3 and ozone[16–17]. This technique makes it possible to obtain vertical profiles of the atmospheric gas concentrations from ground, air or space platform. Differential absorption lidar have used CO_2 laser operating in the 9–11 μm spectral band and solid state lasers in the 2–5 μm band. This region of the spectrum is rich in absorption lines belonging to many different molecules, and CO_2 laser systems are potentially able to monitor a large number of pollutants or trace gas that commonly occurs. The generation of 2–5 μm wavelengths can be done by various non-linear techniques like the optical parametric oscillator techniques (OPO) in solid-state laser. Discretely tunable infrared TEA CO_2 DIAL system operating in 9–11 μm is used for the range resolved measurement of low vapor pressure organic liquid dispersed on the ground and in vapour chambers at standoff distance of 1–2 km [18–20]. There are various advantages of utilizing laser in the 9–12 μm spectral region:

- The CO_2 gas laser technology is already well developed.
- Many important industrial organic molecules can be characterized uniquely by their fingerprint spectra in the 9–12 μm regions, which is important when mixtures of several gases are present simultaneously.

- The 9–12 μm regions are normally a relatively transparent atmospheric window, allowing multiple kilometer ranges.
- CO_2 laser can be tuned over a significant part of the window, although they are only line tunable rather than continuously tunable.

In the DIAL technique, two wavelengths are required: one wavelength is tuned to match a peak (maximum) of an absorption line of the gas of interest, while the second nearby wavelength is tuned to a region of low absorption. When two-wavelength lidar signals propagate through the atmosphere, a dual-channel receiver detects the backscattering at both wavelengths. Comparison of the two return signals gives a direct measure of the concentration of the absorbing gas, and its position in space is determined using the range resolution of the lidar. In this way, it is possible to measure the concentration and spatial distribution of a large number of atmospheric gases.

21.6.1 CO_2 DIAL

DIAL setup designed and assembled at Laser Science & Technology Centre, Delhi comprises a compact pulsed tunable TEA CO_2 Laser (M/S OptoSystems Ltd., Russia) emitting ~ 50 mJ energy (max energy at peak wavelength) and tunable in the wavelength range of 9.2–10.8 μm (with pulse width of 2 μs) as a transmitter. This fully computer controlled laser with controllable PRF (from 10 – 100 Hz) can fire laser pulses as per a predefined program allowing selection of λ_{on} and λ_{off} wavelengths corresponding to the absorption spectrum of the suspected target molecule. The receiver system comprises a 200 mm beam reducing telescope (0.1X) together with focussing optics (Φ = 25.4 mm, focal length = 15 mm, ZnSe lens) placed at its exit port which focuses the received backscattered radiation on a liquid nitrogen cooled MCT detector (active area ~0.1 mmx 0.1 mm, from M/S Laser monitoring systems, USA). An optical filter in the 9–11 μm band is employed in the receiver optics chain to cut down incoming optical radiation outside the desired band. A National Instruments, 12 bit, 10 MS/s DAQ card is used for acquiring the lidar return signals. The CO_2 laser beam was folded at 90 degree to be sent into the desired direction by the firing mirror@45 degree positioned at the centre of entrance aperture of the receiver telescope. This allowed a coaxial configuration for the transmitter and receiver. The experimental scheme of the CO_2 DIAL setup is shown in Fig. 12.12. The whole system was mounted on a trolley as shown in Fig. 21.13 to make the system transportable. The lidar system was used in a topographic target configuration for measuring the absorption at λ_{on} and λ_{off} wavelengths for diethyl ether (DEE). Aerosols of diethyl ether were enclosed in a perspex chamber of dimensions (40 cm × 40 cm × 40 cm) shown in Fig. 21.12, with openable windows of size 8 cm × 8 cm (cut at a height of 15 cm from the bottom of chamber) for laser beams to pass through. The perspex chamber mounted on a metallic frame was placed at various distances from the laser transmitter. An atomiser (M/s TSI Inc. USA) was used to create aerosols of the chemical species (DEE, liquid at room temperature). Dry nitrogen in standard gas cylinders was used as the carrier gas for generating the aerosols. The windows were kept closed initially for aerosols to be trapped inside the enclosure. During laser firing (and DIAL sensor testing) the windows were opened to allow laser beams to pass through.

In order to test the DIAL sensor, some preliminary laboratory experiments with this assembly using the chemical, DEE were carried out. Aerosols of DEE were released in the perspex chamber. The chamber was positioned on a metallic frame at a distance of 60 m in the laser beam path so that the beam interacts with DEE molecules and gets reflected from the hard target at distance of 100 m. We have used λ_{on} –9.2824 μm (9R18) and λ_{off} –10.2071 μm (10R26) for detection of DEE chemicals. Firstly, the laser beam at both wavelengths with equal energy of 7 mJ was transmitted through the chamber without DEE aerosol inside and the return signals at both wavelengths were recorded as shown in Fig. 21.14(a). It was observed that return signals at both wavelengths were approximately equal, which means no absorption was taking place inside the chamber. Subsequently, the chamber was filled with aerosolized diethyl ether and return signals once again measured as shown in Fig. 21.14(b) at both wavelengths. The return signal strength at λ_{on}(9.2824 μm) showed significant reduction of about 280 mV from its original value due to absorption by DEE molecules whereas no significant change was observed at λ_{off} (10.2071 μm). A small volume of 40 ml diethyl ether was used at the time for creating aerosols in the experiments.

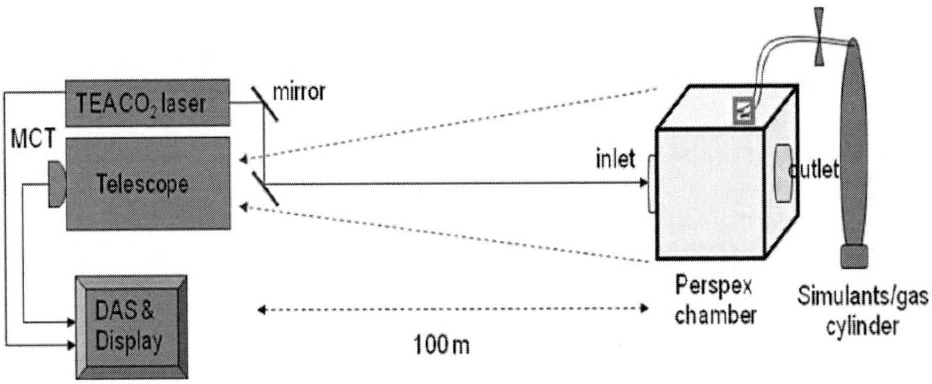

Figure 21.12 Experimental scheme of CO_2 DIAL setup

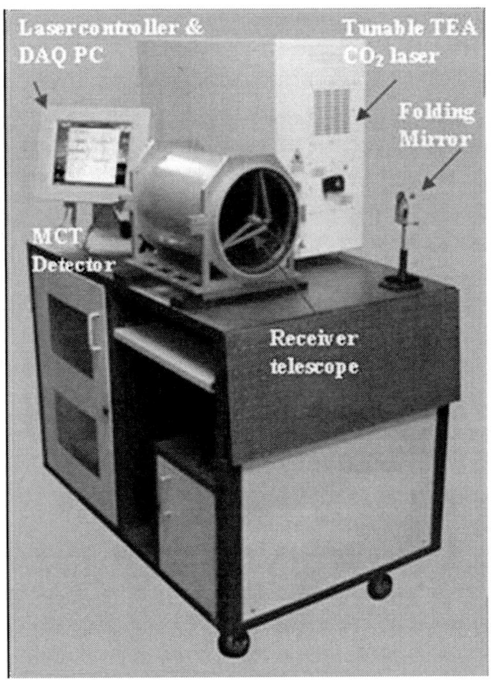

Figure 21.13 Photograph of trolley mounted CO_2 DIAL system

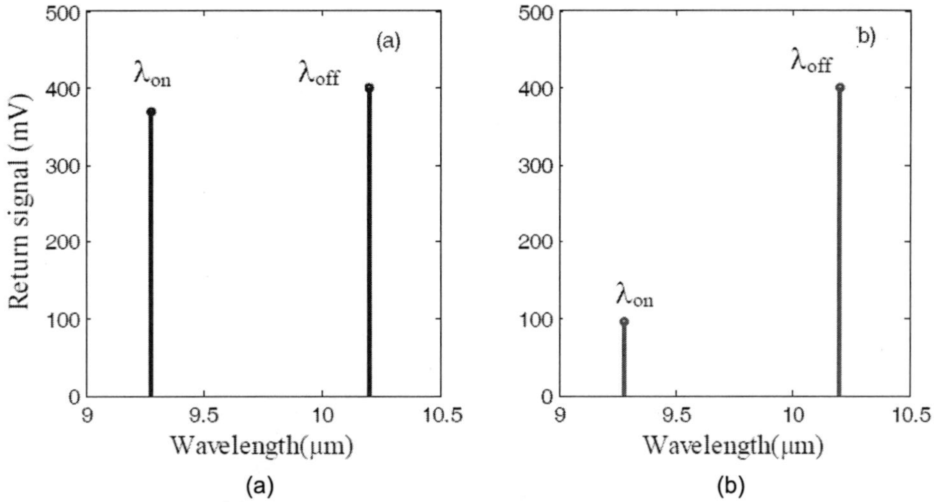

Figure 21.14 Return signals recorded at λ_{on} (9.2824 μm) and λ_{off} (10.2071 μm) wavelengths. (a) Signals measured in the absence of DEE aerosol in the chamber (b) Signals measured in the presence of DEE aerosols in the chamber

21.6.2 OPO DIAL

At LASTEC, a DIAL prototype system operating in the mid–IR wavelength band has also been designed and developed. This system uses a 6 mJ compact pulsed (pulse width ~ 4 ns) optical parametric oscillator (OPO), tunable in the wavelength range of 3000–3500 nm and driven by a 200 mJ Nd:YAG laser, as the DIAL transmitter (OPOTEK Inc., USA). This fully computer controlled laser with a prf of 10 Hz can fire laser pulses as per a predefined program allowing selection of ON and OFF wavelengths corresponding to the absorption spectrum of the suspected target molecule. The receiver system comprises transmissive ZnSe optics (Dia 25 mm) which focuses the received backscattered radiations on a liquid nitrogen cooled indium antimonide detector (active area 0.1 mm × 0.1 mm). A 3–5 micron optical filter is employed in the receiver optics chain to cut down incoming optical radiations outside the desired band. A National Instruments 12 bit, 10 MS/s DAQ card is used to acquire the lidar data for analysis. The lidar system was used in a topographic target configuration for measuring the absorption at λ_{on} and λ_{off} wavelengths for atmospheric trace gases and various chemical agents. Figure 21.15 (a) and (b) shows the photograph of the trolley and tripod mounted OPO DIAL system. Lidar master controller software was developed in-house to communicate with laser source (to fire wavelength sequence), energy meter (to measure pulse-to-pulse energies), and data acquisition card. Figure 21.16 shows the display panel of a lidar controller. Atmospheric methane was considered as a trace gas to be measured using this system. Two wavelengths namely online (3314 nm) and offline (3318 nm) were used for detecting methane. Return signals at both wavelengths were measured and presented in the right side of the controller panel. It was observed that the transmitted energies at these two wavelengths were not the same. Hence, return signals were normalized with respect to transmitted energies. 10 pulses were averaged for each wavelength (left side figure). Strong returns from the topographic target located at 60 m were received at both wavelengths, which showed the discernible changes in signal strength. The difference in signals was measured by taking the signal strengths from the topo-target (60 m). Significant changes in signals are attributed to absorption of atmospheric methane over path length. Its background concentrations were also determined and found to be 2.23 ppm, which is close to reported values of ambient concentration of methane under Delhi weather conditions.

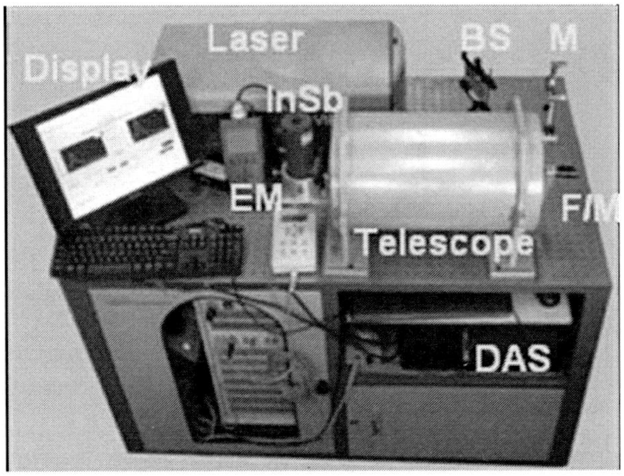

Figure 21.15(a) Trolley mounted OPO DIAL system

Figure 21.15(b) Tripod mounted OPO DIAL system

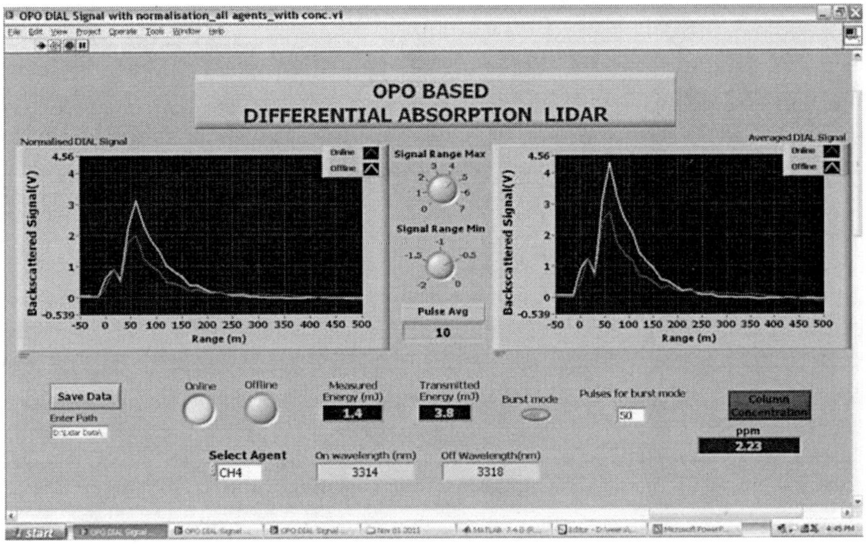

Figure 21.16 Differential absorption lidar master controller panel

21.7 Standoff Detection of Explosives

Detection of explosives is based on a wide variety of technologies that focus on either the bulk explosives or traces of explosives. Bulk explosives can be detected indirectly by imaging characteristic shapes of the explosive charge, detonators, and wires or directly by detecting the chemical composition or diamagnetic properties of the explosive material. Trace detection relies on vapours emitted from the explosive or on explosive particles that are deposited on the nearby surfaces.

The problem of detecting explosives at significant standoff distances remains one of the most difficult and the most important challenges for security specialists. Bulk detection techniques suffer from an intrinsic $1/r^4$ fall-off of the signal intensity with distance. Trace detection suffers from the fact that available explosive particles are too low in number or the available vapour plumes are too dilute for detection at appreciable standoff distances[6]. Vapour pressures are approximately 10 ppb for TNT and 10 ppt for RDX and PETN, for example. In terms of torr, for example, the typical values of vapour pressures for TNT, RDX, PETN, NG, EGDN are 5.8 $\times 10^{-6}$ torr, 4.4 $\times 10^{-9}$ torr, 1.4 $\times 10^{-8}$ torr 3.1 $\times 10^{-4}$ torr and 0.07 torr, respectively. Composite materials such as PBX (plastic bonded explosives) further suppress these small vapour pressures. Another challenge is the matrix interference, that is, the complexity of composition of suspicious materials.

The best hope for standoff detection is the analysis of particulates absorbed on surfaces by using a laser to cause these explosive particulates to emit characteristic (spectral) radiation [21–23]. In such an application, the signal would be expected to fall as $1/r^2$ (instead of $1/r^4$) if the signal from the particulates is emitted isotropically. The challenge is in the near real-time detection of traces of explosives in ambient conditions, in situ, which are not only concealed (by a 'shoe bomber', for example) but also placed in sealed/plastic containers, at standoff distances of more than 100 meters. Further, there should not be any requirement of collection or preparation of the sample of analyte for its detection and identification.

21.7.1 Resonance Raman scattering

In Raman spectroscopy, the explosive sample is irradiated (at frequency ν_0) by the intense laser beam, and the scattered light is usually observed in the direction perpendicular to the incident beam. The scattered light is of two types: one, called Rayleigh scattering, is strong and has the same frequency (ν_0) as that of the incident beam; and the other type, called Raman scattering, is very weak (10^{-5} of the incident beam) and has frequencies $\nu_0 \pm \nu_m$, where ν_m is the vibrational frequency of the molecule under observation. The $\nu_0 - \nu_m$ and $\nu_0 + \nu_m$ are called Stokes and anti-Stokes lines, respectively, with the former being more intense than the latter. Well-defined Raman peaks obtained enable the simultaneous identification of many individual explosive substances without the need of separation procedures. Thus, this technique is highly effective in the detection of IEDs.

The major limitations of Raman scattering are that it is a weak effect and background fluorescence can be a problem. The use of powerful lasers can lead to undesirable photodecomposition of the explosive material and absorption of the laser can lead to fluorescence that can obscure the Raman signals. This low Raman signal makes it difficult for the identification of target molecules in the dispersed phases such as the solution or the vapor phase. The problem of small Raman backscattering cross-section can be overcome if one uses a UV pump laser to excite the sample to an actual excited level instead of a virtual level. When the exciting radiation falls within the frequency range of the molecule's absorption band in the visible or ultraviolet spectrum, the radiation may be scattered by two different processes: resonance fluorescence or the resonance Raman scattering. Resonance fluorescence differs from the resonance Raman scattering in that the absolute frequencies of the fluorescent spectrum do not shift when the exciting radiation frequency is changed, so long as the latter does not move outside the absorption band. The absolute frequencies of the resonance Raman scattering, on the other hand, would shift by exactly the amount of any shift in the exciting frequency, just as do those in the normal Raman scattering.

Advantages of resonance Raman spectroscopy

The use of resonance Raman scattering (RRS) provides a significant improvement in sensitivity and selectivity as compared to spontaneous (normal) Raman spectroscopy. The scattered intensities in the case of resonance Raman scattering processes are two to six orders of magnitudes higher than in normal Raman scattering process. Besides the enhancement of the return signal due to the resonance phenomenon, UV Raman cross-section is further enhanced because of the $(1/\lambda)^4$ dependence of the scattered radiation. Many explosives have optical absorption bands in the UV, and therefore, will have some Raman peaks enhanced by resonance. A careful choice of excitation wavelength from a tunable UV laser will enhance the Raman spectrum from the target molecules as compared to the other molecules present there. There is no strong absorption of water in UV. Therefore, UV Raman detection is possible for compounds in aqueous solution (liquid explosives, for example). Use of solar-blind UV laser wavelength (< 300 nm) allows a Raman system to be used in daylight without interference from ambient light. A careful choice of excitation wavelength can enhance the Raman spectrum from the target material but not from other materials.

Limitations of resonance Raman scattering (RRS):

(i) The use of laser line of the same frequency as an electronic transition can cause a strong fluorescence. (ii) Fluorescence interference, either from the analyte itself or from the matrix, is a major obstacle in resonance Raman spectroscopy. (iii) The ratio of RR scattering to fluorescence varies with molecules of different materials. (iv) Another problem is that the use of a high-energy short-wavelength (UV) laser could cause some of the samples to ignite unless care is taken to choose spot diameter size.

21.7.2 Fluorescence rejection by deep UV resonance Raman scattering

A major advantage of deep UV resonance Raman spectroscopy is that no fluorescence interference exists when excitation is provided at wavelengths below about 250 nm [24]. A typical Raman spectral range of 4000 cm^{-1} occurs in less than 30 nm above the excitation wavelength at 250 nm. Independent of excitation wavelength, no known material fluoresces below about 280 nm. This provides complete spectral separation of Raman and fluorescence emission bands, resulting in high signal-to-noise ratio and low detection limits.

21.8 Future Perspective

It is well understood that LIDAR plays a major role in standoff detection of various chemical, biological and explosive (CBE) materials. Three major techniques namely, differential absorption, ultra violet laser induced fluorescence and Raman spectroscopy are being used for this purpose. The Raman technique can detect most CBE materials only up to few meters, whereas the other two methods can detect chemical and biological agents up to a few kilometers. With the advent of new compact broad band tunable lasers, fast response detectors and electronics components, it will be easy to develop compact and portable standoff laser based sensors for these applications.

Bibliography

1. R. M. Measures. 1992. 'Laser remote sensing-fundamentals and applications.' Kreiger Publishing Company, Krieger Drive, Malabar: Florida, 32950.
2. V. A. Kovalev and W. E. Eichinger. 2004. 'Elastic LIDAR: Theory, practice and analysis methods.' Hoboken, NJ: Wiley,
3. C. Weitkemp. 2005. *LIDAR: Range resolved optical remote sensing of the atmosphere.* Springer.
4. L. Szinicz. 2005. 'History of chemical and biological warfare agents.' *Toxicology* 214: 167–181.
5. C. R. Prasad, P. Kabro and S. Mathur. 1999. 'Tunable IR differential absorption lidar for remote sensing of chemicals.' *Proc. SPIE.* 3757: 87–95.
6. J. R. Simard, G. Roy, P. Mathieu, V. Larochelle, J. McFee and J. Ho. 2004. 'Standoff sensing of bioaerosols using intensified range gated spectral analysis of laser induced fluorescence.' *IEEE Transactions on Geoscience and Remote. Sensing* 42, 4: 865–874.
7. J. C. Carter, M. Angel, M. L. Snyder, R. E. Whipple and J. G. Reynolds. 2005. 'Standoff detection of high explosive materials at 50 meters in ambient light conditions using a small raman instrument.' *Appl. Spectrosc.* 59, 6: 769–775.
8. S. Veerabuthiran, A. K. Razdan, M. K. Jindal, D. K. Dubey and R. C. Sharma. 2011. 'Mie lidar observations lower tropospheric aerosols and clouds.' *Spectrochimica Acta Part A: Molecular and Biomolecular Spectroscopy* 84: 32–36.

9. J. D. Klett. 1981. 'Stable analytical inversion solution for processing LIDAR returns.' *Applied Optics* 20: 211–20.
10. W. Carnuth and R. Reither. 1986. 'Cloud extinction profile measurements by LIDAR using Klett's inversion method.' *Applied Optics* 25: 2899–2907.
11. S. D. Mayor, S. M. Spuler and B. M. Morley. 2005. 'Scanning eye-safe depolarization LIDAR at 1.54 microns and potential usefulness in bioaerosol plume detection.' *Proc. SPIE.* 5887: 137–48.
12. E. Yee, P. R. Kostenluk, G. Roy and B. T. N. Evans. 1992. 'Remote biodetection performance of pulsed monostatic LIDAR system.' *Applied Optics* 31: 2900–2913.
13. K. Baxter, M. Castle, S. Barrington, P. Withers, V. Foot, A. Pickering and N. Felton. 2007. 'UK small scale UVLIF LIDAR for standoff BW detection.' *Proc. of SPIE.* 6739: 67390Z–1 to 67390Z–10.
14. O. Steinvall, P. Jonsson and F. Kullander 2007. 'Performance analysis for standoff biological warfare agent detection LIDAR.' *Proc. of SPIE.* 6739: 673912–1 to 673912–14.
15. S. Veerabuthiran and A. K. Razdan. July 2011. 'Parametric studies of fluorescence lidar for detection of biological warfare agents in the atmosphere.' *Journal of Battlefield Technology* 14, 2: 5–12.
16. Y. Zhao, W. A. Brewer, W. L. Eberhard and R. J. Alvarez. 2002. 'Lidar measurement of ammonia concentrations and fluxes in a plume from a point source.' *Journal of Atmosphere and Oceanic Technology* 19: 1928–1938.
17. S. L. Jain, B. C. Arya and Arun Kumar. 2003. 'Measurement of liquid water content and trace species in the atmosphere using differential absorption lidar.' *Proceeding of SPIE.* 4893: 121–131.
18. C.B. Carlisle, J. E. Van der Laan, L. W. Carr, P. Adam and J. P. Chiaroni. 1995. 'CO_2 laser based differential absorption LIDAR system for range resolved and long range detection of chemical vapor plumes.' *Applied Optice* 34: 6187–6201.
19. S. Veerabuthiran and A. K. Razdan. November 2010. 'Differential absorption lidar for standoff detection of chemical warfare agents: simulation studies.' *Journal Battlefield Technology* 13, 3: 11–18.
20. F. M. D'Amico, R. G. Vanderbeck and R. G. Warren. 1999. 'Range resolved frequency agile CO2 lidar measurements of smokestack vapor effluents.' *Proceeding of SPIE.* 3855: 128–133.
21. T. A. Parpar, I. Levy, H. Schubert and A. Rimski-Korsakov (Eds.). 2006. 'Remote detection of explosives by enhanced pulsed laser photo dissociation/laser-induced fluorescence method in 'stand-off detection of suicide-bombers and mobile subjects.' *Springer* 59–68.
22. V. M. Korenskii and S. Vartapetov. July 2006. 'Remote detection of narcotics and explosives by fluorescent lidar.' *Presented at the 23rd International Laser Radar Conference (ILRC)* 24–28 Nara: Japan. Paper ID: 2P–5.
23. R. J. Lacey, I. P. Hayward, H. S. sands and D. N. Batchelder. 1997. 'Characterization and identification of contraband using resonant raman spectroscopy.' *Proc. SPIE.* 2937: 100–05.
24. 'Fluorescence rejection in resonance raman spectroscopy using a picosecond-fated intensified CCD camera'; Poster Presentation on Sep 28, 2006 at the 33rd Federartion of Analytical Chemistry and Spectroscopy Societies (FACSS) Meeting, Sep 2006.

List of Contributors

1. **C. Figueira de Morisson Faria,** Department of Physics and Astrophysics, University College of London.
2. **B. B. Augstein,** Department of Physics and Astronomy, University College London.
3. **J. Wu,** PhD Student, Department of Physics and Astronomy, University College London.
4. **Tahir Shaaran,** PhD Student, Department of Physics and Astrophysics, University College London.
5. **Alex Brown,** Associate Professor, Department of Chemistry, University of Alberta.
6. **Ryan R. Zaari,** Post-doctoral fellow Varganov Group Department of Chemistry, University of Nevada.
7. **Debabrata Goswami,** Department of Chemistry, IIT Kanpur.
8. **Amartya Bose,** PhD Student, Theoretical Chemistry, University of Illinois at Urban Champaign.
9. **Manabendra Sarma,** Assistant Professor, Department of Chemistry, IIT Guwahati.
10. **Satrajit Adhikari,** Physical Chemistry Department, Indian Association for the Cultivation of Science, Jadavpur.
11. **M. K. Mishra,** Professor, Department of Chemistry, IIT Mumbai.
12. **Yoshiaki Teranishi,** Assistant Professor, Institute of Physics, National Chiao Tung University.
13. **Hiroki Nakamura,** Professor, Institute of Molecular Science, National Chiao Tung University.
14. **Sheng H. Lin,** Institute of Molecular Science, National Chiao Tung University.
15. **V. V. Suran,** Department of Physics, Uzhgorod National University.
16. **I. I. Bondar,** Department of Physics, Uzhgorod National University.
17. **Man Mohan,** Department of Physics and Astrophysics, University of Delhi.
18. **Rinku Sharma,** Associate Professor, Applied Physics Department, Delhi Technological University.
19. **Vivek Kumar,** Research Scholar, Physics Department, Indian Institute of Technology, Delhi.
20. **A. K. Shukla,** Associate Professor, Department of Physics, Indian Institute of Technology, Delhi.
21. **Kishore T. Kapale,** Associate Professor, Department of Physics, Western Illinois University.
22. **Sulakshana Thanavnthri,** Professor, Physics Department, Naugatuck Valley Community College, Waterbury, CT.
23. **Jonathan P. Dowling,** Department of Physics and Astrophysics, Louisiana State University.
24. **Aranya B. Bhattacherjee,** Associate Professor, Department of Physics, Atma Ram Sanatan Dharma College, University of Delhi.
25. **Priyanka Verma,** Department of Physics, Motilal Nehru College, South Campus, University of Delhi.

26. **S. Aithal,** Principal, Srinivas Pre-University College, Pandeshwar, Mangalore.
27. **P. S. Aithal,** Director/Principal, Srinivas Institute of Management Studies, Mangalore.
28. **N. G. Bhat,** Professor and Head, Department of Chemistry, Srinivas Institute of Technology, Valachil, Mangalore.
29. **Smruti Rath,** National Institute of Science and Technology, Odisha.
30. **S. K. Tripathy,** Professor, Department of Electronics and Communication, Nation Institute of Science and Technology, Orissa.
31. **Tarun Kumar,** Assistant Professor, Ramjas College, University of Delhi.
32. **Sonam Mahajan,** Research Student, Department of Physics and Astrophysics, University of Delhi.
33. **A. K. Gupta,** Director, Instruments Research and Development Establishment, Dehradun.
34. **Ajay Mishra,** Senior Scientist, Instruments Research and Development Establishment, Dehradun.
35. **Maria Farooqui,** Researcher, Instruments Research and Development Establishment, Dehradun.
36. **Nimish Dixit,** Deputy Director, Instruments Research and Development Establishment, Dehradun.
37. **Ashok N. Kaul,** Dean, Research and Consultancy, D.I.T University, Mussoorie Road, Dehradun.
38. **Anil Kumar Maini,** Director, Laser Science and Technology Centre (LASTEC), Defense Research and Development Organization (DRDO), Ministry of Defence, Metcalfe House, Delhi.
39. **Varsha Agarwal,** Scientist 'E'. Laser Science and Technology Centre (LASTEC), Defence Research and Development Organization (DRDO), Ministry of Defence, Metcalfe House, Delhi.
40. **Manoj Sharma,** Scientist 'D', Laser Science and Technology Centre (LASTEC), Defence Research and Development Organization (DRDO), Ministry of Defense, Metcalfe House, Delhi.
41. **R. S. Solanki,** Scientist 'D', Laser Science and Technology Centre (LASTEC), Defence Research and Development Organization (DRDO), Ministry of Defense, Metcalfe House, Delhi.
42. **Gagan Preet,** Scientist 'C', Laser Science and Technology Centre (LASTEC), Defense Research and Development Organization (DRDO), Ministry of Defense, Metcalfe House, Delhi.
43. **M. Kundu,** Scientist SE, Basic Theory and Simulations Division, Institute for Plasma Research.
44. **P. K. Kaw,** Institute for Plasma Research, Bhat, Gandhinagar, Gujarat.
45. **Anil K. Razdan,** Additional Director, Laser Science and Technology Centre (LASTEC), Defence Research and Development Organization (DRDO), Ministry of Defence, Metcalfe House, Delhi.
46. **S. Veerabuthiran,** Scientist "D", Laser Science & Technology Centre, Metcalfe House, Delhi.